Physics of the Invisible Sun

Instrumentation, Observations, and Inferences

Frontinspiece photo: Spectacular picture of solar corona, as observed during the total solar eclipse of August 21, 2017, from Oregon, USA, made by digital image composition of 287 eclipse images (Continuum: 161, emission lines Fe XIV: 32 On-band and 32 Off-band, Fe XI: 31 On-band and 31 Off-band). Charged particles are the proxy of the 'shape' of magnetic field lines in the solar corona, where the ions, such as Fe XI and Fe XIV populate different magnetic structures with different temperatures. The free coronal electrons, on the other hand, scattering the photospheric light (white-light continuum) towards the observer are omnipresent, and therefore provide a complete view of the solar magnetic field. The composite of these images displays the large scale structure of the coronal magnetic field and the distribution of Fe ions. (By permission of Miloslav Druckmüller.)

Physics of the Invisible Sun

Instrumentation, Observations, and Inferences

Ashok Ambastha

CRC Press
Taylor & Francis Group
Boca Raton London New York

CRC Press is an imprint of the
Taylor & Francis Group, an informa business

CRC Press
Taylor & Francis Group
6000 Broken Sound Parkway NW, Suite 300
Boca Raton, FL 33487-2742

International Standard Book Number-13: 978-1-138-19744-2 (Hardback)

Visit the Taylor & Francis Web site at
http://www.taylorandfrancis.com

and the CRC Press Web site at
http://www.crcpress.com

To my wife
Kumud

Contents

Preface

O my Lord, sustainer of all that lives, Your real face is covered by Your dazzling effulgence. Kindly remove that covering and exhibit Yourself to Your pure devotee. O Pushan, O Sun, thou offspring of Prajapati, Remove Thy rays, gather up thy effulgence, So that I may see that which is Thy most auspicious effulgence. **Isopanishad, Verses 15-16**

In the past, several books have been written on various aspects of the physics of the Sun that are either at a very popular level, or highly technical and theory oriented with lots of mathematical details. This book aims at providing the essence of challenges and the exciting developments that have taken place in the last couple of decades in solar instrumentation and observing capabilities in various wavelengths, in optical (visible) and invisible windows, of the electromagnetic spectrum. These developments have led to inferences that unravel many hitherto unknown phenomena and helped in new discoveries in the frontier areas of solar physics. These involve the invisible outer layers of the solar atmosphere, and the deep solar interior hidden beneath the visible "surface", the photosphere. Our aim in this book is to update these recent developments by elaborating on the instrumentations, observations and deriving physical inferences. It is expected that this book will be useful to general readers interested in the Sun, as well as to the students of physics, space sciences, solar physics and astronomy.

Chapter 1 provides a brief chronological sequence about the Sun as derived historically from only the very limited observations of visible (optical) disk of the Sun, its global properties inferred from the general laws of physics, the extended solar system, birth and evolution of the Sun, and the Sun as a star in the Milky Way galaxy. The photosphere is only a very thin outer 'skin' of a much larger Sun extending inwards to the hidden interior, and upwards to outer atmospheric layers which are much too faint to be directly observed. The photosphere and its features are easily observed using even modest aperture solar telescopes suitably aided with optical solar filters. Dark absorption lines, known as Fraunhofer lines, are observed embedded in the solar photospheric spectrum. These spectral lines are essentially the 'thumb prints' of the elements present in the solar atmosphere. The photons emitted continuously from the Sun are the messengers carrying information about Sun's structure and dynamics. Collecting these photons, and using known principles of physics, one can determine physical conditions that exist in the photosphere.

Chapter 2 is devoted to ground-based optical observations, instruments and analysis techniques that are generally used. Ground-based solar optical observations require careful site surveys over extended periods to select suitable sites with good atmospheric "seeing", low dust and water vapour content, and large number of cloud-free days in a year. Various types of telescope configurations, dome designs, and instruments aided with active and adaptive optics capabilities, are the next step to optimally exploit the diffraction-limited resolutions along with the atmospheric "seeing" of the site. High altitude sites, surrounded by large bodies of water, appear to provide better quality of observations. Imaging and spectroscopic methods are used for observing the photosphere and chromosphere, their associated features, and energetic transient activities on a regular basis to synoptically monitor the Sun. This is achieved by blocking the light of the brighter photosphere using special optical filters centred on specific photospheric and chromospheric spectral lines. For example, a chromospheric H_α filter enables one to routinely observe the faint "red" chromosphere above the photosphere that is otherwise observed only at the solar limb just before (and after) the totality of a solar eclipse. Using the technique of artificial occultations, it is possible to obtain images of the near-corona in certain visible coronal emission lines from high altitude sites with very low dust content.

Chapter 3 discusses the outer layers of solar atmosphere, observed features at various scales, optical observations and instruments in chromospheric and white light corona, and transient energetic phenomena, such as flares and prominence eruptions. Also discussed are the discoveries of helium and the million degree temperatures of solar corona deduced spectroscopically through the detection of visible forbidden lines from highly ionized species imprinted in solar coronal spectrum.

Magnetic field, combined with velocity (flows) and an overall solar differential solar rotation, are believed to be the key factors to bring variability and solar activity at widely ranging temporal and spatial scales. Although the energy output from the Sun's interior has remained nearly constant over the last many millions of years, there is an interesting 11-year solar activity cycle that operates in the Sun in terms of appearance and disappearance of sunspots and rise/fall of energetic, explosive phenomena. Recent observations of solar irradiance from space over the past couple of decades have revealed variations with the solar cycle which are minute but not negligible. The 11 year sunspot cycle or 22 year solar magnetic cycle is believed to be driven by the giant dynamo operating inside the Sun in the convective region. Understanding of the active and explosive Sun is one of the fundamental problems which have essentially led to modern solar research. Chapter 4 is devoted to Zeeman and Hanle effects used in solar magnetic field measurements at photosphere, chromosphere and corona, solar activity and magnetic cycles, and dynamo models.

Although it is possible to observe the outer, visible layer of the Sun using specially designed instruments routinely, the deep solar interior is hidden under the gaseous, but opaque photosphere. Our knowledge of the Sun's interior has been traditionally derived from theoretical models based on the global properties of the Sun. Inside the Sun (and other stars), energy is produced continuously by the thermonuclear fusion process which releases high-energy photons and neutrinos. Detection of solar neutrinos produced in the hot and dense central core provides an important tool to infer about the *solar interior*. In Chapter 5, we describe the solar standard models constructed based on theoretical formulations and numerical modelling, energy production in the central core, and its transport through the radiative and convective zone, detection of solar neutrinos as a probe and test of the reliability of solar standard models, the neutrino problem and its eventual resolution.

Chapter 6 is devoted to yet another important tool for probing the solar interior that became available subsequent to the discovery of the so-called *5-minute oscillations* of the solar photosphere in the 1960s. These photospheric oscillations were interpreted to result from the superposition of millions of modes of oscillations trapped within cavities of different depths in the interior of the Sun. Accurate determination of the amplitude and frequency of these global modes of solar oscillation is termed *helioseismology*, which has made it possible to probe the structure and dynamics of the solar interior. Helioseismic observations over long temporal periods from a network of ground-based observatories and from space have successfully revealed physical properties, such as density, temperature, flows, and rotation rate, as functions of depth in the solar interior. Discovery of an interface layer, *tachocline* located in a narrow region between the differentially-rotating convection zone and the almost uniformly-rotating radiative interior, is a major contribution of helioseismology. Also discussed are some aspects of 'local helioseismology' at small scales of active regions, which have helped in understanding the birth of sunspots, far-side imaging of the Sun, *etc.* Along with solar neutrinos, helioseismology has led to major refinements in solar models, and the standard model of particle physics.

Chapter 7 deals with the Sun as observed in longer wavelengths, *i.e.*, IR and radio regimes. Of these, infrared instrumentation is more difficult due to the observational and technical issues, but holds a lot of promise in the measurement of chromospheric magnetic fields and helioseismology. IR radiations can reach to the ground-levels only in certain bands due to molecular absorption by atmospheric water vapour, CO_2, *etc.* The intense instrumental and surrounding emissions in IR overwhelm the emission coming from the source, that need to be discarded. We discuss the new instruments for detection, imaging and polarimetry in IR. Then, we move on to the longer wavelengths of the radio window that is accessible from the ground. But, it became available only in the early 20th century when Karl Jansky in the 1930s detected radio waves coming from the densest part of the Milky Way. After World War II radio emission from the Sun was serendipitously discovered using military radars.

This discovery provided a method to study the outer solar atmosphere that had so far not been accessible by the optical solar telescopes.

We describe the thermal and non-thermal radio emissions and bursts from the Sun, detection and imaging techniques, radio diagnostics of coronal magnetic fields, radio heliographs and spectrographs, propagation and monitoring of solar transients in the interplanetary medium. The limitation of spatial resolutions in the radio window had been a major problem which has been overcome by the advent of aperture synthesis and long base-line radio interferometry. A brief introduction to solar MHD formulation is also provided in Chapter 7 due to its importance in waves and oscillations of the coronal plasmas, storage of energy in active regions, and its release in transient phenomena. These are being observationally confirmed and appear to be promising diagnostic tools for determining the physical properties, such as the density, temperature and magnetic field strength of the magnetized coronal plasma. Developments in this field form the basis of *coronal seismology*, which is emerging as a new discipline in solar physics. These are also briefly covered in this chapter.

Chapter 8 discusses solar astronomy involving short wavelengths, such as UV, EUV and X-rays. This is a relatively new science as it had to await the access to space by rockets and satellites as these high-energy photons are absorbed in the higher layers of Earth's atmosphere. Solar instrument payload on the Skylab space-mission in 1973 gave a major boost to X-ray solar observations. It has given rise to dramatic discoveries using specially designed instruments for detecting and *digital imaging* over these wavelengths. The hot corona at several million-degrees emits EUV and X-ray photons. Non-thermal processes associated with energetic transient events lead to sudden, many-fold enhancements in soft and hard X-ray emissions, and even produce γ-rays. Recent developments in instrumentation in these wavelengths has made it possible to observe the corona and the plasma processes occurring over the entire solar disk, and not just at the limb. Large dark areas having very low X-ray emissions, called the *Coronal Holes*, and spectacular, large-scale eruptions in the corona, the CMEs, have been discovered only since the advent of the space era. The question of high temperatures of the solar coronal plasma and its heating mechanism has remained an outstanding problem in solar physics for several decades. Some mechanisms have been proposed to dissipate the energy as the continuous source of coronal heating as supported by recent observational evidence.

Chapter 9 addresses the ever-expanding solar corona that is the source of solar wind; a radially moving outward flow of plasma from the Sun that streams past the Earth and beyond to the outermost boundaries of the heliosphere. The activity cycle and explosive phenomena on the Sun influence and disturb the vast *heliosphere* consisting of the interplanetary space between the Sun and planets, and the planetary magnetospheres. For example, flares and CMEs cause disturbances in the solar wind that shake the Earth's (and other planets') magnetic field and pump energy into the radiation belts. The enhanced electromagnetic radiations and high-energy particles released during the course of explosive solar flares ionize the Earth's upper atmosphere. The stormy *space weather* can physically damage satellites, change their orbits and shorten mission lifetimes. The excess radiation can pose a threat to astronauts operating in space stations. Shaking the Earth's magnetic field, the geomagnetic storms produced by CMEs encountering the magnetosphere can cause current surges in power lines that can destroy equipment and disrupt communication and electric power transmission, particularly in high latitude regions.

In Chapter 10, we discuss the recent major instruments being developed for ground-based and space-borne telescopes, which are coming up as a part of national and international collaborations. The future directions of newer technology and suggestions for observing from vantage locations in space are described. As we become more dependent upon satellites in this space era, we increasingly feel the importance of space weather and the need to predict it. The space weather and geomagnetic activity have important practical consequences, and their forecasting lies in continuously observing the Sun. A new thrust is thus on *heliophysics*, encompassing the Sun, interplanetary space ranging from the inner regions all the way to the outer limits of the solar system. This requires a synergy between the ground-based programs and space missions.

The material for this book has grown out of my course of lectures given to research scholars and graduate students during the past few years. I have made many composites of images from GONG, SDO, and SOHO using IDL, and GIMP, and art-work of line diagrams using a vector graphics application, InkScape. I would like to convey my gratefulness to several colleagues for sending or giving permission for use of their figures. In particular, I am indebted to H.M. Antia, Sarbani Basu, R. Ramesh, Shibu Mathew, Bill Livingston, Markus Aschwanden, David Hathaway, Jorgen Chistensen-Dalsgaard, Dean Pesnell, D.J. McComas, and Miloslav Druckmüller. I have benefited greatly by the solar images made available from Global Oscillations Network Group (GONG), SOHO (MDI, LASCO, and VIRGO), and SDO (HMI, AIA), which are gratefully acknowledged. Many images of the island observatory, telescopes, and observational sequences from my host institute, Udaipur Solar Observatory (USO), have been used. I would like to express my gratitude to my colleagues at USO, Udaipur and Physical Research Laboratory (PRL), Ahmedabad, for their direct and indirect support for this project.

Finally, I would like to express thanks to my family; and in particular, to my wife, Kumud, for her forbearance and support without which it would have been difficult to complete this work.

Udaipur (India) Ashok Ambastha

Author

Prof. Ashok Ambastha obtained his B.Sc. in 1972, and M.Sc. in 1974 from Pt. Ravishankar University, Raipur. He was awarded his Ph.D. degree in 1981 for his research in galactic dynamics at the Physical Research Laboratory (PRL), Ahmedabad (India). After that, he pursued postdoctoral research on plasmas trapped in magnetic mirrors. Prof. Ambastha joined Udaipur Solar Observatory (USO), PRL, Department of Space, Government of India, as a Research Associate in July 1983. He contributed extensively to the formulation and consolidation of USO's scientific programs serving in various academic positions, superannuating as Senior Professor in 2014, but continuing as Honorary Scientist. His research experience spans over 40 years in Plasma Physics, Galactic Astrophysics, and Solar Physics. He worked at the Solar Physics Group, NASA Marshall Space Flight Center, Huntsville, Alabama (USA) during 1992-1994 as an NRC-Senior Resident Research Associate. He was a Councillor (1998-2001) and the Secretary (2004-2006) of the Astronomical Society of India (ASI). He served as a Member of the COSPAR-URSI-SCOSTEP committee of the Indian National Science Academy, Delhi, during 2008-2012, and on the editorial board of the *Journal of Astrophysics and Astronomy*, Indian Academy of Sciences, Bengaluru, during 2010-2016. He is a member of the International Astronomical Union (IAU), ASI, and the National Academy of Sciences, Allahabad. Prof. Ambastha led several total solar eclipse expeditions to many countries for studies of the structure and polarization of the solar corona. He has more than 100 research publications in reputed national and international scientific journals and has edited four books/proceedings.

Chapter 1

Introduction

1.1 Ancient observations of the Sun

The Sun has been an object of fascination and worship all over the world for several millennia. Ancient Chinese astronomers are believed to have brooded over solar eclipses and sunspots to divine future events for their Emperor. Babylonians were perhaps the foremost of astronomers from the ancient times, apart from those in the Indian subcontinent, far east, and Europe to some extent. But, surprisingly, no astronomical records are available from ancient Egypt. Babylonians began systematic observations of the sky as early as in 700 B.C. They studied the motions of Sun, Moon and planets with a particular aim of recording ephemerides, and recorded numerous lunar and solar eclipses.

Many cultures around the world have interesting myths about the Sun, reflecting its importance in all our lives. In ancient Egypt, *Re* was known as the Sun god and the creator. *Re* was portrayed with a hawk head, wearing a fiery disk like the Sun on his head, and took many forms. *Re* cried one day, and humans were made from his tears. For the Aztecs in central Mexico, *Tonatiuh* was a Sun god. Aztecs believed that four suns had been created in four previous ages. They died at the end of each cosmic era. *Tonatiuh* was the fifth Sun, and the present era is still his.

In Greek mythology, *Apollo* was the god of the Sun, logic, and reason. He was the son of *Zeus* (Jupiter) and *Leto* (Letona). The Romans also believed in *Apollo* as the god of light, music, and healing. Chinese believed that there existed ten suns that appeared in turn in the sky during the Chinese ten-day week. Once, all the ten suns decided to appear all at once, making life on Earth difficult to survive. So, *Yi* shot nine suns, and only the Sun that we see today remained in the sky. One of the oldest Japanese religions, Shinto, considered *Amaterasu* as the Sun goddess.

Hindu sacred writing *Rig Veda* composed over 3,000 years ago described many gods and goddesses. The Vedas are full of hymns describing *Surya* as the Sun God, or *Aditya*, representing as the source and sustainer of all life on earth. Surya is imagined as riding a chariot drawn by seven horses. The Vedic scriptures refer to the Sun as the storehouse of inexhaustible power and radiance, The origin of the worship of the Sun in India is thus several centuries old. Some aspects of prehistoric astronomy in India are described in [201].

FIGURE 1.1: The Konark Sun Temple, built in 13th-century CE, is located near Puri, Odisha state of India. Credit: Alokprasad84 - https://commons.wikimedia.org/w/index.php?curid=12664454

Many Sun temples were built across sub-continental India over different centuries, enshrining the Sun God as the principal deity: Dakshinaarka Temple (13th century A.D.) in Gaya, Bihar; Suryanaar Koyil, Kumbhakonam in Tamilnadu, South India; Suryanarayanaswamy temple, Arasavilli, Andhra Pradesh (7th century built by Kalinga king); Konark on the East Coast of India; Modhera Sun Temple (built in 1026 A.D.) in Gujarat (Western India); Surya Pahar in Assam, North Eastern India; and Bhramanya Dev Temple at Unnao, near Jhansi in Central India. Ruins of ancient Sun temples are found at Martanda near Srinagar in Kashmir, dating back to the first century A.D., and in Multan (presently in Pakistan) built in the 7th century A.D.

The grandest, and still well preserved, Sun temple amongst these is located at Konark, near Puri, Odisha (India). It was built in the 13th century (Figure 1.1). What remains of this temple complex is a 30 m high chariot with immense wheels, and horses, all carved from stone, which was over 61 m high originally. The structures and elements of this temple have intricate artwork, iconography, and themes. The architecture consists of the chariot's twelve pairs of wheels symbolic of the 12 months of the Hindu calendar, with each month paired into two cycles of Shukla and Krishna pakshh of the Moon. The chariot is pulled by a set of seven horses signifying seven days of a week. This temple was so designed that the rays of the Sun would fall on the image of Surya at the time of the equinoxes.

There are other ancient structures around the world, such as Stonehenge. It is a massive prehistoric stone monument located north of Salisbury, England. The biggest of Stonehenge's stones are up to 9 metres tall and weigh 25 tons. This prehistoric structure is estimated to be 5000 years old, around the late Neolithic and early Bronze Ages. Some have suggested that Stonehenge was part of a ritual landscape. However, from its design it is possible to suggest a celestial observatory function, which might have allowed prediction of eclipses, solstice, equinox and other celestial events important to a contemporary religion. These structures are aligned with the rising and setting of the Sun on certain days, called the solstices. These occur twice a year, around 21 June and 22 December, when the Sun appears to reach its highest point above or below the equator.

The German "Stonehenge" near a small German town, Goseck, was radiocarbon dated to 4900 B.C. [278]. First spotted in 1991 from aerial photographs, the giant circular ridge, 75 metres wide, arguably represents the remains of the world's oldest, and the first observatory [53]. It took archaeologists 12 years to realise that this ridge was the remains of an ancient solar observatory. This is an early Neolithic henge structure with entrances orientated to the rising and setting solstices. It was apparently created by Europe's first civilisation, *i.e.*, long before the cultures of Mesopotamia and the pyramids of Egypt. It is one of the best studied of the more than 250 ring ditches identified by aerial surveys throughout Germany, Austria, the Czech Republic, Slovenia and Croatia. It lies nearly on the same latitude as Stonehenge in England.

One of the oldest solar observatories in the Americas was reported to have been uncovered in coastal Peru in 2007 [405]. The ceremonial site provides evidence of sophisticated cults of the Sun operating in South America as early as 2300 years ago. Previously, archaeologists had uncovered 4000-year-old fragments in Peru showing images of a staff god with rays emanating from its head, perhaps like the Sun. The site may have been a place for public rituals and feasts associated with the seasons and the Sun.

1.2 Pre-telescopic observations of eclipses and sunspots

For much of human existence, the Sun going dark in the middle of the day understandably frightened our ancient (and not so ancient) ancestors. Ancient civilisations often relied on mythology to make sense of a seemingly apocalyptic event. In South America, a celestial jaguar was thought to have eaten the Sun. The ancient Chinese believed a dragon was responsible for devouring the bright yellow ball in the sky. Ancient Mesopotamians saw it as a sign that the king was nearing death. In Indian Hindu mythology, the demon Rahu spent his days chasing the Moon and the Sun. Every once in a while, he caught and swallowed them, but since he was only a head, the Sun and Moon slide right back out after only a few moments. It was around the 8th century BCE, that eclipse observations were recorded, and people started realising that there were patterns to it.

Ancient astronomical texts give numerous and precise lunar and solar eclipses from many civilisations, including the ancient European. From a Babylonian diary covering parts of the year 137-136 B.C., there is a description relating to a total solar eclipse. However, few astronomical records have survived from the Greek and Roman past. Many eclipses were recorded during medieval Europe, but mostly without any useful technical details. These ancient eclipse observations have played a major role in our understanding of the rotation of the Earth, and the motion of the Moon [353].

Chinese astronomers recorded solar activity around 800 B.C. in China, and Korea frequently observed sunspots. However, there are no known illustrations of sunspot activity or solar events from these sources. The earliest surviving record of sunspot observation dates from 364 B.C., based on comments by Chinese astronomer Gan De in a star catalogue. From the Chinese court of Han dynasty records dating back from 28 B.C. until the late Ming dynasty in the mid-17th century, Chinese history books are known to have recorded more than 100 sunspots. By 28 B.C., Chinese astronomers were regularly recording sunspot observations in official imperial records. Gaocheng Observatory, established in 1279 A.D. by the famous Chinese astronomer Guo Shoujing, is considered to be the oldest of 27 ancient observatories in China dating back to the early Yuan dynasty (1271-1368 A.D.).

The ancient culture of India has over three thousand years of tradition in mathematics and astronomy in which the Sun has played an important role. The Surya Siddhanta is one of the earliest traditions (siddhanta) in archeo-astronomy of the Hindus dating back to the 3rd century B.C. The early interests in astronomy were in making calendars, and in observing stars near the zodiac as a guide to the motion of the Sun and the Moon. Many leading figures, such as Aryabhata I (476 A.D.), Bhaskara I (629 A.D.), Varahamihira (505 A.D.), Brahmagupta (628 A.D.), Bhaskara II (1114 A.D.), and Jain mathematician Mahaviracharya (9th century A.D.) contributed to the development of Indian astronomy in the early periods. They described lunar and solar eclipses in their Sanskrit text treatises known as Siddhantas, such as Aryabhatiyam, Surya Siddhanta, Brahm Sphuta Siddhanta, *etc.* [27], [215], [355].

Aryabhata correctly insisted that the Earth rotates about its axis daily, and that the apparent movement of the stars is a relative motion caused by the rotation of the Earth, contrary to the then-prevailing view in other parts of the world, that the sky rotated (first chapter, Aryabhatiya). Varahamihira's important contribution is the encyclopaedic Brihat-Samhita, which covers wide-ranging subjects including planetary movements, eclipses, rainfall, clouds, architecture, growth of crops, *etc.* A.B. Keith mentioned 'After Bhaskara, no progress can be recorded in Indian astronomy' [208]. But, the ancient tradition was followed by astronomers from ancient Kerala, and later by Samanta Chandrasekhara Simha (1835-1904) of Odisha [293].

The first clear mention of a sunspot in western literature was by the ancient Greek scholar Theophrastus, student of Plato and Aristotle and successor to the latter, around 300 B.C. [375]. Benedictine monk Adelmus observed a large sunspot on 17 March A.D. 807 that was visible for eight days; however, Adelmus incorrectly concluded he was observing a transit of Mercury [394]. Sunspot activity in 1129 was described by John of Worcester, providing the first surviving sunspot drawings based on the visual sighting of the Sun on

8 December 1128. However, the observations were misinterpreted again as planetary transits [129].

There are many reliable records of total eclipses from that period, which describe (i) complete disappearance of the Sun, (ii) the date, and (iii) the place of observation (stated or inferred). Because of their meticulous record-keeping, ancient astronomers discovered that eclipses followed a systematic time-table of the so-called "*Saros cycle*", a period of ~ 18 years (18 years, 11 days and 8 hours, to be exact). According to Herodotus's Histories, written in 450 BCE, Greek philosopher Thales of Miletus was the first to accurately predict the time and place of a solar eclipse. His exact method is not clear, or if he used the Saros cycle. But, it appears that by the 5th century BCE, eclipses were no longer thought to be the works of demons or gods [286].

Using the written observations from the ancient Babylonians and his own calculations, Egyptian mathematician Ptolemy was able to accurately describe 19 solar eclipses going back centuries. For the next 1,500 years, complex writings and mathematics of *The Almagest* became the definitive tome for predicting the positions of celestial bodies and eclipses up until the time of Galileo.

The first unambiguous mention of the solar corona was by Leo Diaconus, a Byzantine historian. He wrote of the 22 December 968 total eclipse, which he experienced in Constantinople (modern-day Istanbul, Turkey): "at the fourth hour of the day ... darkness covered the earth and all the brightest stars shone forth. And it was possible to see the disk of the Sun, dull and unlit, and a dim and feeble glow like a narrow band shining in a circle around the edge of the disk. –"Leo Diaconus" [175]. It is interesting that in medieval Europe, many significant astronomical observations were made not by skilled astronomers, but by monks in monasteries.

Copernicus was able to observe at least four solar eclipses during his life, often through a camera obscura, an optic phenomenon that occurs when an image is projected through a tiny pinhole, which is then reversed and inverted. But it was an Islamic scholar Alhazen in the 2nd century who was the first to build and use a camera obscura in observing and measuring solar eclipses. In 1543, as Copernicus lay on his deathbed, this research contributed to one of astronomy's greatest revelations: It was the Sun, not the Earth, that was the centre of our solar system.

1.3 Geocentric and heliocentric universe

Early astronomy was essentially centred on observing planetary motions. Greek scholar Claudius Ptolemy (100-160 A.D.) wrote his *Almagest*, formalising the Earth-centred (*geocentric*) model universe. This model considered the Earth fixed at the centre around which all celestial bodies, including planets, Moon, Sun, and stars moved in the sky. The apparent motion of planets in the sky was explained using the concept of circular *epicycles*, whose centres moved around the Earth in circular *deferents*. The geocentric theory was able to explain the observed *retrograde* motion of planets, such as Mars and Jupiter. Retrograde motion is the apparent change of position of a planet from east-to-west in the sky relative to the stars. It is not real motion in that the planet does not physically move backwards in its orbit, but only appears to do so because of the relative positions of the planet and the Earth, as they revolve around the Sun.

The geocentric idea was generally accepted until the 16th century. However, there were still many doubts about Earth's centrality on a fixed position, raised from time to time by many Islamic and Greek scholars. It was in 1543 that Nicolaus Copernicus published his treatise, *On the Revolutions of the Celestial Spheres* describing his Sun-centred (*heliocentric*) model of the universe. According to this, the centre of the universe is located at or near the Sun, and the distance from the Earth to the Sun is imperceptible compared with the distance to other stars. Rotation of the Earth accounts for the apparent daily rotation of the stars, and apparent annual cycle of movements of the Sun is caused by the Earth revolving round it. Apparent retrograde motion of planets is explained to be caused by motion of the Earth from which one observes the universe.

While Copernicus rightly observed that the planets revolve around the Sun, it was Johanes Kepler who correctly defined their orbits. According to his work, published between 1609 and 1619, the planets revolved not in perfect circles, as thought by Copernicus, but in elliptical orbits. In fact, the major contribution from Kepler was his three laws of planetary motion:

- Law of planetary orbit - planets move in elliptical orbits with the Sun at one focus.

- Law of areas - the radius vector, *i.e.*, the line joining the planet to the Sun, sweeps out equal areas in equal times, or dA/dt =constant, where A denotes the area, implying that the planet moves faster at the perihelion as compared to the aphelion, and

- Law of periods - the square of a planet's orbital period T is proportional to the cube of the semi-major axis a of the orbit, or, $T^2 = ka^3$, where k is a constant $= 1$, if T is in years and a is in astronomical units (A.U.), *i.e.*, the average Earth-Sun distance.

In fact, it can be stated that Kepler led Newton (1642-1726) to the Law of Gravitation. Newton used his own Law of Motion $F = ma$, and Kepler's third Law to derive the famous Law of Gravitation $F = Gm_1m_2/r^2$, which gives the gravitational force F between two celestial bodies of masses m_1, m_2 separated by a distance r.

Kepler's theory was able to give a superb map of the solar system, but could not give a scale of their distances. It was Cassini, who in 1672 measured the Earth-Mars distance using observations of Mars from Paris, and French Guiana. Along with Kepler's third Law, he then calculated Earth-Sun distance to be 140 million km, not much off from the actual value.

1.4 The invention of the telescope and discovery of sunspots

The dark features on the Sun, clearly depicting the umbrae/penumbrae of the sunspots, were perhaps first discovered in 1610 immediately after the invention of telescopes more or less simultaneously by Galileo Galilei (Italy), Thomas Harriot (England), the Jesuit Christoph Scheiner (Germany), and Johann Goldsmid (Holland) [374]. However, credit for the discovery belongs to Galileo for making a convincing case that sunspots are indeed solar features. Galileo made a series of sunspot observations and drawings in summer of 1612 (Fig. 1.2a). These were published in 1613: *Istoria e Dimostrazioni Intorno Alle Macchie Solari e Loro Accidenti Rome* (History and Demonstrations Concerning Sunspots and their Properties). A couple of Galileo's telescopes have survived, and are on display in the History of Science Museum in Florence, Italy.

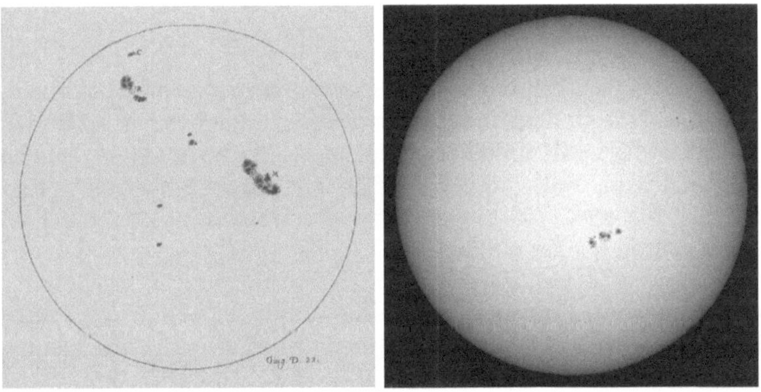

(a) (b)

FIGURE 1.2: (a) Galileo's drawing of sunspots shows umbra-penumbra structures (The Galileo Project `http://galileo.rice.edu/sci/observations/sunspot_drawings.html`), (b) Solar image of photosphere of February 14, 2011, showing sunspots, and the limb-darkening effect (Image courtesy of SDO-AIA/HMI science team.)

One is often told of a story that Galileo became blind by looking at the Sun through his telescope. However, this does not appear to be true. His sunspot observations were initially made near sunrise and sunset, and later by projection. Galileo was reported to have become blind much later at the age of 72, nearly 25 years after his telescopic observations of the Sun, apparently from a combination of cataracts and glaucoma [342].

The physical nature of sunspots remained unclear for centuries after their discovery. Scheiner believed them to be dense objects embedded in the Sun's luminous atmosphere. In the late 18th century William Herschel (1738–1822) suggested that sunspots were opening in the Sun's luminous atmosphere, allowing a view of the underlying, cooler surface of the Sun [175].

Sunspots observations continued in the 17th century. German Johannes Hevelius (1611-1687) and the French Jesuit Jean Picard (1620-1682) were the most active observers of the sunspots. A systematic solar observing program was carried out under the direction of Jean Dominique Cassini (1625-1712) at the newly founded Observatoire de Paris.

1.4.1 Observations of planetary transits from India

The first use of telescopes from India was in 1651, *i.e.*, barely 40 years after its use by Galileo. Many important astronomical expeditions and observations were made from India since the 17th century. For example, Jeremiah Shakerley observed the 1651 transit of Mercury from Surat. In 1689, Father Jean Richaud, a French Jesuit priest, made astronomical observations from Pondicherry in south India, and discovered the binary nature of the bright star Alpha Centauri.

Modern astronomy started taking root in India way back in the latter half of the 18th century mainly initiated by Europeans. Raja Sawai Jai Singh II set out to build a series of huge immovable masonry astronomical instruments at Delhi (1724), at his newly founded capital Jaipur (1734), and later smaller ones at Mathura, Ujjain, and Varanasi (1737). The rare cosmic event of transits of Venus in 1761 and 1769 saw a flurry of astronomical activity. The King of France deputed Guillaume Le Gentil (1725-1792) to observe these transits from Pondicherry. It was important to observe the transits because it could provide the Sun-Earth distance, a very important astronomical parameter.

During the transit of Venus in December 8, 1874, India was again a preferred location. Major J. F. Tennent made observations from Roorkey and Lahore during the 1874 Transit. An Indian astronomer of Orissa, Pathani Samanta Chandrasekhar predicted the 1874 Transit working solely within the Indian Traditional Astronomical system. He also observed the Venus transit of 1874. He was a Siddhantic Astronomer, un-influenced by western Astronomy. His works are summarised by him in 'Sidhanta Darpana', published in 1899, by Calcutta University. The original manuscript of 2500 Sanskrit shlokas was written by him in Oriya script on palm leaves. Chindamani Raghunathachari of Madras observatory, and Ankitam Venkata Narsinga Rao also observed the transit of 1874 from Visakhapatnam. In 1874 Narsinga Rao published a pamphlet on the Transit of Venus.

1.5 Some early solar observatories

The French astronomer Pierre Jules Ceśar Janssen was perhaps one of the first to realise in 1868 that reducing the thickness of air through which solar light passes should be advantageous for making solar observations. He suggested that a high mountain site with very few water molecules and negligible dust content would be helpful in reducing the extinction or absorption of solar light. Janssen was appointed the director of the new astrophysical observatory established in 1875 by the French government at Meudon, and took a bold step to build a solar observatory on top of Mont Blanc at an altitude of 4809 metres in 1893.

A British civil servant, William Petrie and trained surveyor-astronomer Michael Topping established an astronomy observatory in Egmore, Madras, in 1786. In May 1882, the British government astronomer at Madras, India, Norman Robert Pogson proposed the need for observing the Sun. A committee, chaired by Lord Kelvin, decided to establish a solar physics observatory at a more suitable site, and shifted the observatory from Madras to Kodaikanal in 1899. Some of the early instruments are still in operation there [30]!

Since 1858, a daily photographic record of the solar disk began at Kew in England using a solar telescope especially designed for photography. Mount Wilson Solar Observatory (California, U.S.A.) was founded in 1904 by George Ellery Hale who built a 60-foot tower telescope in 1908, and a 150-foot tower telescope in 1912. It was a remarkable solar telescope located at an altitude of 1740 m, based on the concept of high mountains, and combined with a high tower platform to reduce the ground heating effects.

Long time-series of daily sunspots are available from many observatories, such as Royal Greenwich Observatory, Kodaikanal Observatory, (1904-present, [339]), Mt. Wilson Solar Observatory (1917-1985, [181]), Kanzelhöhe Solar Observatory (Hungary), and several other observatories in Russia, Italy, and Belgium. Nearly a century-long continuous record of solar observational data from Kodaikanal Solar Observatory was combined with the data from Mt. Wilson Solar Observatory, to study long-term variations of various solar parameters. More recently, 24-hour coverage of the Sun is being carried out from USAF-International Solar Optical Observing Network (ISOON) and from space-borne instruments.

1.6 Major discoveries post-Galileo: from the 17th to the early 20th century

1.6.1 The vanishing act of sunspots during 1645-1715

One of the most surprising discoveries was that few or no sunspots were observed during a long period of about 70 years from about 1645 to 1715 [317]. Two papers were published by Edward Maunder in 1890 and 1894 [260], [261], where he mentioned a prolonged sunspot minimum. In fact, his work was based on the earlier papers by Gustav Spörer, who had noted that during a 28-year period (1672–1699) the observations revealed fewer than 50 sunspots. This contrasts with the typical 40,000 to 50,000 sunspots seen over a similar 25-year sampling in modern times. This was not due to a lack of observations, as during the 17th century Cassini had carried out a systematic program of solar observations at the Observatoire de Paris along with many others.

Historical reconstructions of sunspot numbers also indicate that the dearth of sunspots during that period was indeed real. This period is now known as the *Maunder Minimum* [111], [113]. A simultaneous decrease in auroral counts suggested their connection with solar activity that also greatly reduced in this time period. The documented occurrence of exceptionally cold winters throughout Europe during those years may be causally related to the reduced solar activity, although this still remains a controversial topic [186].

1.6.2 Discovery of *invisible* radiations from the Sun

It was first shown by Isaac Newton in the 1660s that sunlight can be separated into chromatic components using refraction of sunlight through a prism. William Herschel extended Newton's experiment in 1800 by detecting the temperature rise in thermometers placed beyond the red end of the visible solar spectrum [173]. He concluded that invisible 'rays' existed beyond the red end of the solar spectrum (Figure 1.3), and named it *infrared* radiation. He further suggested that these were fundamentally not different from the visible light, and could not be seen simply because the human eye is not sensitive to them.

Herschel unsuccessfully attempted to detect such radiation beyond the violet end of the visible solar spectrum. Later, Johann Wilhelm Ritter (1776-1810) and William Hyde Wollaston (1766–1828) used an experimental setup similar to Herschel's, but instead of thermometers they placed a piece of photochemical paper soaked in silver chloride beyond the violet end of the spectrum. The blackening of the photo-sensitive paper demonstrated the existence of the invisible radiation at the shorter wavelength side too, the *ultraviolet* radiation [269].

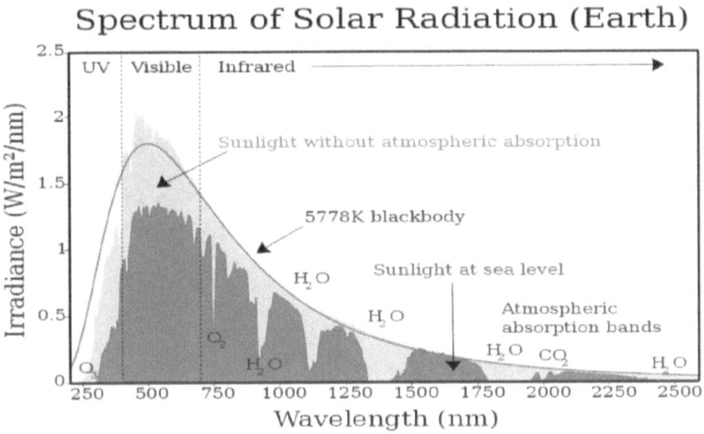

FIGURE 1.3: The solar electromagnetic radiation spectrum: lighter and darker shades show the flux at the top of the Earth's atmosphere, and at sea level, respectively, along with specific absorption bands due to the atmospheric gases. The continuous curve corresponds to a black-body radiation spectrum at T=5778°K. Figure adapted from Nick84 `https://commons.wikimedia.org/w/index.php?curid=24648395`

1.6.3 Solar spectral lines: the birth of spectroscopy

Following the dispersion of light, discovered by Newton's prism experiment, William Hyde Wollaston (1766-1828) investigated refractive properties of various transparent substances [396]. In 1802, he noticed dark lines in the spectrum of the Sun. But Wollaston did not pursue this matter further beyond suggesting that these dark lines marked the boundaries of 'natural colors'.

Joseph von Fraunhofer (1787-1826) rediscovered the 'dark lines' in the solar spectrum in 1817. He pursued these lines, named after him as *Fraunhofer lines* (Fig. 1.4). He used these lines as standard wavelength for determining the refractive index of optical glasses. Similar lines were being observed in laboratory experiments in the spectrum of white light passing through heated gases.

It was being realised by many that the Fraunhofer lines could be used to infer properties of the solar atmosphere. It was in 1859 that Gustav Kirchhoff realised it was possible to deduce the chemical composition of the Sun, and other stars by analysing the spectra of the light they emit. The Fraunhofer absorption lines were also discovered in stellar spectra, which were used to identify the elements present in stars. It was demonstrated that every absorbing atom or ion leaves its characteristic pattern of spectral lines at their specific wavelengths superimposed on the continuum spectrum.

The importance of these spectral lines in spectroscopic diagnostics was being slowly established. For example, the relative strengths of the absorption

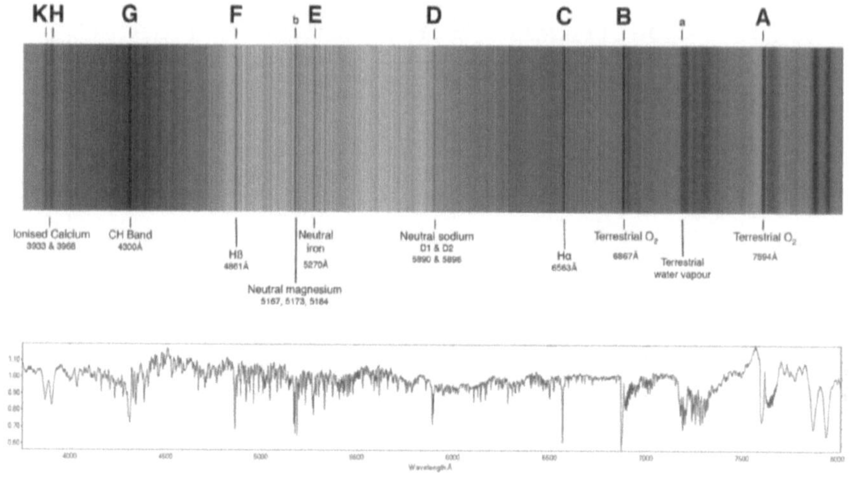

FIGURE 1.4: The visible continuum spectrum of the Sun along with some prominent absorption lines. Image credits- Creator:Paul Luckas.

lines were found to be indicative of the relative abundances of various elements. From the wavelength of the spectral lines in solar and stellar spectra, it became possible to determine the temperature, and velocity of the medium, and their chemical composition.

Similar to the absorption lines, emission lines are also produced due to the hot outer atmospheric layers above the solar photosphere, which are at much higher local temperatures. The emission lines are being used to infer the properties of solar outer layers, as we shall discuss in the later chapters.

This development led to the next major step in solar physics towards the birth of the science of spectroscopy. It revolutionised not only solar physics, but also astronomy as a whole. Even today, most of the information about the Sun, stars and galaxies is obtained through spectroscopy [269].

1.6.4 Photographic observations of the Sun

After centuries of naked-eye observations of the sky, and hand-drawings visually aided by early telescopes, the next big step came with the photography technique. The first successful photograph of the Sun's photosphere was made on 2 April 1845 by the French physicists Louis Fizeau (1819-1896) and Léon Foucault (1819-1868). Although Warren De Le Rue was not the first to photograph a solar eclipse, his eclipse photos of 1860 are still studied by many astronomers today (Figure 1.5a). Even in its infancy, it was pretty clear that photography would lead to a revolution for science. One of the first discoveries

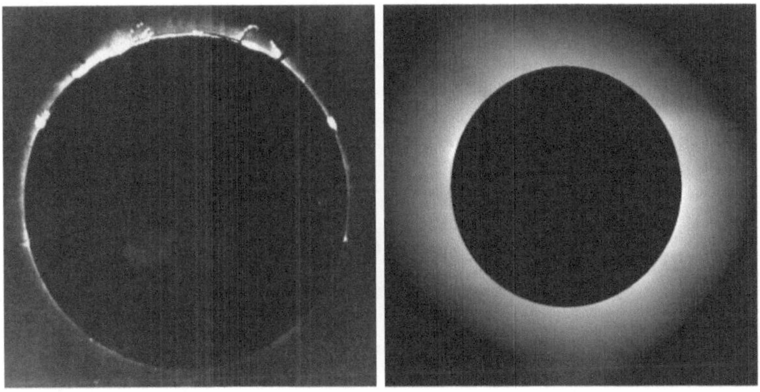

(a) (b)

FIGURE 1.5: (a) The photograph of a solar eclipse, taken by Warren de la Rue in Spain 18 July 1860. (Photo by Ann Ronan Pictures/Print Collector/Getty Images). (b) This eclipse image was taken by the PRL-USO Team on 29 March 2006 from Turkey. Solar eclipse images, like these, taken during 1919 in Brazil, helped prove the prediction of Einstein's theory of general relativity.

from Sun's photographs was the *limb-darkening effect*, a gradual decrease in intensity from the disk-centre to the limb or the edge of the Sun (Fig. 1.2b).

In 1915, Einstein announced his General Theory of Relativity that, in part, said light in space-time does not travel in a perfectly straight line, but bends in relation to an object's gravitational field. This was proved four years later, in 1919, using solar eclipse photographs, such as in Figure 1.5(b), when astronomers were able to observe and accurately measure the Sun's gravity-shifted positions of starlight on their way to the Earth.

1.6.5 Solar differential rotation

Johann Goldsmid a.k.a. Fabricius of Holland in 1611 was the first to publish his results on sunspots' apparent motion in terms of axial rotation of the Sun [275]. Christoph Scheiner, in 1630, noted that sunspots near the equator traversed the solar disk more rapidly than those near the poles.

Early 19th century solar astronomers were increasingly intrigued that solar rotation period obtained by tracking sunspots over the preceding two centuries varied from 25 to 28 days [112]. This puzzle was resolved by Richard C. Carrington (1826-1875) in England, and by Gustav Spörer (1822-1895) in

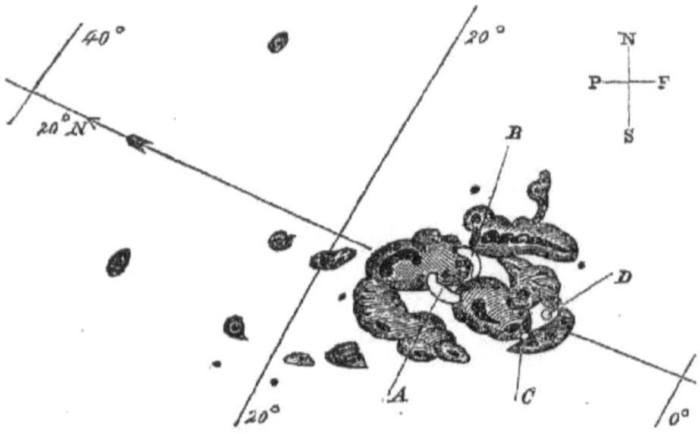

FIGURE 1.6: The drawing of the white light flare by R.C. Carrington on September 1, 1859.

Germany. In 1863, Carrington used sunspots to produce the first measurement of differential solar rotation. It found that the sunspots located at higher latitudes are carried around the Sun more slowly than the sunspots at lower latitudes. Thus, Carrington concluded that the Sun rotates '*differentially*'. This was yet another argument in favour of the gaseous nature of the Sun's outer layers (see also Section 3.3.7, page 61).

1.6.6 Discovery of solar flares

While making sunspot sketches on September 1, 1859/11:18 GMT, a solar flare was first discovered independently by two English amateur astronomers R. Hodgson and R.C. Carrington (Fig. 1.6). Coincidentally, it happened to be a very large and intense flare, a class of flares termed *white light flares*. These are $\sim$ 50% brighter than the solar disk. In fact, such flares belong to a rare class of flares that are bright enough to be seen in the white light solar photosphere.

It was reported that the geomagnetic instruments at Kew were disturbed on September 2, 1859; perhaps the first such report of solar-terrestrial relation! However, Lord Kelvin (William Thomson) 'proved' that the Sun could not possibly affect the Earth in such a way. Unfortunately, his view was to prevail for many decades, hampering the early progress of solar-terrestrial physics.

Nowadays, flares are routinely observed in optical wavelengths at ground-based observatories as sudden chromospheric brightening, using narrow pass-band optical filters centred at the Balmer absorption H_α 656.3nm spectral line.

1.6.7 Discovery of helium and the forbidden lines

Solar astronomers were particularly interested in spectroscopically studying solar prominences which were observed during a solar eclipse as colourful flame-like features jutting out of the Sun's limb (Figure 1.5a). French astronomer Pierre Jules César Janssen noticed a bright yellow spectral line (at 587.49nm) emanating from the chromosphere of the Sun, while observing the solar eclipse of August 18, 1868, through a prism. This observation was made from Guntur, Madras State of British India (now in Andhra Pradesh). This line, independently also observed by Sir Joseph Norman Lockyer, was termed as D3 in solar spectrum as it was observed close to the D1, D2 Fraunhofer lines of sodium. He also noticed that this yellow line actually did not match up with the wavelength of any other known element to that date.

Lockyer along with the English chemist Edward Frankland hypothesised this mysterious yellow emission line to be due to a new element on the Sun, and named it *helium* (derived from *helios* – the Greek word for the Sun). His paper arrived at the French Academy of Sciences on the same day as Janssen's paper, so both jointly received credit for the discovery of helium.

This was the first time a chemical element was discovered on an extraterrestrial body before being found on the Earth. Despite being the second most abundant element in the observable universe, helium is relatively rare, and was discovered only later on Earth.

An important outcome of this eclipse expedition was that it made Janssen realize that the chromospheric spectrum could be observed even during ordinary daylight conditions without a solar eclipse, provided a means could be found to filter out all but that wavelength of visible light. With this idea, Janssen came to invent the spectrohelioscope to better analyse the Sun's spectrum, and he proceeded on a bold project of establishing an observatory on the top of Mont Blanc (4808-m altitude).

From the spectrum of solar corona during a total solar eclipse (TSE) in 1869 another bright emission line was discovered which was initially attributed to a strange element called *coronium.* Its origin remained puzzling for over 70 years, and only after many laboratory experiments in the 1930s, it was identified as a forbidden spectral line corresponding to highly ionized Fe XIV 530.3 nm.

1.6.8 Magnetic field of sunspots - the Zeeman effect

In 1896, Pieter Zeeman discovered in laboratory experiments that individual absorption lines or emission lines in the electromagnetic spectrum of light were split into many lines in the presence of strong magnetic fields. It was attributed to the interaction between the magnetic field and the inherent magnetic moment of the atom, ion, or molecule.

George Ellery Hale in 1907-1908 first noticed the Zeeman effect in solar spectra also. From this fundamental observation, he suggested that sunspots were indeed the seats of strong magnetic fields [155]. This is discussed in later chapters in more detail.

1.6.9 The Evershed flow around sunspots

British astronomer John Evershed first detected that the spectral lines in sunspots showed Doppler shifts. He discovered this phenomenon in January 1909, while working at the Kodaikanal Solar Observatory in India [118]. The Evershed effect, named after him, is the radial outflow of gas across the photospheric surface of the penumbra of sunspots from the inner border with the umbra towards the outer edge (see also Section 3.3.5, page 60). It has been interpreted as a siphon flow along magnetic flux tubes driven from high pressure to low pressure end [272].

The discovery by Evershed of flows in sunspots, and that by Hale, just a year earlier of sunspot magnetic fields using the Zeeman effect in sunspots, arguably stand as the two most important contributions that seeded the development of a whole new subject of solar magnetohydrodynamics (MHD).

1.7 Determination of the Sun's general properties

The basic physical quantities related to the Sun, such as its mass, radius, luminosity, temperature, and distance from the Earth, *etc.*, can be determined from the fundamental laws of physics, combined with simple observations, as listed in Table 1.1.

1.7.1 Distance and diameter of the Sun

From simple observational considerations, the Sun's visible disk's average angular diameter θ is known to be around 32 arc-minutes. On the other hand, the average Sun-Earth distance $D_{\odot}$ (defined as the astronomical unit, or A.U.) has been determined historically from trigonometric parallax of Venus, as observed during its transits over the Sun's disk.

Even before the transit of Venus was used, the calculation of A.U. was attempted much earlier, in 1672, using the principle of parallax for simultaneous observations of Mars by Giovanni Cassini (in Paris) and Jean Richer (in French Guiana). They compared the apparent position of Mars relative to the background stars, and obtained a value of A.U. that differed by only $\sim 7\%$ from the modern value.

From transits, the distance from Earth to Venus is first measured. Then, Kepler's third law of planetary motion, relating the orbital period of a planet to its relative distance from the Sun, is used to determine the Sun-Earth distance. In 1771, the French astronomer Jérôme Lalande, based on the analysis of observations of Venus transits that occurred in 1761 and 1769, calculated a value for A.U. that was just 2% higher than its modern value.

Since 1961, a more direct method of radar signal has been employed that uses the time it takes a radar echo to return from another object. The average Sun-Earth distance, thus measured, is $D_\odot = 149,597,870.7$ km.

The distance from the centre of the Sun to its visible surface, the photosphere, is defined as the solar radius, although it is difficult to precisely define the edge of the Sun, as it is a gaseous sphere. In astronomy, it is used as a measure of the size of stars. It varies very slightly from equator to poles due to its rotation producing an oblateness of the order of 10 parts per million. The mean diameter of the Sun can be calculated as $d_\odot = D_\odot \times \theta \approx 1.49 \times 10^8 \times (0.5 \times 2\pi/360) \sim 1.3 \times 10^6$ km.

1.7.2 Mass and mean density of the Sun

Let us consider the Earth (having a mass $m_\oplus$), moving around the Sun (mass $m_\odot$) in a circular orbit, with velocity (v) at an average Earth-Sun distance ($D_\odot$). From Newton's Law of Gravity, a planet is held in orbit around the central massive star due to the gravitational force $F_g = (Gm_\oplus m_\odot)/D_\odot^2$ acting on it. Also, due to its orbital motion, a centripetal force acts on Earth, given by $F_c = m_\oplus a = m_\oplus(v^2/D_\odot)$, where a is the acceleration due to Earth's circular motion around the Sun.

These two forces must balance out for equilibrium, *i.e.*, $F_c = F_g$, which implies that $m_\odot = v^2/(GD_\odot)$. But, we have v = (the distance travelled by Earth in its orbit) /(Period of Earth's orbit) $= 2\pi D_\odot/P_\oplus$. Therefore, one can obtain the mass of the Sun from the resulting equation: $m_\odot = (v^2/GD_\odot) = (4\pi^2 D_\odot)/(GP^2)$. As $D_\odot$, P (one year for Earth) and G ($=6.67 \times 10^{-11} m^3 kg^{-1} s^{-2}$) are known, we can calculate mass of the Sun, $m_\odot = (1.989 \pm 0.003) \times 10^{30} kg$, which is nearly 99.8% of the mass of the entire solar system. From the mass and diameter of the Sun, one can also calculate its mean density, $\rho_\odot$, which turns out to be 1408 kg/m^3, which is around $0.255 \times \rho_\oplus$.

One must bear in mind that there is a minute change in solar mass due to continuous mass loss through the following processes: (i) loss due to radiation $\sim 4 \times 10^9 kg/s$ (derived from Sun's luminosity), and (ii) loss due to the solar wind expanding away from the Sun at the rate of another $10^9 kg/s$.

One can calculate the escape velocity v_e of a parcel of mass m moving out from solar surface from the condition that its kinetic energy must be equal or greater than the potential energy, *i.e.*, $mv_e^2/2 = Gmm_\odot/r_\odot$, where $r_\odot$ is Sun's radius. Thus, the escape velocity $v_e = \sqrt{2Gm_\odot/r_\odot}$ =618 km/s, and the gravitational acceleration on the Sun, $g_\odot = Gm_\odot/r_\odot^2$= 274 m/s^2.

1.7.3 Effective temperature, luminosity, and the solar constant

Effective temperature, T_{eff} is an estimate of the surface temperature of the star, which is assumed to radiate nearly as a black-body. The continuum emission of the stars (or the Sun) correspond essentially to the thin outer visible layer, known as the photosphere. Black bodies at higher temperatures radiate more intensity (or flux) at all wavelengths, while, the spectral shape of the radiation profile depends only on its temperature.

The effective temperature of the Sun can be determined by the peak wavelength λ_{peak} of its output radiation using Wien's displacement law for black-bodies, *i.e.*, $\lambda_{peak}T$ = constant = 2.8978×10^{-3}. This relation provides the effective temperature of the Sun $T_{eff} = T_{\odot} = 5700°K$, using its $\lambda_{peak} = 510 \times 10^{-9}$ m.

According to Stefan's law for a black body, the surface emissivity is given by σT^4, where σ is the Stefan-Boltzmann constant. Then, the effective temperature is related to the luminosity, $L_{\odot}$, of the star and its radius $r_{\odot}$ as:

$$L_{\odot} = (4\pi r_{\odot}^2)\sigma {T_{eff}}^4 = 3.844 \times 10^{26} W, \tag{1.1}$$

where, we have used T_{eff} = 5700° K, and Stefan-Boltzmann constant $\sigma = 10^{-8} W m^{-2} K^{-4}$.

This may be used to calculate the radiation flux outside the Earth's atmosphere (at the distance $R_{\odot}$= 1 A.U. from the Sun) integrated over all wavelengths, or the total solar irradiance (T.S.I.). This quantity, traditionally termed the *solar constant*, has been measured since the late 1970s directly, using radiometers aboard several satellites, as $S_{\odot} = 1367 \pm 4 W m^{-2}$ (see Section 4.9, page 109). It may be noted, however, that it is NOT a constant, as it has been varying from the early stages of the Sun over different time scales. Recent space observations also indicate that on the average, there is a variation of around 0.1% from the minimum to the maximum activity phases of the Sun.

1.7.4 The age of the Sun

A method for inferring the age of the Sun is based on radioactive dating of the oldest solar system objects, such as chondritic meteorites. The problem is, however, in precise relationship between the age of the meteorite, and that of the Sun. The other difficulty is to define the zero time for the Sun. From stellar evolution theories, the age of the Sun is usually referred to zero age main sequence as 4.57 ± 0.18Gyr.

1.7.5 The elemental abundances

Spectroscopic measurements of the solar photosphere provide the relative abundances, in mass, of elements on the Sun. The relative metal abundance $(Z/X) = 0.0233 \pm 0.0014$, where X is hydrogen abundance. This result is, in

TABLE 1.1: Some general properties of the Sun

Property	Value	Remark
Angular diameter θ	32 arc-min	from observations
Distance $R_\odot$	149.6×10^6 km (1 A.U.)	Kepler's law & parallax of Venus
Radius $r_\odot$	6.96×10^5 km ($109 r_\oplus$)	from geometry : d = 2r = $R\theta$
Mass $M_\odot$	1.99×10^{30} kg $\sim 333,000 M_\oplus$	from Kepler's 3rd law,
Av. Density $\rho_\odot$	1408 kg/m^3 (0.255 $\rho_\oplus$)	
Temperature $T_\odot$	$5800°K$	from Wien's law $\lambda_{peak} T_{eff} = const.$
Solar Constant $S_\odot$	1367 $Watts/m^2$	from space observations
Luminosity $L_\odot$	3.85×10^{26} Watts	from inverse-square law: $L = 4\pi R^2 S$
Average Mag. field	1–3 Gauss	polarimetry & Zeeman effect
Composition	74 % H, 24 % He, 2% other (by mass)	from spectroscopy
Solar rotation	28 days (at equator) 37 days (near poles)	from observations
Present age	4.57×10^9 yrs	meteorites and theory

general, consistent with meteoritic abundances. However, there is a remarkable exception that solar lithium abundance is depleted with respect to the meteorites.

1.8 The extended solar system

The solar system consists of a variety of objects, including our Earth, inner planets like Mercury and Venus, the asteroid belt object, minor planets, the outer Jovian planets, such as Jupiter, Saturn, Neptune, Uranus, and even comets occasionally making their transit around the Sun, and returning back to where they came from. Mercury is only 0.39 A.U. from the Sun, while Jupiter is located at around 5.5 A.U., and Pluto, now removed from the list of planets, is farther out at 39.2 A.U. Our solar system extends much beyond Pluto, to other similar minor planets such as Eris, Makemake and Haumea, that are part of *Kuiper belt objects*, extending from $\sim$ 30 A.U. to 50 A.U. (Figure 1.7).

Beyond the Kuiper belt, at about 80-200 A.U. is the region where solar wind from the Sun meets with the interstellar medium, known as the *termination shock* (Section 9.6). But this is not yet the boundary of our solar system. Even beyond, there is a theorised cloud, called the *Oort cloud* of icy objects, where Sun's gravity still operates. The Oort cloud objects are not directly

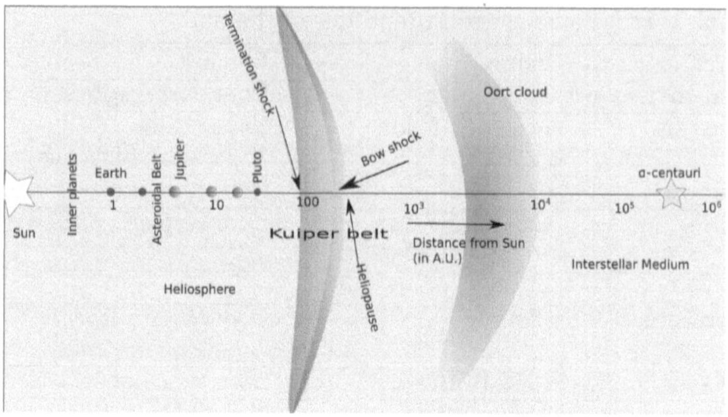

FIGURE 1.7: Extended solar system.

observable at this immense distance of ~100,000 A.U. (1.87 light-years). However, the long-period comets, that visit the Sun from time to time, are believed to originate from this region. The influence of Sun's gravity extends to a distance of about 2 light-years, *i.e.*, nearly half the distance from the Sun to the nearest star: *Proxima Centauri*, within which objects would probably be orbiting the Sun, and hence, still be a part of the extended solar system.

1.9 Birth and evolution of the Sun

A spiral galaxy like our Milky Way contains stars, stellar remnants, and diffuse interstellar medium (ISM) of gas and dust components. Higher density regions of the interstellar medium form molecular clouds, or diffuse nebulae, where star formations are observed to take place. Molecular clouds, also called stellar nurseries, are a type of interstellar cloud, the density and size of which permit the formation of molecules, most commonly molecular hydrogen. Such clouds possess low temperatures of the order of a few tens of degrees K, and densities of a few 10^2 particles per cm^3, masses ranging from a few to over a million solar masses, and diameters from 20 to 200 parsecs (1 parsec ≈ 3.26 light-years). They are located mostly in the disks of spiral galaxies, like Milky Way, and irregular galaxies.

The Sun and the rest of the solar system are believed to have formed from a rotating cloud of gas, and dust called a solar nebula, about 4.5 billion years ago. As the nebula collapsed because of its overwhelming gravity, it spun faster and flattened into a disk. Most of the material was pulled toward the centre to form our Sun, which accounts for 99.8% of the mass of the entire solar system.

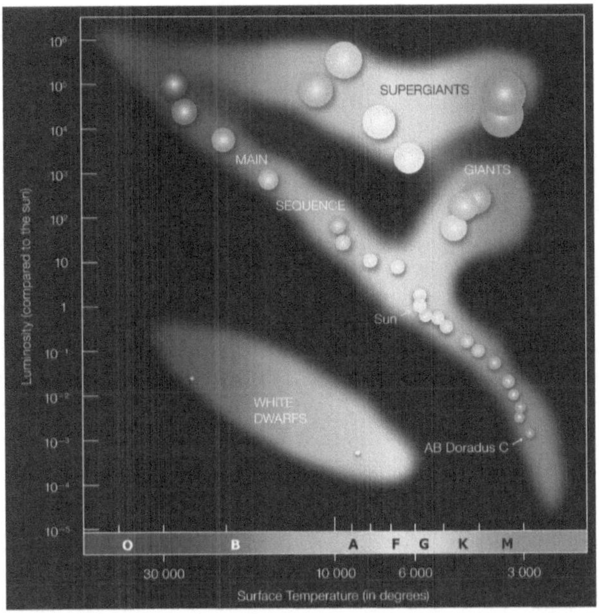

FIGURE 1.8: Classification of stars: The Hertzsprung-Russel diagram. By ESO- `https://www.eso.org/public/images/eso0728c/`

Most stars are classified by their temperature (spectral type) from hottest to coolest as follows: O, B, A, F, G, K, M. These categories are further subdivided into subclasses from hottest (0) to coolest (9). The scatter plot of stars with their spectral class or the effective surface temperatures against luminosities or magnitudes, shows well-defined, distinct groups in which the stars are populated. This is the Hertzsprung-Russel diagram, or H-R diagram shown in Figure (1.8). The stars are clearly grouped in: (1) a *Main Sequence of stars* extending from the top left to bottom right, (2) a horizontal branch of giants and supergiants at the top right corner, and (3) white dwarfs at the bottom side of the main sequence. From the location of a star on the HR diagram, its luminosity, spectral type, colour, temperature, mass, age, chemical composition and evolutionary history are known.

Spectral lines of stars can show different characteristics within the same spectral type, or temperature. Therefore, another type of classification system for stars is devised using the luminosity. The differences in spectral lines among stars having the same spectral type are a function of the radius of the star. This results in different luminosities (L), related to the absolute magnitude (M_V) of a star. Luminosity is proportional to the fourth power of the temperature (T^4), according to the Stefan-Boltzman law assuming that stars radiate as black-bodies. Thus, two stars with similar effective temperatures, but different luminosities must differ in size, and belong to different luminosity classes within their spectral type, as determined from their spectra. There is a wide variation in stellar luminosities ranging from one million times more luminous to a ten-thousandth of the luminosity of the Sun.

The collapse of molecular clouds results in the proto-star phase, where nuclear burning in the central compressed core of the star begins. It then evolves towards the main sequence. This journey of proto-stars to the main sequence depends on their mass: low mass proto-stars contract more slowly. For example, a proto-star with a mass of 0.5 $M_{\odot}$ would take around 150 million years to contract, and evolve to the main sequence, while a 15 $M_{\odot}$ proto-star would take only around 60,000 years to do so. Similarly, the duration that a star spends on main sequence depends on the ratio of mass and luminosity, or, the amount of hydrogen in the star, and the rate at which it is consumed. The more massive a star is, the shorter is its main-sequence life time.

Like all stars, the Sun will some day run out of its energy as hydrogen in the core is consumed resulting in a helium core, and hence, thermonuclear reactions would cease. Therefore, the outward thermal pressure opposing the inward gravitational pull would slowly diminish, and when the Sun starts to die, it will swell up as a red giant, so big that it will engulf Mercury, Venus, and even Earth. It is predicted that the Sun is a little less than half-way through its lifetime, and will last another 6.5 billion years before it finally shrinks down to end up as a white dwarf.

1.10 The Sun as a star in the Milky Way

The Sun is a modest, average sized star, which is located at the centre of our solar system, holding planets and other members of the solar system together by its gravity. The Sun, and everything that orbits it, is located in the *Milky Way galaxy* consisting of a central bulge, 5 major spiral arms, over 400 billion stars, dust and gases. Our Sun has got no special position amongst all these stars. In fact, it is located far out at around 28,000 light-years (or, 8.6 kilo-parsecs) from the Milky-Way centre, near the outer edge, in a local spiral arm called the Orion Spur. In size and temperature too, our Sun is an average star compared to many other stars.

Along with all the members of the solar system, our Sun orbits the centre of the Milky Way galaxy, moving with an average velocity of ~828,000 km/hr, and takes around 230 million years to make one complete orbit around the Galactic Centre (G.C.), the location of a humongously massive black hole. Notwithstanding these modest aspects about the Sun, the fact remains that the Sun is our closest star, and the governor of our solar system. It is a great deal easier to study its surface in detail, that is not achievable for any other star. In fact, our next nearest star, Proxima Centauri, is about 250,000 times as far, therefore, it is difficult to resolve its surface even with the largest telescopes.

1.11 Summary

The Sun has played a major role in the development of mathematics and physics through the past several centuries. Sun's interior and exterior provide for an ideal cosmic laboratory for atomic, nuclear and neutrino physics, high temperature plasma physics, magnetohydrodynamics (MHD), and general relativity.

The Sun had been observed with a lot of interest even before the invention of the telescope in 1610. Solar eclipses, planetary transits and sunspots were the early subjects of careful expeditions and observations from the 17th century onwards. Some major discoveries were made, such as the Maunder Minimum, invisible radiations from the Sun, photographic observations of eclipses, solar differential rotation, and solar flares. With the advent of spectroscopy, another dimension was added, which led to the discovery of Fraunhofer absorption lines in the solar visible spectrum, helium and the forbidden lines in coronal spectrum, presence of solar magnetic field by Zeeman splitting and the Evershed flow around sunspots using the Doppler effect of spectral lines, *etc.*

The photosphere is only a very thin outer 'skin' of a much larger Sun extending downwards to the hidden interior, and upwards to outer atmospheric layers which are much too faint to be directly observed. The photosphere and its features are easily observed using even modest aperture solar telescope suitably aided with optical solar filters. The spectral lines are essentially the 'thumb prints' of the elements present in the solar atmosphere. The photons emitted continuously from the Sun are the messengers carrying information about Sun's structure and dynamics. Collecting these photons, and using known principles of physics, one can determine physical conditions that exist in the photosphere. From the very limited observations of the visible (optical) disk of the Sun, its global properties, such as mass, diameter, distance, effective temperature, elemental abundances, age, were inferred from the general laws of physics. The extended solar system goes out much farther than the outer planets. The Sun as a star, its birth and evolution, its place in the Milky Way galaxy are other major topics of interest.

The Sun is the source of all forms of energy which sustain Earth's weather and climate. There are several manifestations of the Sun which unfold to us at different spatial and temporal scales, such as the quiet Sun, dynamic Sun, active Sun, and even stormy and explosive Sun. The interplanetary space, in which the planets are immersed, and even planetary environments are affected by these various facets of the Sun. Study of these aspects has developed as an important field, *space weather*, which has a bearing on many applied fields related to space sciences, satellite applications, communications, *etc.*

Chapter 2

Observing the Visible Sun

2.1 Introduction

The Sun radiates energy over the entire electromagnetic spectrum from the shortest wavelength γ-rays to the longest radio waves. The Sun's emission peaks in the visible, at the yellow wavelength, with a continuum extending from about 400 to about 750 nm. Most of the solar energy is contained in the visible and near-infrared parts of the spectrum (Figure 1.3).

Although the Sun, and other stars, are not perfect black bodies, their continua radiations are very closely approximated by the blackbody spectrum. One can define an effective temperature T_{eff} of the star as the temperature of a blackbody of the same size as the star, that would emit the same total power (see Section 1.7.3, page 18).

The Sun's photosphere, as seen in the visible light, defines the Sun as a sphere with only a very limited aspect of a much more extended Sun both within and outside it. It hides underneath it the interior which includes the central core, radiative, and convective zones. Above it, the very extended atmosphere includes the usually invisible chromosphere, transition region and the corona. The physics of its outer atmosphere is much different than the physics involved in the solar interior and the photosphere.

Solar radiation propagates from the Sun's photosphere, through the interplanetary medium, nearly unhindered to the Earth. But, as it passes through the Earth's atmosphere, it is absorbed partly, or completely, at different wavelengths by a number of atomic and molecular processes. The high energy EUV, and X-ray photons from the Sun lead to the ionization of the atmosphere, resulting in producing the Earth's ionosphere, while, the UV component is absorbed by the ozone layer, and re-emitted as heat, thus heating the stratosphere. A part of this heat is radiated back to outer space, while the rest is transmitted downwards to the Earth's surface. The consequent heating of various layers of Earth's atmosphere leads to atmospheric convection currents and turbulence.

The low-energy photons at the longer wavelengths, *i.e.*, the infrared component of solar radiation, are also absorbed by atmospheric molecules, such as CO_2 and water vapour. In fact, only a fraction of flux in the near IR range can manage to reach ground-based observing instruments, particularly, at high altitudes. Thus, the Sun is observable from ground only through the optical and radio wavelength windows, and to some extent, in the near-IR from high altitude sites. In this chapter, we discuss solar optical observation techniques and instruments employed for analysis and diagnostics. We shall discuss radio observations in Sections 7.4.1 and 7.5.6.

2.2 Ground-based solar optical observations

Due to the Sun's proximity, it is possible to resolve the surface features on the Sun's outer visible layers even with modest aperture telescopes, whereas the distant stars appear only as point sources. However, a major limitation in observing arises due to the Earth's atmosphere. Significant absorption and scattering effects can result due to the presence of atmospheric gaseous and dust particles, reducing the intensity of light, degrading the image quality, and even changing the wavelength distribution of the radiation. Therefore, an observing site should be as free of atmospheric dust and cloud coverage as possible. High-altitude mountain sites with very low dust and water vapour contents located above the inversion layer are more suitable for observing the Sun or night time astronomical objects.

A major concern about observing through the atmospheric layer arises from the fact that it is not static and stratified uniformly. It continuously changes due to air parcel motions resulting from thermal turbulence at short temporal and spatial scales. In particular, the ground first gets heated during the day, which then produces thermal convection and turbulence in the air. This results in a continuously changing optical refractive index of the air parcels crossing through the line-of-sight from the Sun to the solar telescope. The atmospheric conditions thus potentially limit (i) the maximum resolving power of a telescope, and (ii) the detectability of the faintest features. Therefore, the image of a point-source, such as a star, would appear to 'dance' randomly around its mean position along with changing intensity, resulting in the 'twinkling' of stars. This effect is more pronounced for stars located at lower altitudes.

For meeting the science goals which require long exposures, these atmospheric scintillations lead to severe degradation by blurring of images. This effect also operates on extended sources, like the Sun, and is a major concern for observing the Sun at high resolutions. A similar effect operates on radio waves too, not in the Earth's atmosphere, but while traversing through the interplanetary plasma medium. The effect is called the *interplanetary scintillations*, and occurs due to the density fluctuations in the interplanetary plasma.

The spatial resolution of fine scales structures in extended objects gets limited due to the 'blurring', or reduction in contrast, when the image is

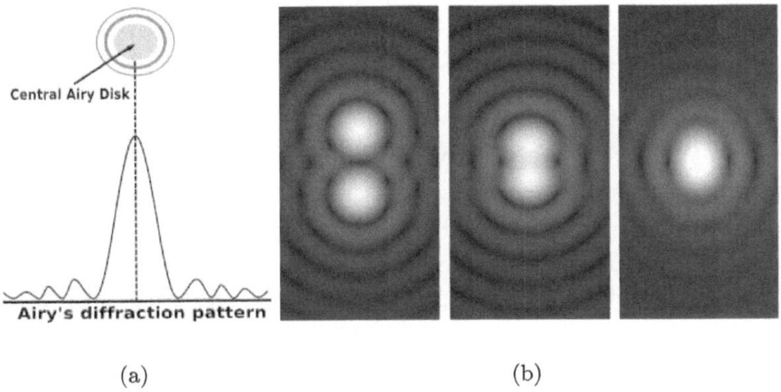

FIGURE 2.1: (a) Airy diffraction pattern for a point source, and (b) diffraction patterns for two point sources passing through a circular aperture. Points far apart can be distinguished, but sources closer than the Rayleigh criterion are difficult to resolve distinctly.

acquired over a time longer than the time over which atmospheric fluctuations occur. Furthermore, it is required that the S/N ratio of the light flux, or the intensity at the wavelength of interest should be adequate. In order that the telescope could achieve its expected spatial resolution to the extent possible, the site's 'seeing' condition should be good.

2.2.1 Diffraction-limited spatial resolution of a telescope

When a point source, like a star, is imaged by a telescope, a circular diffraction pattern, and not a sharp point image, results at the image plane. This arises due to the diffraction of the plane wave-front arriving from a point source, located at infinite distance, by the telescope's aperture. This is known as the 'Airy pattern' in which the central bright spot of the image contains around 84% of the light, with the intensity of surrounding bright rings rapidly decreasing (Figure 2.1a).

The radii of the minima are given by the zeros of the Bessel's function of order one. The size of the Airy's disk limits the resolution of the telescope. Two point sources can only be resolved, if the Airy's disks corresponding to them are adequately and distinctly separated (Figure 2.1b). The Rayleigh criterion for resolution requires that the central maximum of one source should lie on the first minimum of the other, and defines the theoretical resolution as the first minimum of Airy's disk, α in radians:

$$\alpha = \frac{1.22\lambda}{D}, \tag{2.1}$$

where λ is the wavelength of the light and D is the diameter of the telescope's aperture. This is usually referred as the 'diffraction limited' resolution of a telescope, valid for small aperture (say, $D \lesssim 20$ cm) telescopes.

2.2.2 Atmospheric 'seeing'

The diffraction limit of larger aperture telescopes encounters an additional limitation caused by the *atmospheric seeing* of a site. The effective spatial resolution is then called 'seeing limited', which is independent of the size of the telescope beyond a certain aperture. In fact, optical telescopes are less seeing limited, when they are installed at high altitude mountain sites, or flown in balloons, and 'not seeing limited' when used from satellites located outside the atmosphere. Of course, the telescope must be structurally and optically free from any distortions and aberrations in order to obtain the best results.

A quantitative measure of atmospheric seeing is given by the Fried Parameter r_0 which is defined for long exposures [323]. It corresponds to the aperture of an imaginary diffraction limited telescope, which has the same resolving power as a large, optically perfect telescope used under the condition of the actual seeing. As the seeing is a constantly varying parameter at a site, r_0 depends on the atmospheric stability of the observing site, time of the day, and also on the observing wavelength λ. For example, the site test results for the Canary Islands have frequently shown $20cm \leq r_0 \leq 30cm$, and at times r_0 even as large as 50 cms.

A one meter aperture telescope has a diffraction limit of ≈ 0.1 arc-sec at a wavelength of 500 nm. However, its resolving power gets seeing limited due to smaller r_0 values even at better sites. Thus, exceptionally good and stable 'seeing' conditions are required to improve the resulting resolutions. Another major requirement for setting up a large ground-based solar optical telescope is ascertaining the suitability of the site for observability over a large number of days in a year, or the duty cycle, and number of hours on a given day. Therefore, careful site surveys over long periods are important pre-requisite for solar site selection.

2.3 Site survey and selection

A good solar site should possess good 'seeing' conditions, which implies not only good image quality, but also good atmospheric parameters. To meet and ascertain these considerations, extensive site survey programs are carried out at potential solar sites with suitable instruments including meteorological and weather stations.

A potential site for extensive survey is usually selected based on: (i) annual sunshine hours, (ii) altitude of the site, (ii) location in or near a large lake, (iii) an island located in sea/ocean, *etc.* It is generally believed that lake sites reduce the degrading effect caused by atmospheric turbulence in the immediate surroundings of the telescope. In fact, the beneficial effect of lakes on daytime solar seeing was already reported over a century ago by John Evershed.

The quality and stability of atmospheric seeing are monitored at the potential site using various instruments, such as weather station, dust counter, sky brightness monitor, and solar scintillation monitor or *seeing monitors*. Various parameters are collected over long periods, such as the cloud coverage, wind speed, wind direction, relative humidity, pressure and temperature.

In order to determine an estimate of the near ground temperature gradient, temperature sensors are placed close to the ground, and also at various heights above it. A dust counter monitors the aerosol particles counts and sizes to evaluate the transparency of the site. A sky brightness monitor is used to measure the brightness away from the solar disk at different wavelengths.

A solar scintillometer determines the amount of atmospheric turbulence at and near ground level. Such measurements are useful in determining the quality of the atmospheric seeing required for high resolution solar observations. Commercial solar scintillation monitors are available, off-the-shelf, for daylight seeing analysis. These instruments instantaneously determine the seeing in arc-sec, and its variation over the day. Special seeing instruments were designed [38], which use the image of the Sun's limb (Solar Differential Image Motion Monitor, or S-DIMM), or measure the localised variations of refractive index of the atmosphere that produce intensity fluctuations (SHAdow BAnd Ranging, or SHABAR).

The CalTech survey conducted visual observations at several sites in California after which the Big Bear Solar Observatory(BBSO) site in California, U.S.A., was selected ([409]). For a better quantitative understanding of the behaviour of the optical seeing quality, solar scintillometers were used to test the quality of a number of lake sites [39]. They reported mountain sites to have systematically more scintillation than the lake sites. The lake sites had low scintillation over nearly the entire day, whereas it increased during the day on the mountain sites. This behaviour obviously resulted from much lower heating at the lake sites during the day as compared to the mountain sites.

From the experience of BBSO, a similar site survey was conducted at many lake sites in Rajasthan (India) during 1972-75, and the island site of the Udaipur Solar Observatory (USO), Udaipur, India, was selected (Figure 2.2). It was established in 1975 in the middle of a large artificial lake. From these site surveys, it is believed that lake/ocean sites, combined with high altitude, offer several benefits, such as temperature stabilization as a large water body reduces convection and produces laminar air flows around the telescope. There are many existing solar observing sites located around the world based on these considerations.

FIGURE 2.2: The island site of Udaipur Solar Observatory, Udaipur, India. The 50 cm-aperture Multi-Applications Solar Telesecope (MAST) that became operational in 2015 is housed under a collapsible dome seen at the right.

2.3.1 Building and dome designs

Apart from the site selection, the other important aspects are the structural designing of observatory building, dome, and the design of the telescope. To avoid the surrounding air turbulence developing as the Sun heats the ground, solar telescopes are usually installed on top of high towers. Special white paints are used on the external walls of the observatory building to reduce unwanted heating, and avoiding generation of thermal turbulence around the site.

Open, collapsible, or retractable domes are often used for housing solar telescopes, so that the thermal turbulence of air produced in the observing floor of the telescope are flushed out, and not allowed to build up around the observing platform. For instance, the Dutch Open Telescope (DOT) is built on an open framework to allow the wind to pass through the structure and provide cooling around the telescope's main mirror. A collapsible dome is used for housing the 50 cm-aperture Multi-Applications Solar Telescope (MAST) of the Udaipur Solar Observatory, India (Figure 2.3a).

Sufficient care is needed to avoid direct sunlight falling on the telescope's mechanical structures and the optical elements, which may cause their heating and create turbulence locally and within the telescope itself. This is usually achieved by employing a large shield at the front end of the telescope, having a diaphragm to allow the Sun's rays to reach the primary aperture. This shield covers the telescope support structure from direct sunlight throughout the observing period while tracking the Sun.

(a)

(b)

FIGURE 2.3: (a) The collapsible dome of MAST, showing the open structure of the solar telescope on the top floor. (b) The sunlight is sent downwards to the backend instruments, imager, spectrograph and polarimeter, on separate optical benches located below on the observing floor.

A major difficulty, specifically related to large solar telescopes, is the large heat load which is produced at the prime focus. The heat load problem becomes particularly more severe in telescopes having short focal lengths, in addition to large apertures. This generates convection of air leading to internal thermal turbulence within the telescope body, and also produces distortions (and even burnout damage) of the optical elements involved. This requires rejection of a large part of sunlight out of the telescope, using a heat stop placed before the prime focus.

Another option which has been used to reduce the telescope's internal heating is to use vacuum or inert gases, like, a helium filled interior through which the sunlight is directed. This reduces or even eliminates the air turbulence inside the telescope. However, the option of vacuum does not work for large telescope apertures (over 1 metre) as the pressure difference at the entrance window of the vacuum tube becomes too large and produces stresses. Therefore, more recent solar telescope designs use completely open structures, and the hot air is flushed out of the optical path. Active cooling of the mechanical structures, and the dome are done to minimise the temperature difference between the air inside and outside the telescope.

2.4 Solar optical telescopes

The main component of any telescope is the primary 'objective' lens (in a refracting telescope), or a mirror (in a reflecting telescope). Lens objectives, normally doublets to correct for the chromatic aberration, are generally preferred in smaller aperture solar telescopes. The light thus collected is then passed through a suitable 'analyser' such as a photometer, imager, spectrograph, or a polarimeter. Finally a 'detector' records the analysed product simply visually, or on photographic plate, digital recorder, *etc.* Currently, large format electronic CCD devices have replaced the old, traditional detectors.

It is preferable to use a reflecting (mirror), instead of refracting (lens) primary or objective component in a large telescope, which eliminates many problems. There are various configurations of reflectors having advantages over refractors (Figure 2.4). Firstly, only one (reflecting) surface of the mirror needs to be ground with precision, unlike both the surfaces in a lens. The other major difficulty involved is the weight of a large aperture lens, whereas the weight of a large mirror can be minimised by using a honeycomb structure for the rear part of a reflecting surface. A lens needs to be supported at the edges, which can develop distortions due to the stress caused by its weight. On the other hand, a reflecting surface can be supported from behind which avoids weight-related distortions. Furthermore, it is possible to eliminate it even if some distortions result in the mirror's shape by thermal and mechanical effects as the telescope tracks the Sun, by designing an active system of pressure pads at the rear.

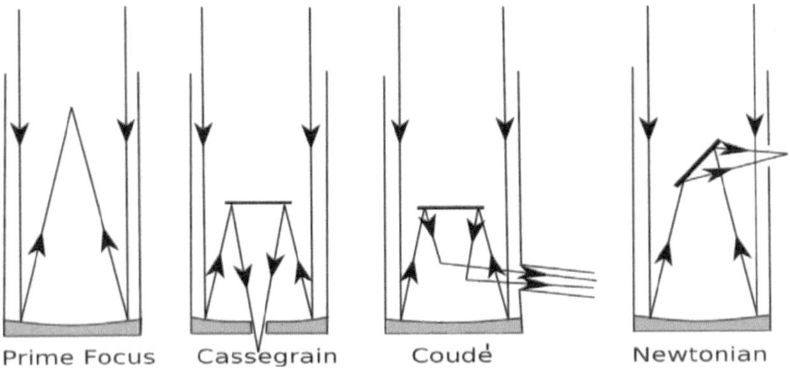

FIGURE 2.4: Schematics of various types of telescope optics.

Traditional solar telescopes used main optical elements of modest apertures with very long focal lengths to achieve large magnification of the solar image, and also to reduce thermal load at the prime focus. However, this is very difficult to operate for larger aperture telescopes, or for very long focal length instruments that require unmanageably large mechanical structures. Therefore, preferably 'fixed' optical platforms are used where sunlight is diverted for analysis. Thus, solar telescopes are broadly divided into two classes: fixed, and steerable telescopes, that essentially correspond to long and short primary focal length, respectively.

2.4.1 Telescope mounts

Professional telescopes require rigid and stable mounts, along with tracking ability to compensate for the effect of the rotation of the Earth around its axis. In 'equatorial' mounting, the telescope can move around its polar axis (to be aligned parallel to the Earth's rotation axis). There is also a declination axis at right angles to the polar axis to point the telescope at the desired source in the sky, such as the Sun, or any other star. The telescope is then driven around its polar axis to track the source without any significant adjustment needed in the declination axis. However, as the required polar and latitude alignments of the telescope are never perfect, a finer adjustment in both the axes is usually required, which can be carried out manually or electronically using a servo-controlled guider.

Equatorial mounts are only suitable for modest aperture telescopes. Large telescopes are associated with massive optical components, and mechanical structures associated with large torques. The weight of the telescope can cause bending of the telescope and its axes. A solution is to support the declination axis at both ends, using a fork mounting. This fork mount is then mounted on a disk or a horse-shoe. However, some residual flexure problems remain due to the variation of stresses on the axes as the telescope moves with time of the day.

With the availability of fast computers for telescope control, it has become possible to drive around the two axes simultaneously for tracking the source. Larger telescopes now usually adopt *alt-azimuth* designs, where the main support axes are always vertical and horizontal. The stresses on the axes do not change much with the direction the telescope is pointing. In addition, an 'active' optics arrangement is usually employed to adjust for the primary mirror shape to maintain sharp images during the time scales of a few minutes. This allows correction for small tracking errors, and for changes in mechanical structures due to slight bending due to gravity as the telescope is driven.

2.4.2 Heliostats, coelostats, and siderostats

The difficulty involved in large focal length telescopes can be alleviated by having a *heliostat* arrangement as the light feed, where only a single mirror is moved to track the Sun. The rest of the equipment is mounted on a stationary platform. The mirror is placed at a height along the Earth's polar axis in a system called polar heliostat, and the sunlight is reflected by this mirror and is diverted onto an objective lens or a mirror. The drawback is that this system produces a rotating solar image with time, which needs to be corrected. There are various ways to compensate for the rotation of the image, either by rotating the observing table itself, or by using an image rotator.

The examples are the 150-foot tower telescope at Mt. Wilson Solar Observatory, southern California, U.S.A., and the Richard B. Dunn Solar Telescope (DST), located at Sunspot, Sacramento Peak, New Mexico, U.S.A.

DST from top to bottom is a single instrument, where the tower portion rises 41.5 metres above the ground level, while the rest is 67 metres underground (Fig. 2.5a,b). The telescope's entire optical system, along with the 12 metre-diameter observing floor, is suspended from the top of the tower by a mercury float bearing. The entire optical and mechanical structure of the telescope weighs over 250 tons. The DST is equipped with a wide range of instruments. On the observing room floor, a 51-centimetre-diameter image of the Sun is produced for detailed studies.

To overcome the problem of image rotation produced in a single mirror tracker, heliostat, a *coelostat* arrangement is often used. This arrangement employs one mirror to track the Sun that reflects the sunlight to another 'fixed' mirror, which then directs the light to the objective lens or mirror system to form the solar image. As the face of the tracking mirror lies in and rotates about the polar axis, a coelostat configuration gives a non-rotating image in a fixed direction.

A 'siderostat' arrangement uses a single mirror that is alt-azimuth mounted, and the movement of the mirror is transformed through a mechanical gear arrangement into an equatorial motion. In all these systems the

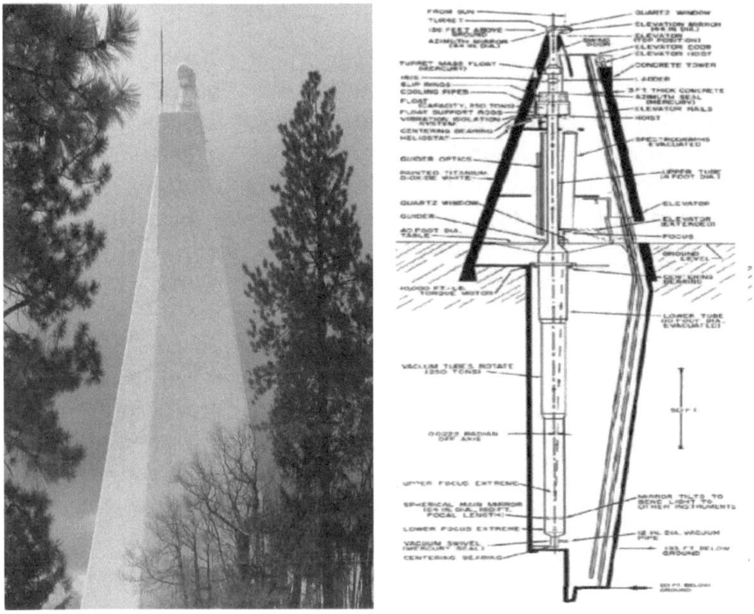

(a) (b)

FIGURE 2.5: (a) The Richard B. Dunn Solar Telescope (DST) at Sacramento Peak Observatory, New Mexico, U.S.A. The turret is seen on the top of the tower, which tracks the Sun. (b) The schematic of the optical path of the DST. Figures received with thanks from SacPeak Observatory.

advantage is that the image plane is fixed. Therefore, large and heavy back-end instruments can be placed on separate platforms than the telescope itself.

To track the Sun in all these mirror based systems, the rotation rate of the tracking mirror is required at once in 48 hours. In both heliostat and siderostat arrangements, the solar image rotates once in 24 hours around its centre, while in a coelostat the image does not rotate and is fixed. Therefore, coelostats are generally preferred, but have the disadvantage of using two mirrors.

2.4.3 Active and adaptive optics

Night time astronomy deals with very faint objects, which necessitates use of large telescope apertures and long integration times in order to collect adequate numbers of photons. The Sun is a relatively immensely bright object, and usually it poses problems related to the extreme of heat loading which needs partial rejection, and/or reduction of intensity by use of neutral density optical filters.

There are, however, observational requirements based on certain science goals that necessitate observing the Sun in very narrow wavelength bands around the spectral lines of interest. For observing dynamic solar transients, it is also required to observe at high temporal rates. In addition, simultaneous observation and analysis by multiple back-end instruments of the telescope requires diverting the sunlight and thereby reducing the intensity in each of the instruments. These observing requirements, and the need for high spatial resolutions, eventually make even the large aperture solar telescopes photon-starved.

To some extent, the problem of a shortage of photons in modest solar telescopes is obviated by longer integration times, but the atmospheric degradation and inaccurate tracking can make very long integrations futile. Therefore, large aperture telescopes in solar physics become unavoidable. There are several such ongoing solar projects that have proposed large apertures. It has become possible to build large aperture telescopes due to improvements in telescope design, combined with fast computers. Modern solar telescopes such as DKIST propose to employ large apertures of several metres (Section 10.2.1 on page 253).

While large aperture telescopes collect more photons than a small aperture telescope over the same time interval, they are usually not able to achieve the expected higher spatial resolutions, which is limited by the 'seeing' quality of a ground-based site due to atmospheric turbulence (Sections 2.2.1 and 2.2.2). For example, even a 10-m telescope, such as the Keck telescope, located at a very good site having exceptional seeing does not fare better than a 20-cm telescope unless it is aided with *active optics* to correct for the telescope's mirrors, and *adaptive optics* to compensate for atmospheric turbulence.

The reflecting surface needs to be distorted to match the distorted wavefront caused by the turbulent atmosphere at a very fast rate. It is not possible to adjust the shape of the primary mirror so rapidly. However, the overall motion of the image can be removed by tilting the secondary mirror. The waveform distortion due to effects of atmospheric seeing is corrected by re-imaging the telescope's exit pupil onto a much smaller 'active' mirror (membrane) which can be distorted on the short time scales involved. The wavefront distortions are monitored by using a reference image of the object being observed [36].

Adaptive optics employ a small deformable (membrane) mirror having tens or hundreds of piezoelectric crystals attached to its back. These crystals act as tiny actuators in order to counteract or cancel changes in the wavefronts coming from the object, resulting from the irregularities in Earth's atmosphere. These actuators can make rapid adjustments in μm sizes at the rate of several hundred times per second.

Active/adaptive optics

Active optics involve continuous adjustment of a primary mirror shape on fast time scales of a few seconds to minutes to maintain the sharpest image. On the other hand, adaptive optics aim at correcting the image for a degrading effect of atmospheric turbulence at much faster time scales of 10-100 millisec.

- Option: correcting before light reaches the detector
- Correction of wavefront distortion:
    1. Imaging of entrance pupil ($4 > 100$ points)
    2. Thin deformable mirror ($3 > 100$ actuators)
    3. Fast sampling (> 1000 Hz)
- Applicable on small fields of view (few arcmins): mainly for large long-focus telescopes (disk subfields)

2.4.4 Off-line image restoration

This process is based on the idea of correcting/restoring images after the acquisition and intermittency of turbulence ($< 1s$), where brief moments of random image stability every few seconds are available at the observing site. The method consists of

- obtaining short image burst sequences at high cadence (around 2 to 100 images/sec),
- applying sharpness measures, such as gradient and contrast enhancement using FFT, wavelets, and numerical filters, and
- selecting image(s) with the highest sharpness.

For imaging the Sun, an additional difficulty is posed as it is an extended object, and often the best resolution is available only in limited parts of the solar image. Also relative motion (shift) of different patches may occur at some parts of images. This requires 'de-stretching' techniques based on local pattern matching. In order to take fast bursts of images, one requires a shutterless, high speed imager, and fast data storage capability.

2.4.5 Some notable solar sites and observing strategies

We briefly list here some of the notable solar sites, along with observing strategies adopted for their telescopes.

- *Observing from high altitude (mountain) sites using tower platforms*: the 150-foot solar tower at Mount Wilson (Alt. 1740 m, completed in 1912, 54 m high tower, aperture 30 cm), the Swedish solar telescope at La Palma, Canary Islands (Alt. 2360 m, 28 m tower, aperture 98 cm).
- *Observing from small island sites surrounded by large water bodies (lakes or ocean sites)*: Big Bear Solar Observatory, USA (Alt: 2060 m, D=1.6 m); Udaipur Solar Observatory, India (Figure 2.2) (Alt: 600 m, collapsible dome, D=50 cm); Huairou Solar Observatory, China (vacuum, D= 60 cm); GREGOR (D=1.5 m) and VTT, Tenerife, Canary Islands; Dutch Open Telescope, La Palma (open clamshell).
- *Using vacuum, or inert gas filled telescope to reduce internal instrumental turbulence*: THEMIS (sealed telescope tube with low pressure helium, D=90 cm), New Vacuum Solar Telescope (NVST D=98.5 cm) of Fuxian Lake Solar Observatory (FSO), China.
- *Telescopes interfaced with active/adaptive optics*: GREGOR (156 wavefront sensors, 256 actuators, 2000Hz, resolution 0.08 arc-sec); Goode Solar Telescope, BBSO (3 deformable mirrors, 2 arc-min corrected FOV).

2.4.6 Advantages of ground-based telescopes

In recent years, several space-based solar telescopes have provided excellent quality of data. However, the ground-based solar observations remain important due to their many advantages. It is possible to use very large, and heavy instruments at ground-based sites. The limitation posed by atmospheric degradations can be reduced by using active and adaptive optics.

Continuous regular monitoring of the Sun can be carried over periods of decades from ground-based sites. They can be deployed with large, multiple detectors for fast temporal cadence (a few seconds), for simultaneous multi-wavelength observations for on-line and off-line analysis. Telemetry of large volumes of data is not a concern that is faced in space-based instruments.

There is more flexibility in maintenance and upgradability. It is possible to have a network of ground-based stations spread geographically across longitudes with identical solar instruments for continuous solar observations.

2.5 Solar observations: the analysis techniques

The Sun (or other cosmic objects) is observed by collecting the photons using a telescope. Imaging and analysis of the collected photons are carried out using suitable back-end (focal plane) analysers. Ground-based optical telescopes routinely observe the solar photosphere, chromosphere and even the solar corona from some high altitude sites. There are broadly four types of astronomical analysers:

- photometers for brightness measurements, and imagers,
- spectrometers for spectral measurements,
- polarimeters for measuring the state of polarization and deducing magnetic fields,
- interferometers for high angular and spectral resolution measurements.

Often neutral density (ND) filters are needed to observe the white light Sun in continuum, with a constant attenuation across the visible wavelengths. These are used to reduce the intensity of light by reflecting or absorbing a portion of it. There are other classes of optical filters commonly used to image the Sun in certain desired wavelengths. These filters selectively transmit light in a particular range of wavelengths, or wavelength bands, while blocking the remainder.

Absorption and interference filters are the two most common types of such filters in use. Absorption filters work by reducing the incident light through absorption of specific wavelengths. These are usually broadband filters having pass-bands of around 10 nm, which are essentially coloured glass plates with optical quality surfaces.

2.5.1 Interference filters: Fabry-Pérot (FP) etalon monochromators

Interference filters reflect and destructively interfere with unwanted wavelength, rather than absorbing. They allow the passage of only a narrow passband of around 1–2 nm around a desired spectral line of interest. Modern interference filters are based on the Fabry-Pérot interferometer, designed in the late 1800s by Alfred Pérot and Charles Fabry.

Interference filters are being used as narrow passband filters, and consist of a glass plate of a fixed spacing and semi-reflective mirror coating on both sides. Multiple partial reflections between the two surfaces of the plate produce interferences, and the wavelength is defined by mirror spacing. Fabry-Pérot interferometers consist of two glass plates (or, wedges) with air spacing. They may be tuned in wavelength by thermal or piezo control (Figure 2.6).

Finesse of an etalon is defined by the spacing of peaks, $\Delta\lambda$, relative to the peak width, $\delta\lambda$, and is given by: $\mathcal{F} = \Delta\lambda\ /\delta\lambda$, which increases with reflectivity R of the mirrors, from typically 10, up to > 100. One order can be easily selected by using a pre-filter centred at a desired wavelength, and one can achieve very narrow passbands of 0.02-0.06 nm at much lower costs.

In order to obtain near-simultaneous observations at photospheric and chromospheric heights, two voltage-tuneable lithium-niobate Fabry-Pérot

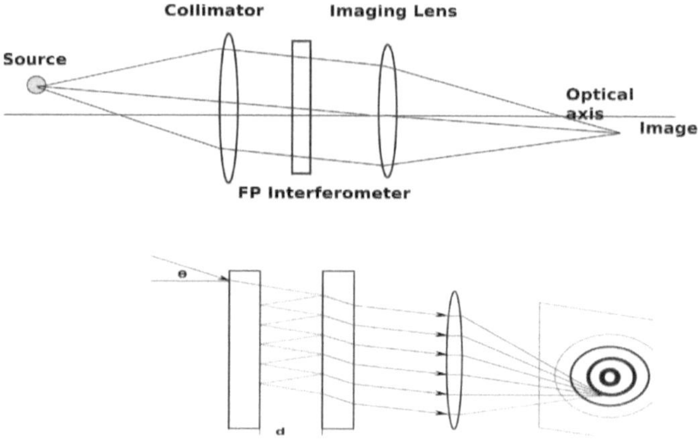

FIGURE 2.6: (top) Using a filter which is a resonant cavity (thickness = nλ), coatings or order-sorting filter can be used to isolate order. (bottom) Loci of a constant wavelength in the focal plane are circles, such that $n\lambda = 2d\mu cos(\theta)$, where n=interference order, d=gap, μ=refractive index, λ=wavelength, and θ=angle of incidence.

etalons are being used as narrow band imagers at the Multi-Application Solar Telescope, Udaipur (India) (Figure 4.4, page 91). A set of interference blocking filters is employed to make the selection of corresponding wavelength bands. Both of the etalons are used in tandem for photospheric observations in Fe I 617.3 nm, and chromospheric observation in H_α656.3 nm spectral lines, whereas only one of the etalons is used for the chromospheric Ca II 854.2 nm-line [259].

2.5.2 Lyot filter based monochromators

In 1933, Bernard Lyot [92] in France, and in 1938, Öhman in Sweden designed another class of optical filters with very narrow passbands of around 0.1 nm which have been used extensively in solar observations. These filters use the properties of birefringence of calcite and quartz crystals. These are generally called Lyot filters or polarization-interference monochromators, and consist of an alternate sequence of polarizers and birefringent crystals [252]. Each plate of a Lyot-Öhman filter is 1/2 the thickness of the preceding plate.

In each plate, the ordinary and extraordinary polarization components of a light beam experience different refractive indices, and thus have different phase velocities. Therefore, the polarization state of an arbitrary wavelength will in general be modified and the intensity will be attenuated in a subsequent polarizer. The phase retardation is an integer multiple of 2π, the polarization state keeps the same and the loss is very small. The separation and narrowness of the transmission peaks depend on the number of elements, plate thicknesses and orientations.

The largest plate sets the bandwidth and the smallest sets the Free Spectral Range (FSR). A birefringent crystal rotates the plane of polarisation by an amount proportional to the thickness divided by the wavelength. A linear polarizer is placed in the emergent beam to pick out those wavelengths for which n is integer.

Birefringent filters are essentially based on natural crystals having high optical purity, and hence, they are usually very expensive. These filters produce instantaneous 2-dimensional images at a much faster cadence as compared to the time consuming process involved in scanning an image by the slit used in spectroheliographs.

Birefringent filters are generally confined around a given wavelength, or are tunable over a limited wavelength range by slow, mechanical rotation of polarizers. The acceptance angle (field of view) of a birefringent filter is rather limited. Solar beams faster than $f/15$ create non-uniformity over the field of view. Universal Birefringent Filters (UBF) covering a wide spectral range from about 420 nm to 700 nm have been designed, which allow the observation in any specific spectral line using blocking filters around their wavebands.

In the case of birefringent filters, transmission drops steeply to become nearly zero beyond the passband, hence a much cleaner passband is obtained. The more economic FP etalons, on the other hand, usually do not have a sharp cut-off, and the passband falls exponentially, with considerable amount of continuum leaks through etalon filters. Therefore, Hα images taken through a Lyot filter and through an FP-type etalon filter show chromospheric features quite different in appearance.

2.5.3 Magneto-optical filters

The magneto-optical filter is an extremely narrow band filter made of a cell containing sodium or similar kinds of vapour [4]. The cell is placed in a strong magnetic field in making use of magneto-optical effects in the vapour. The filter has high throughput and wavelength stability, and secure absolute wavelength standard. They provide a possibility of making a filter with a very narrow passband (less than $30m\mathring{A}$), and a high transmission (up to nearly 100% apart from losses arising from the glass cell, lenses and polarizers. They have been utilised for imaging [241] and velocity fields measurements in helioseismological observations [364], [326].

2.5.4 Michelson interferometers

The Michelson interferometer is a device that produces interference between two beams of light. Each part is made to travel a different path. After traversing these different path lengths, the two parts of the light are brought together where they interfere with each other according to their path length difference. The interference pattern can be seen on a screen. They are slit-less, therefore, allow imaging. However, finesse is only around 2. The Global Oscillations Network Group (GONG) uses an instrument based on a polarising Michelson interferometer having a path difference of about 30,000 waves. This is called a Fourier Tachometer [40]. It provides very high precision measurements of solar velocity imaging for helioseismological observations.

2.6 Solar spectrographs

Spectroscopy is a fundamental tool that allows to determine the chemical compositions, physical properties, and radial velocities of astronomical sources. An astronomical spectrograph is an important 'analyser' instrument for any astrophysical and solar studies. The basic components of all astronomical spectrographs are (i) a slit at the focal plane of the telescope on which the object's light is focussed, (ii) a collimator, which receives the diverging beam of light and gives out parallel light, (iii) a dispersing element (a prism or a grating), and (iv) a camera to focus the resulting spectrum on to a detector. These are illustrated in Figure 2.7.

The first spectrographs used photographic paper as the detector. More recent spectrographs use electronic detectors, such as CCDs which can be used for visible, near IR and UV light, while, different techniques are used for γ-ray, X-ray, and radio spectroscopy. Types of spectrographs according to dispersing element and geometry are as follows:

(i) By type of dispersing element:

- Grating (transmission or reflection)
- Prism (rare, except as a cross-disperser) - Grism = adds a grating on a prism for zero-deviation transmission dispersion
- Narrow-band imaging
- Interferometry

(ii) By geometry:

- Long-slit or multi-slit
- Aperture of multi-fiber
- Integral field units: lenslets or fiber bundles - tunable imagers (*e.g.*, Fabry-Pérot)

A diffraction grating is the principal device used in a spectrograph to generate wavelength-dependent interference patterns at UV/Optical/IR (and

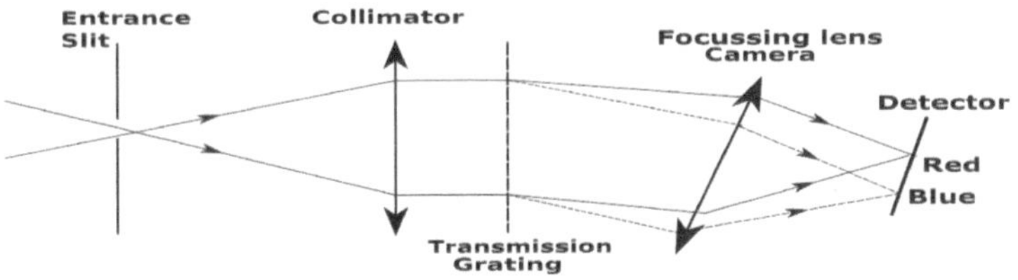

(a)

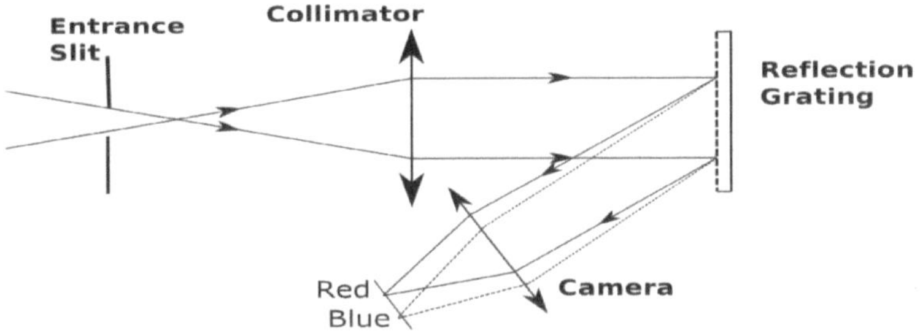

(b)

FIGURE 2.7: Schematic of an astronomical spectrograph incorporating (a) a transmission grating, and (b) a reflection grating. The slit is in the focal plane of the telescope. The grating is conjugate with the primary mirror. The detector is in the focal plane of the camera.

even X-ray) wavelengths. A diffraction grating acts as a set of equally spaced slits in an otherwise opaque screen, where each slit can be considered as radiating Huygens' secondary wavelets. The amplitude at any point on the image side of the slit is obtained by summing the amplitude contributed by each set of secondary wavelets.

Most practical gratings are reflection gratings (Figure 2.7b), not composed of a screen with equally spaced slits (as in transmission gratings), but of alternating reflecting and non-reflecting strips, made by ruling on a reflecting surface. It should be noted that ruling gratings is not an easy task as typical required spacing tolerance is of the order of ~nm.

The disperser in the spectrographs is ruled with a number of grooves per mm (typically 100-1000). At the camera image location, the wavelength λ of light depends upon the incidence angle θ at the grating relative to the incoming beam, and the angle of diffraction ϕ of the camera with the normal to the grating (Fig. 2.8).

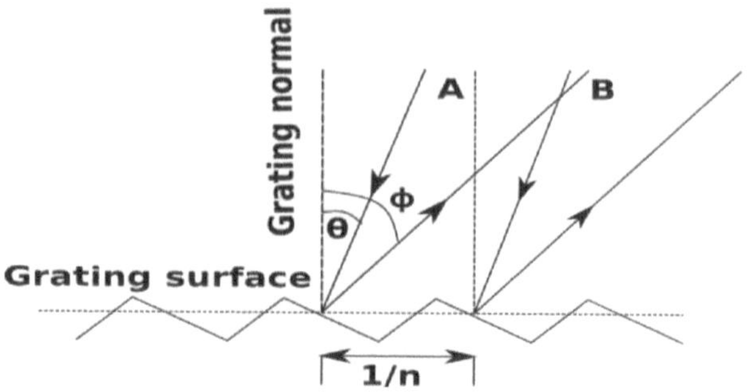

FIGURE 2.8: Grating relation.

Spectrograph's dispersion is defined as the distance on this detector plane in order to change wavelengths by a certain amount. In general, the greater the projected number of grooves/mm (*i.e.*, as seen along the light path), the higher would be the dispersion, keeping all other parameters equal.

The light reflecting (or transmitted) from adjacent grooves, say, A and B will interfere constructively, if the difference in path length is an integer number of wavelengths. The path difference is $d\sin\theta + d\sin\phi$, (where d is the groove spacing on the grating). The relationship governing these parameters is called *the grating equation*:

$$n\lambda = d\,(\sin\theta + \sin\phi), \tag{2.2}$$

where n is the order at which the grating is being used. The angle of incidence, θ, is the same for all wavelengths, so from this relation, it is evident that the angle of diffraction, ϕ, must then be a function of wavelength, *i.e.*, $\sin\phi = n\lambda/d - \sin\theta$.

As n and λ both occur in the relation 2.2, the orders cannot be separated. For example, 600 nm in the first order falls at the same place as 300 nm in the second order. The problem in separating the orders is overcome by using appropriate *blocking filters*, usually placed just after the light has passed through the slit. For example, if we were to observe in 600-800 nm in first order, there will be overlapping light of 300-400 nm in second order, and 200-266.6 nm in third order, *etc.* As the atmosphere transmits very little light below 300 nm, the third or higher orders pose no difficulty, but 300-400 nm light needs to be blocked. For this purpose, blue cut-off filters may be used, which pass all the light longer than a particular wavelength, say, 600 nm and higher, but cut off the light in the blue.

The angular spread (or dispersion) of a given order n with wavelength, for a given incidence angle θ, can be obtained by differentiating the grating equation (2.2):

$$\frac{d\phi}{d\lambda} = \frac{n}{d\ \cos\phi} = \frac{(\sin\theta + \sin\phi)}{\lambda\ \cos\phi}. \tag{2.3}$$

In the Littrow arrangement, one has $\phi \approx \theta$, therefore, angular dispersion $d\phi/d\lambda$ is given by

$$\frac{d\phi}{d\lambda} = \frac{2\tan\phi}{\lambda}. \tag{2.4}$$

A typical grating spectrograph is used with small angle of incidence, where light enters and leaves the grating nearly perpendicularly, and in low orders, where n is usually 1 or 2, in order to avoid overlapping wavelengths from different orders. To achieve high dispersion, the only way is to make d very small, *i.e.*, using a grating with a large number of groves per mm.

Gratings have a blaze angle that results in their having maximum efficiency for a particular value of $n\lambda$. Blazed gratings are made by cutting the grooves at a certain angle such that most of the reflected light from the grating is directed in a particular direction, while in other directions or orders the reflected light is decreased. Such blazed gratings have been found to be very useful for working in specified orders.

High dispersion is useful only when the spectral lines to be studied are very narrow and the spectrograph slit is of the right width to resolve the fine solar features, or when line profile studies are made. The plate scale of a telescope can be described as the angular size (arc-sec) corresponding to the size in mm at the focal plane of the telescope. For example, the plate scale, for a telescope with the diameter of the primary mirror (or lens) D, and effective focal length, F, (or $f - number\ f/\# = F/D$), can be obtained using a simple expression $\tan\alpha = D/F$, where α is the angle subtended by the aperture D of the primary mirror at the prime focus.

Angle α is small for long focal-length telescopes, so $\tan\alpha \approx \alpha$, which expressed in radians, and converted into arc-sec (where 1 radian = 206265 arc-sec), gives $\alpha = 206265D/F$. Now, the plate scale is the ratio between the angle α and diameter D of the telescope, in arc-sec/mm, so one has:

$$p_s = \frac{206265}{D \times f/\#} = \frac{206265}{F} \tag{2.5}$$

in arc-sec per mm at the focus of the telescope. For example, a 20 cm, $f/30$ telescope yields an image scale of 0.03 mm/arc second. The solar seeing is rarely better than 1 arc-second, therefore, to keep the slit width less than 0.03 mm is of no advantage.

In the simplest terms, we may define resolution of an optical system as the smallest detectable adjacent points in an image, which do not appear to touch each other. In case of a solar spectrograph, the resolving power depends on the image scale of the telescope as well as the slit width and is given by $R = s/(2000\ d)$. Here s is the slit width in mm, d the diameter of solar image in mm, and 2000 arc seconds is taken as the mean angular diameter of the solar image.

2.6.1 Spectroheliograph

In 1891, George Ellery Hale invented an instrument to photographically record the monochromatic images of the solar chromosphere. Independently, around the same time, H. Deslandres of the Meudon Observatory, France, and John Evershed, India, constructed spectroheliographs. A spectrograph disperses the Sun's light into a spectrum of colours, interspersed with numerous dark Fraunhofer absorption lines, which are in fact images of the slit. The image of the Sun is formed on the slit of the spectrograph, and produces a spectrum of the part of the Sun corresponding to the slit. As the image (or the slit) is moved, the spectrum of other parts of the Sun is obtained.

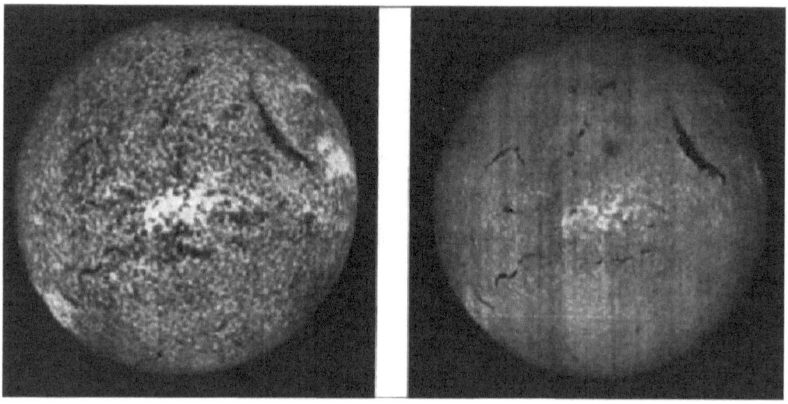

FIGURE 2.9: Spectroheliograms obtained at Meudon Observatory in (left) Calcium-K, and (right) H-alpha on March, 21, 1910.

In a spectroheliograph, a region of interest on the Sun, or the whole solar disk can be imaged in a particular spectral line by simply scanning, and staggering a series of slit images. For example, to image the Sun in H_α 656.3 nm line, the image of the Sun is moved across the slit, and to isolate this particular spectral line, a second slit is placed in the focal plane of the spectrograph. Then by letting a photographic plate move in this plane, at the same speed as the solar image on the first slit, the image of the Sun in that line is built up. In this manner, it is possible to obtain a monochromatic image of the Sun in any desired wavelength (Fig. 2.9).

Spectroheliographs have the advantage that the bandpass can be varied and made very small to achieve high spectral purity. However, there is the disadvantage that spectroheliographs are relatively time consuming in making an image of the Sun. Also, image degradations may occur due to seeing variations and guiding errors. Until 1938, however, nearly all the monochromatic observations of the chromosphere were made through spectroheliographs.

Spectrographic technique is being used in measurements of Doppler velocities using shifted line profile measurements [234], and spectro-polarimetry technique related to Zeeman polarization of magnetic sensitive lines. We shall discuss these instruments in Section 4.1.

Spectroscopic diagnostics

The purposes of spectroscopy are: (i) to measure accurate wavelengths of emission and absorption lines, (ii) to measure the relative strengths and/or equivalent widths of emission or absorption lines, (iii) to measure shapes of emission or absorption lines, and (iv) to measure the spectral energy distribution of the continuum radiation. These measurements enable a lot of astrophysics, as follows:

- Line shifts: radial velocity
- Line broadening: temperature, turbulence
- Emission or absorption lines: density and temperature gradients
- High solar intensity allows high wavelength dispersion: high-spectral resolution at high cadence

2.6.2 Slit-less spectrograph

If the slit and collimator lens are not used, and the dispersing element is placed to receive the parallel beam of light before the telescope objective, the images of the object are dispersed. This is used to obtain 2-dimensional spectral information about extended sources, such as solar chromosphere, corona, planetary nebulae, *etc.* The slit-less spectrum consists of images of the object at each of the emission wavelengths, displaced by the dispersion of the instrument (Figure 3.10).

2.6.3 Echelle spectrograph

An alternative method to achieve high dispersion is by operating in a high order for a very large angle of incidence with a low number of grooves per mm as in echelle spectrographs designed typically for $\tan\phi \sim 2$ or greater. A typical echelle grating may have $\sim$ 80 grooves/mm, so, d $\sim 25\lambda$ or for visible light, order n must be of the order of 50. Usually this method involves a fine ruled grating for the initial dispersion, which is crossed dispersing with a prism or another grating.

Thus, an echelle spectrograph uses two diffraction gratings, rotated 90 degrees with respect to each other and placed close to one another. Therefore, an entrance point and not a slit is used and a 2D CCD-chip records the spectrum. Usually one would guess to retrieve a spectrum on the diagonal, but when both gratings have a wide spacing and one is blazed so that only the first order is visible, and the other is blazed so that a lot of higher orders are visible, one gets a very fine spectrum nicely folded onto a small common CCD-chip.

Echelle spectrum thus consists of a number of spectral orders arranged side by side on the detector. The small chip also means that the collimating optics need not to be optimized for coma or astigmatism, but the spherical aberration can be set to zero.

2.7 Ground-based coronagraphs: artificial solar eclipses

The Sun's outer atmosphere, the corona, is extremely faint; around a millionth of the bright photosphere. The stray light in the Earth's atmosphere, produced by scattering of photons coming from the photosphere, is usually much brighter than the faint corona. Therefore, it is visible from the ground only during the rare natural process of total solar eclipses, when one can obtain spectacular images extending to large distances (Figure 2.10a). These images show that the corona is dominated by global solar magnetic fields, and both open and closed field structures are seen.

Observations made during the times of total solar eclipses have contributed remarkably to understanding the structure and dynamics of the corona. However, total solar eclipses are rather rare cosmic events lasting over short durations ranging from a few seconds to the maximum of around 7 minutes. One needs to complete the observations over this short duration of the totality, which are often hampered by inclement weather. They also involve great planning, costs, logistic difficulties, and travel often to remote locations of the narrow track of totality.

This difficulty was overcome to some extent in 1931 by the French solar astronomer, Bernard Lyot who came up with the brilliant invention of an instrument called 'coronagraph' [92]. One needs to select a suitable high-altitude, dust-free site having as little scattering as possible. From such sites this instrument can observe the inner corona by creating artificial eclipses by means of an occulting disk to cover the photosphere. The Lyot coronagraphs are based on an internal occulter.

These coronagraph requires very accurate optical alignment and mechanical stability of the telescope failing which the stray light makes it impossible to see the faint corona. Lyot's design was later followed up and refined by Max Waldmeier at the ETH/Zürich, and Donald H. Menzel (1901-1976) at

(a)

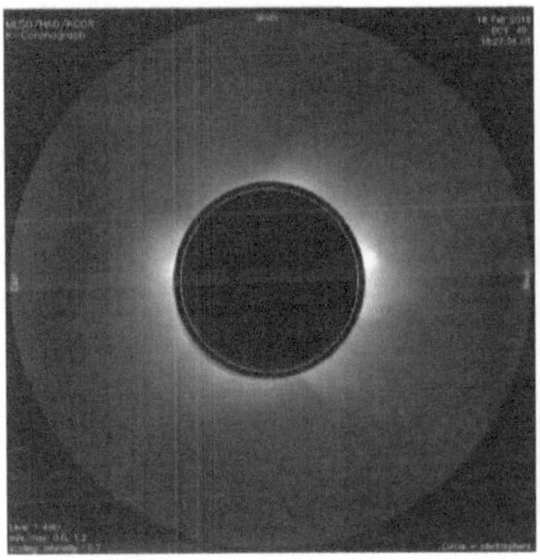

(b)

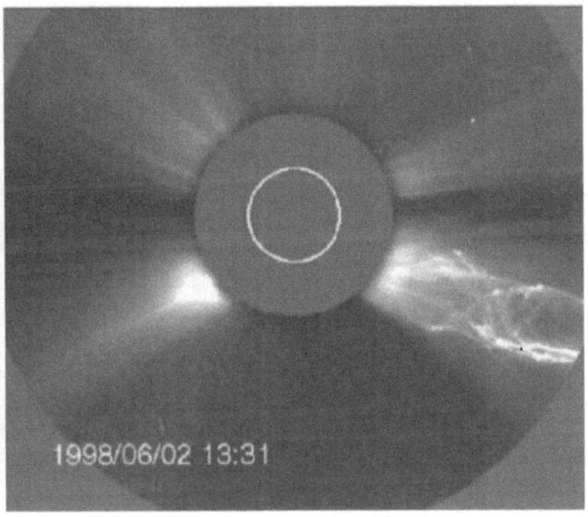

(c)

FIGURE 2.10: (a) White light corona as observed during the solar eclipse of August 21, 2017, constructed by composition of 287 eclipse images. (By permission of Miloslav Druckmüller.) (b) Coronal image taken on February 18, 2016, by artificial occultation of the Sun's disk from the high altitude MLSO-K coronagraph. (c) Coronal image taken on June 2, 1998, by artificial occultation of the Sun's disk from SOHO-LASCO space coronagraph. It shows a coronal mass ejection in progress. White circle represents the size of the Sun's photosphere. (Courtesy of SOHO/LASCO consortium. SOHO is a project of international cooperation between ESA and NASA.)

Harvard College Observatory [188]. At many ground-based observatories, such as Pic du Midi (CLIMSO), MLSO-High Altitude Observatory, Hawaii (MK3 K-coronameter), and Sacramento Peak Observatory, images of the inner E- and K-coronae are being made regularly using this approach (Fig. 2.10b).

With the development of space-borne instrumentation, coronagraphs are also being operated from space (Fig. 2.10c) that use external occulters. The Large-Angle and Spectrometric Coronagraph (LASCO) onboard the Solar and Heliospheric Observatory (SOHO) has been continuously observing the solar corona and dynamic activities, such as coronal mass ejections, since 1996. However, total solar eclipse observations continue to attract solar astronomers, and the public at large for 'a life time experience'.

2.8 Summary

Ground-based solar optical observations require careful site surveys over extended periods to select suitable sites with good atmospheric seeing, low dust and water vapour content, and a large number of cloud-free days in a year. Various types of telescope configurations, dome designs, and instruments aided with active and adaptive optics capabilities, are the next step to optimally exploit the diffraction-limited resolutions along with the atmospheric seeing of the site. High altitude sites, surrounded by large bodies of water, appear to provide better quality solar observations. Imaging and spectroscopic methods are used for observing the photosphere and chromosphere, their associated features, and energetic transient activities on a regular basis to synoptically monitor the Sun. This is achieved by blocking the light of the brighter photosphere using special optical filters centred on specific photospheric and chromospheric spectral lines. For example, a chromospheric H_α filter enables one to routinely observe the faint "red" chromosphere above the photosphere that is otherwise observed only at the solar limb just before (and after) the totality of a solar eclipse. Using the technique of artificial occultations, it is possible to obtain images of the near-corona in certain visible coronal emission lines from high altitude sites with very low dust content.

Chapter 3

The Solar Atmosphere

3.1 Introduction

The solar outer atmosphere begins at the *photosphere*, which is the visible surface of the Sun. It consists of three main irregular layers that are defined by the temperature of the plasma contained within them. Above the photosphere lies the around 1000 times rarefied *chromosphere* at a temperature range of 6000-15,000 K, and depth of 2,500 km. The chromospheric density drops rapidly over several orders of magnitude, from a high of $2.0 \times 10^{-4} kg/m^3$ to a low of $1.0 \times 10^{-11} kg/m^3$ from its bottom to the top.

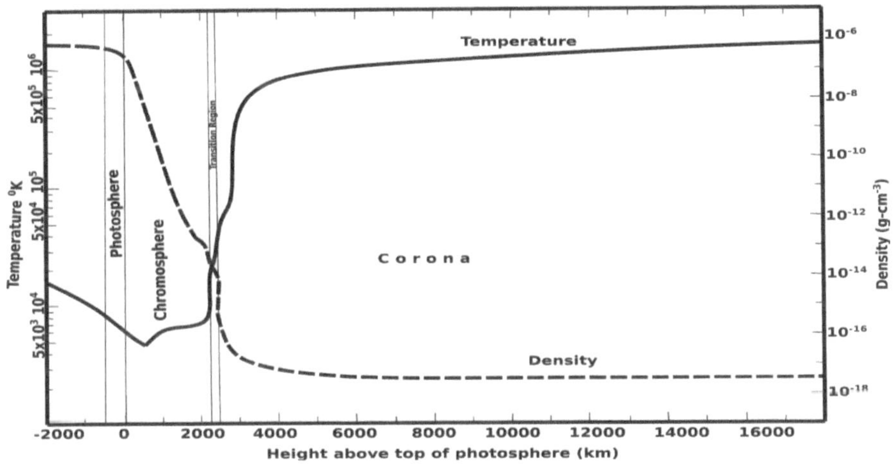

FIGURE 3.1: Temperature and density variation of the solar atmosphere with altitude.

Above the irregular chromospheric layer begins the *transition region* (TR). It has a sudden increase of temperature from around 10^4 to 10^6 K, within a relatively very thin depth of ~500 km. Finally, there is the even more tenuous *corona* at a high temperature upwards of 10^6 K. The solar atmospheric models describe the altitude-dependence of temperature T, density and pressure P (Fig. 3.1).

In these models, all horizontal variations are generally neglected, which is not a valid assumption particularly at higher altitudes. Table 3.1 lists some solar and terrestrial densities for a comparison.

TABLE 3.1: A comparison of densities

Mean density of entire Sun	$1.41 \times 10^3 kg/m^3$
Interior (centre of the Sun)	$1.6 \times 10^5 kg/m^3$
Surface (photosphere)	$10^{-6} kg/m^3$
Chromosphere	$10^{-9} kg/m^3$
Low corona	$10^{-13} kg/m^3$
Mean density of Earth	$5.51 \times 10^3 kg/m^3$
Sea level atmosphere of Earth	$1.2 kg/m^3$

The photosphere is usually the only layer of solar atmosphere possible to visually observe from the ground. The other layers of solar atmosphere are not visible from the ground, except in some special circumstances, such as total solar eclipses, or with the help of special instruments, such as narrow passband filters, and spectrographic means, discussed in the previous chapter.

3.2 The radiative transfer equation

For constructing atmospheric models of the Sun, it is important to learn about how the radiation propagates from the core outwards through the various zones in the Sun's interior, and then from the photosphere through the outer layers. This process is mathematically described by the radiative transfer equations.

The propagation of radiation through a medium is affected by absorption, and scattering processes contributing to its opacity, and also by the production of energy by the emission process (Figure 3.2). The equation of radiative transfer simply states that as a beam of radiation travels through a medium, it loses energy to absorption, gains energy by emission, and redistributes energy by scattering. Equations of radiative transfer have applications in a wide variety of subjects including optics, astrophysics, atmospheric science, and remote sensing.

The equation of radiative transfer describes these interactions mathematically, such that the change in specific intensity over a path-length ds is given as:

$$dI_\lambda = -\alpha_\lambda I_\lambda ds + \varepsilon_\lambda ds = -I_\lambda d\tau + S_\lambda d\tau \tag{3.1}$$

or

$$\frac{dI_\lambda}{d\tau} = -I_\lambda + S_\lambda, \tag{3.2}$$

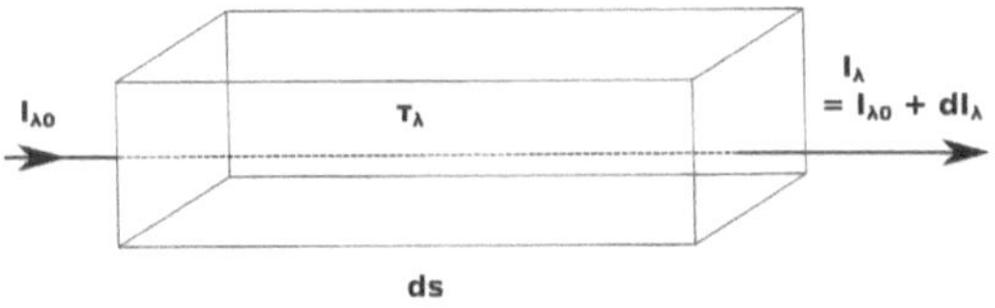

FIGURE 3.2: Propagation of radiation through a box containing a medium of opacity τ_λ.

where $d\tau = \alpha_\lambda ds$ is optical depth of path length ds, absorption coefficient $\alpha_\lambda = \kappa_\lambda \rho$, κ_λ is opacity or mass absorption coefficient, and $S_\lambda = \varepsilon_\lambda / \alpha_\lambda$ is the source function. The intensity will decrease or increase with increasing τ, provided if $S_\lambda < I_\lambda$, or $S_\lambda > I_\lambda$, respectively.

Equation (3.2) is the radiative transfer equation in parallel ray approximation, and needs a small modification of multiplying $dI/d\tau$ by μ (cosine) for applying it to non-vertical rays. However, this simple equation provides interesting insights from its solutions. The physics of the medium is hidden in ε_λ and κ_λ, *i.e.*, S_λ and τ. These quantities depend on the temperature, pressure, chemical abundances, and wavelength of radiation.

In absence of any source of emission, this may be directly integrated to give the usual extinction law: $I_\lambda = I_\lambda(0)e^{-\tau}$. The optical depth τ=0 corresponds to no reduction in intensity, *i.e.*, the medium is transparent, τ=1 gives an e-fold reduction of intensity, and $\tau \to \infty$ gives $I_\lambda \to 0$, such that no radiation reaches the observer, and medium is optically completely opaque.

The layer, where $\tau = 1$, is defined as the surface, or photosphere, from where photons appear to originate. It radiates nearly as a black-body at the effective temperature, $T_{eff} \sim 5700$°K, with an average temperature of 6500°K at the bottom (just above the convection zone) reducing to 4400°K, at the top. A variety of spatial and dynamical features are observed on the photosphere at different scales.

One can solve the radiative transfer equation (RTE) by multiplying with an integrating factor $exp(\tau)$, as follows:

$$\frac{d}{d\tau}(I_\lambda e^\tau) = e^\tau I_\lambda + e^\tau \frac{dI_\lambda}{d\tau} = e^\tau S_\lambda, \tag{3.3}$$

where Equation (3.2) has been used. Equation (3.3) can be integrated over τ to obtain

$$\int_0^{\tau_\lambda} \left[\frac{d}{d\tau}(I_\lambda e^\tau)\right] d\tau = \int_0^{\tau_\lambda} (e^\tau S_\lambda)\, d\tau = [e^\tau S_\lambda]_0^{\tau_\lambda}. \tag{3.4}$$

Using the boundary condition, with $I_\lambda(\tau = 0) = I_{\lambda 0}$, and rearranging, onc obtains

$$I_\lambda = I_{\lambda 0} e^{-\tau_\lambda} + S_\lambda(1 - e^{-\tau_\lambda}). \tag{3.5}$$

The first term on the R.H.S. is the amount of radiation left over from the intensity entering the box (medium) after it passed through the optical depth, τ_λ, and the second term gives the contribution of the intensity from the radiation emitted along the path.

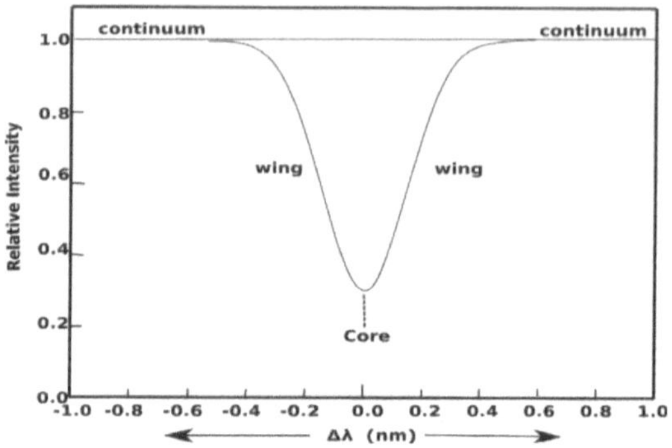

FIGURE 3.3: Spectral profile of a photospheric absorption line showing the relationship between relative intensity and depth in photosphere. The wings of the spectral line form deeper in the photosphere than the core.

3.2.1 Formation of absorption and emission lines

Let us consider now a limiting case of Equation (3.5). In the optically thin medium ($\tau_\lambda << 1$), one can write $I_\lambda = I_{\lambda 0}(1-\tau_\lambda)+\tau_\lambda S_\lambda = I_{\lambda 0}+\tau_\lambda(S_\lambda - I_{\lambda 0})$. From this, it follows that:

- If $I_{\lambda 0} > S_\lambda$, *i.e.*, the intensity of radiation from deep layers is larger than the source function from top layers, the second term in the R.H.S., $\tau_\lambda(S_\lambda - I_{\lambda 0})$ will make a negative contribution which is proportional to the optical depth. Therefore, absorption lines would result at the corresponding wavelengths λ on the continuum intensity I_λ (Figure 3.3). Examples are stellar (solar) photospheres.

- If $I_{\lambda 0} < S_\lambda$, *i.e.*, there would be a positive contribution to $I_{\lambda 0}$, and emission lines would appear on the continuum background intensity I_λ. Examples are solar UV, EUV and X-ray emission lines from hotter and fainter upper layers of the Sun.

3.2.2 The thermodynamic equilibrium (TE)

In the other extreme limiting case of Equation (3.5) of optically thick medium ($\tau_\lambda >> 1$), one has $I_\lambda = S_\lambda$. This is the case of thermodynamic equilibrium (TE), when a single black-body temperature is defined everywhere, *i.e.*, radiation from the object follows Planck's function, $B_\lambda(T)$, or, $I_\lambda = S_\lambda = B_\lambda(T)$.

In the simplest thermodynamic state, described by one single temperature T everywhere, the particles possess a Maxwellian distribution. The states of ionization, and excitation are given by the Saha and Boltzmann equations for the same temperature. In this situation $dI/d\tau = 0$, so that $S_\lambda = I_\lambda = \varepsilon_\lambda/\alpha_\lambda = B_\lambda$. The radiation field is homogeneous, and the isotropic black-body radiation is given by the Kirchhoff-Planck function

$$S_\lambda = B_\lambda(T) = \frac{2hc^2}{\lambda^5} \cdot \frac{1}{exp(hc/\lambda k_B T) - 1}. \tag{3.6}$$

In such an equilibrium, there is no temperature gradient. This clearly is not a realistic description anywhere on the Sun. Therefore, a more useful concept for the Sun is local thermodynamic equilibrium (LTE). It implies that within a localised, finite region a single temperature can be used to describe the particle temperature, population of atomic states and the local ratio between emission, and absorption. Each layer of the Sun's atmosphere is defined by its own temperature.

In the interior of the Sun the photon mean free path is very small, so that where photon is emitted and where it is absorbed have nearly the same temperature, and thus it can be assumed to have LTE. Therefore, the source function in radiative transfer equation can be replaced by the Kirchhoff-Planck function.

In the outer solar atmosphere, LTE is again a useful approximation for the continuum radiation in the visible and infrared wavelengths, and for the weak spectral lines as well as for the wings of most spectral lines. The deviations from the LTE increase with increasing altitude from the solar surface, as the effective IR, UV and visible brightness temperatures are very different. The electrons are still Maxwellian, and the temperature is given by the electron temperature $(T = T_e)$.

3.3 Optical observations of the photosphere

The yellow, visible luminous disk is the outer skin, or the photosphere, of the Sun. It can be observed using eyes (under special protection from the intense glare), and modest optical telescopes. This layer is defined by the optical depth $\tau = 1$ of the solar atmosphere, which rapidly decreases over just a few hundred kilometres.

The momentum conservation equation, in hydrostatic equilibrium, gives $dP = -\rho g dr$. Using the equation of state, $P = \frac{\rho \Re T}{\mu}$, one obtains

$$\frac{dP}{dr} = \frac{\Re T}{\mu}\frac{d\rho}{dr} = -\rho g \Rightarrow \frac{1}{\rho}\frac{d\rho}{dr} = -\frac{\mu}{\Re T}g \tag{3.7}$$

or

$$\rho = \rho_0 exp(-\frac{\mu g}{\Re T}r), \tag{3.8}$$

considering T, μ and g as constants over the photospheric layer. Here, $\Re = 8.317J/(mol.K)$ is the gas constant, and mean molecular weight in a plasma, $\mu = (n_p m_p + n_e m_e)/(n_p + n_e)$. In a fully ionized plasma, one has $n_p = n_e$, and $m_p >> m_e$, therefore, $\mu \approx 0.5$. However, presence of He and small traces of heavier elements in solar plasma leads to a slightly larger value, $\mu \sim 0.6$, throughout most of the solar interior except very close to the photosphere, where due to low temperatures, ionization is lower.

Thus, the density falls off exponentially at the scale-height $H = \frac{\Re T}{\mu g}$ by a factor of $1/e = 0.37$. Considering that $g = 274m.s^{-2}$ at solar surface (photosphere), the density scale height, a measure of thickness of this layer turns out as $H \sim 130km$, assuming T corresponding to the temperature minimum of the photosphere. On the other hand, in the million-degree hot fully ionized outer coronal plasmas, and the deep solar interior, the scale height is as large as $\sim 10^5 km$.

3.3.1 Limb darkening

Limb darkening is an optical effect where the central part of the Sun's disk appears brighter, and the line of sight reaches down to the optical depth τ=2/3, *i.e.*, hotter, deepest photospheric layer (at T_{eff}=6400 K) (Figure 3.4a). As one moves to the edge or limb of the image, one progressively sees the cooler, upper photospheric layers, along a slanted path towards the limb (towards temperature minimum region T_{eff}=4400 K), because $dT/dh < 0$. Thus, there is a limb darkening, or a decrease of solar intensity with the angle θ (Figure 3.4b).

A good approximation for the limb darkening is given by the following relation:

$$I(r) = I(0)\left[1 - u\left(1 - \sqrt{\frac{a^2 - r^2}{a^2}}\right)\right] \tag{3.9}$$

where a is the radius of the solar disk, r is the radial distance from the disk centre, and the limb-darkening coefficient, u is given by

$$u = \frac{I(centre) - I(limb)}{I(centre)}. \tag{3.10}$$

This can be written as:

$$I(\theta) = I(0)[1 - u(1 - \mu)], \tag{3.11}$$

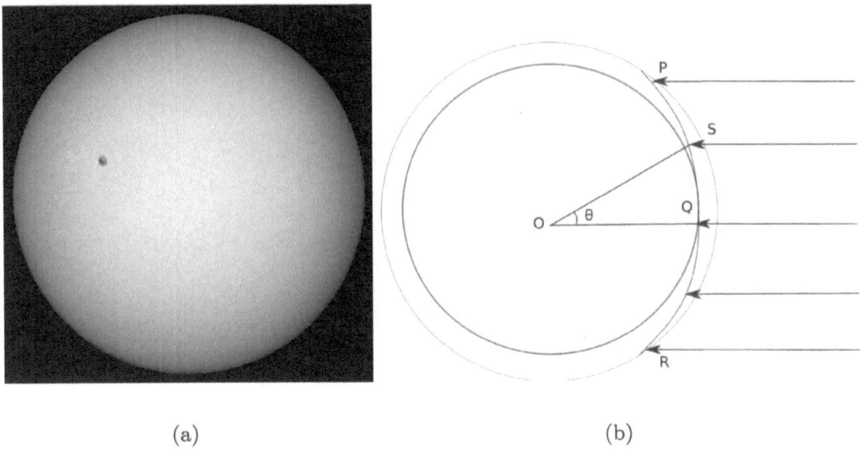

FIGURE 3.4: Geometry of the Sun's image. (a) A solar photospheric image (g-band 430.5nm) clearly displaying limb darkening (USO filtergram), and (b) schematics of observing higher layers as one moves from disk-centre to the limb.

where, θ is the angle between the normal to the Sun's surface and the line-of-sight to the observer, and $\mu = \cos\theta$.

The limb-darkening coefficient, u, is strongly wavelength dependent; limb darkening decreases with increasing wavelength [174]. For example, at $\lambda =$ 320 nm, one gets u=0.95, while at $\lambda = 600$ nm, u= 0.56, reducing even further at larger wavelengths, making the intensity profiles nearly flat from disk center to the limb. Due to the high opacity in far-infrared at 9000 nm, radiation comes only from the outer-most layers of the photosphere both at the limb and disk-center, and the limb-darkening curve in far infrared is almost flat. Thus, observations of solar limb darkening are useful in determining the temperature structure of the Sun's atmosphere.

An opposite effect of *limb-brightening* is observed in the images of higher layers of the Sun's atmosphere, from the height where the temperature minimum occurs, *i.e.*, $dT/dh \sim 0$, and temperature begins increasing with height. The change from limb-darkening to limb-brightening, in fact, provides a means to locate the height of the temperature minimum layer.

3.3.2 The granules

The simplest solar imaging observations in white light continuum through a good solar telescope, under good atmospheric seeing conditions, show a granular pattern covering the entire solar photosphere. Granules are polygonal

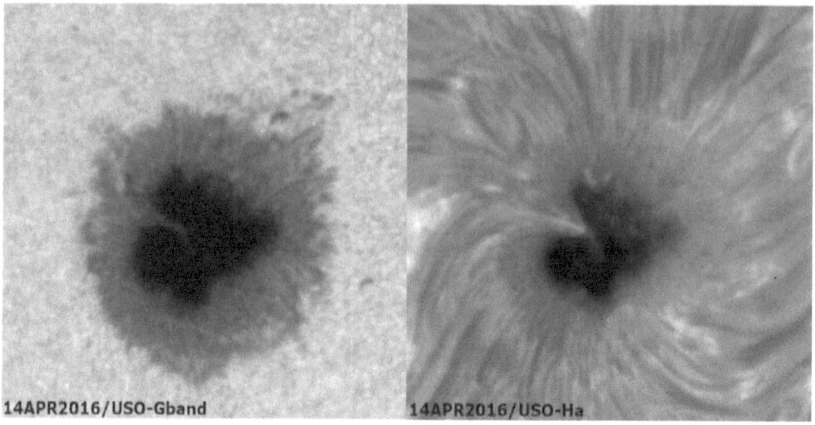

FIGURE 3.5: A high-resolution (left) solar photospheric filtergram taken in the g-band (430.5nm) of solar spectrum shows the sea of granulations, and a single large sunspot, with dark umbra surrounded by a fainter penumbra, and (right) the corresponding chromospheric image in H_α656.28 nm showing the dark umbra, surrounded by penumbra, and a super-penumbral pattern. (USO Filtergrams)

cells in 'quiet' photosphere, and provide the evidence of convection. They are bright upward-moving hot parcels of gas elements having mean cell size of ~1.76 seconds-of-arc [324], (Figure 3.5). This suggests that the granules represent the convection in a very shallow layer (less than 2000 km) below the solar surface. The central upflow velocity is ~0.4 km/s, surrounded by a horizontal outflow of ~ 0.25 km/s, and a small downflow at the granular boundaries.

The granules are separated by thin inter-granular lanes of dark, cooler downward-moving material. As they cool down, the granular material sinks back along the dark inter-granular lanes. Quiet photosphere granules are ~ 30% brighter than the inter-granular lanes, which implies a temperature difference of ~ 400°K.

The granules continuously evolve: coalesce, expand, fragment, and explode with an average lifetime of ~ 6 minutes; the largest granules last longest [362]. The granules are very small and low contrast structures, therefore, their observation requires a relatively large telescope with good spatial resolution, and from a site of good 'seeing', or low atmospheric turbulence.

3.3.3 Mesogranules and supergranules

Apart from granules, even larger scale flow structures, such as *mesogranules* (5-10,000 km) [288], and *supergranules* (average size ~ 30,000 km) are present on the Sun [234]. Mesogranulation was found as late as in 1981 as a network of 'active' granules with a mesh size of 10 arc-second (7000 km). Their vertical velocity is ± 150 m/s with an rms speed of 60 m/s. The horizontal velocity component varies around 500 m/s, reaching up to 1 km/s. The mesogranulation patterns have not been successfully connected to convection cells with a distinct scale.

The supergranules are discernible more clearly in solar Doppler velocity images, with a mean life-time of 36 hours. They are also observed in solar images taken in CaK, magnetograms, *etc.* Indication of these patterns was first obtained in 1956. In 1962 Leighton and his coworkers identified a cellular pattern with a typical cell diameter of 1.6×10^4 km, a mean spacing of cell centers of 3×10^4 km (40 arc-second), and lifetimes of $\sim 120\times$ granules' lifetime. These larger scale sizes of convection are more deep rooted than the granules.

The fluid flows observed in supergranules appear to carry magnetic field bundles to the boundaries of these cells, where they produce the bright chromospheric networks, observed in CaII K and H_α images.

There are also patterns with even larger length scales of the order of over 10^5 km, called *giant cells*, on the solar surface. No typical patterns of the giant cells have yet been established, but they may play a role in the differential rotation of the Sun.

The granules, mesogranules, and supergranules are generally interpreted as manifestations of convection at different scale sizes, with an associated overshoot into the upper regions of the solar atmosphere.

3.3.4 Pores and sunspots

Embedded within the sea of granules are seen, at times, dark structures, such as tiny pores, and large sunspots. Pores are essentially very tiny sunspots of 1000-3000 km size with no developed penumbra. Well-developed sunspots show a dark interior (umbra), surrounded by a lighter area (penumbra) (Figure 3.5). They display rapid change in appearance and numbers.

Dark sunspots are in fact the most prominent, and easiest to observe features on the Sun. They usually occur in groups, and appear dark due to their relatively lower effective temperatures than the surrounding medium. According to the Stephan-Boltzmann law, black-body irradiance is proportional to T^4. Therefore, the ratio of the brightness of a typical sunspot (say, T=4100 K) to that of the surrounding disk (T= 5800K) would be $(4100K/5800K)^{1/4}$ = 25%.

Sunspots mostly occur in a latitude belt between $40°N - 40°S$ around the solar equator, which is also termed the *activity belt.* They first emerge on the solar photosphere as small pore-like structures, and may eventually grow into larger, well-developed sunspots with strong magnetic fields that may reach up to 4000 G.

Sunspot lifetimes may range from less than a few hours to several weeks. They evolve continuously as they are born and grow in size. On the Sun, they make a transit from the east to the west limb, and may reemerge after a full rotation of the Sun, although with very much changed configuration.

Their sizes range from the resolution limit of a telescope to around 100,000 km! But usually not more than $\sim 1\%$ of the Sun's visible hemisphere is covered by sunspots at any given time. There are also prolonged periods when no sunspots are observed on the Sun.

Sunspots and pores possess enhanced magnetic fields as shown by photospheric magnetic field maps. The bright faculae observed in the photospheric images around the sunspots are also the areas of enhanced magnetic fields; however, their field is concentrated in much smaller scale flux bundles than in the sunspots.

3.3.5 The Evershed flow in sunspots

A radial flow is observed at the border between the umbra and the penumbra with speeds varying from around 1 km/s to a maximum of around double this in the middle of the penumbra. It falls off to zero at the outer edge of the penumbra. This is known as *Evershed flow*.

The dynamics of the flow have not been fully understood yet. Although the Evershed effect seems to stop abruptly at the outer edge of the penumbra, the outflow itself must continue, though tracing its path has proved difficult.

Theoretical and observational studies suggested that much of the continuing flow may follow magnetic field lines that go below the visible surface of the Sun, at or just beyond the edge of the penumbra. The model calculations show that the flow acts like a siphon which is driven along a magnetic flux tube by the pressure drop between the endpoints of the tube [276]. This effect is common to every spectral line formed at temperatures below 10^5 °K. This fact would imply a constant downflow from the transition region towards the chromosphere.

Larger sized sunspots often show a superpenumbra in chromospheric images, extending far beyond the penumbral boundary. Along the chromospheric fibril structures of superpenumbrae (Figure 3.5), an inward flow, termed *inverse Evershed flow* of the order of 20 km/s has also been observed.

3.3.6 The solar coordinate system

The rotation axis of the Sun is given by (i) *the angle of inclination* i between the ecliptic plane and the equatorial plane, and (ii) *the angle of the ascending node* α of the Sun's equator, *i.e.*, the angle counted in the ecliptic between the vernal equinox direction, and the direction where the solar equator cuts the ecliptic from below. It is to note that the Earth's precession shifts the equinox direction by 50 arc-second per year. Therefore, α increases by the same rate. Thus, the epoch must be given with coordinates related to the equinox.

Carrington in 1863 determined the angles $i = 7.25°$ and $\alpha(1850) = 73.67°$. The Greenwich sunspot data from the period 1874–1976 imply $i = 7.12° \pm 0.05°$. Lines of longitude and latitude on the Sun wobble by the angle i as Earth orbits around the Sun during the year, because the orbital plane of the Earth is not exactly aligned with the equator of the Sun.

On a solar image, *heliographic latitude* of a feature is defined by ψ as the angular distance from the solar equator. The polar angle or co-latitude $\theta = \pi/2 - \psi$ is used too. As there is no physically unique or fixed reference for defining the longitude ϕ on the solar surface, Carrington devised a method by dividing time into intervals of 27.2753 days, which are called *Carrington rotations.* Thus, there are 13.38 rotations in one year or 365 days. Carrington defined Rotation number '1' to have started on November 9, 1853. At the time of commencement of a new rotation, *Carrington longitude* is assigned to the center of the solar disk as $\phi = 0$. These heliographic coordinates, ψ and ϕ are measured in a system rotating with the Sun.

The Carrington longitude of the central meridian is 360° at the beginning of the Carrington rotation, and decreases to zero at the end. The longitude of a feature on the Sun (such as a sunspot) remains approximately constant in the Carrington coordinate system. For more details one may refer to [360].

Carrington rotations are related to the motion of the Earth around the Sun, *i.e.*, the same point at the solar equator is toward the Earth after one Carrington rotation. This is called the *synodic period*, whereas, the 'true' rotation period with respect to the stars is the *sidereal period* of 25.38 days.

3.3.7 Solar differential rotation

Carrington inferred from following the sunspots' positions on the Sun's disk that the Sun rotates *differentially.* He noticed that sunspots located at higher latitudes are carried around the Sun more slowly than the sunspots at lower latitudes according to the relation

$$\Omega(\psi) = 14.25 - 2.75 \sin^{7/4} \psi \tag{3.12}$$

degrees per day, where ψ is heliographic latitude.

The development of spectroscopic techniques in the second half of the 19th century provided another method to determine Sun's differential rotation independent of sunspot observations. This method, first carried out by Hermann Vogel in 1871, was based on the measurement of the wavelength shift of solar spectral lines using Doppler effect between the approaching and receding solar limbs. By the late 1880s Nils Dúner (1839-1914) demonstrated that the Sun's polar regions rotate about 30% more slowly than its equator.

The advantage of the spectroscopic method is that it is not restricted to only the latitude belts where sunspots are present, or on their appearance.

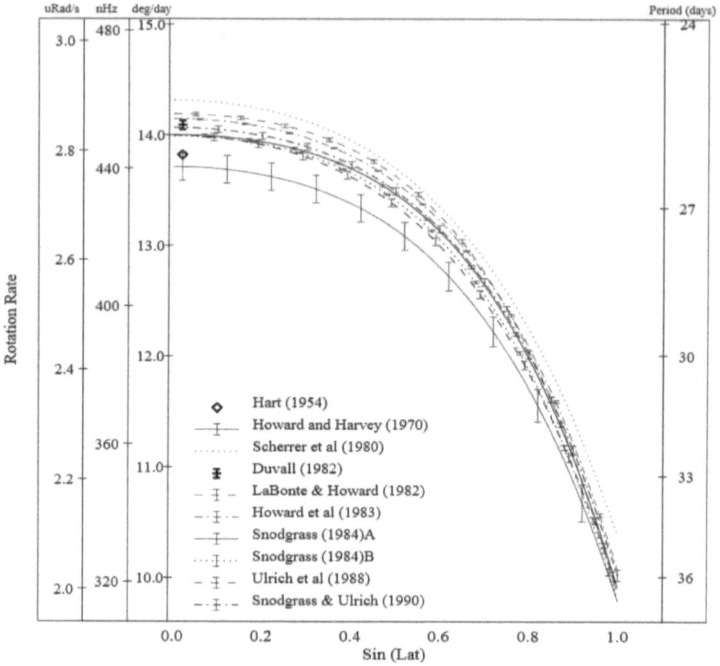

FIGURE 3.6: Plots of differential rotation curves using spectrographic measurements by various researchers. With permission from [37].

These results demonstrated that sunspots rotate at nearly the same rate as the Sun's photosphere (Figure 3.6).

The more modern approach is to relate the photospheric rotation rate with latitude as:

$$\Omega(\psi) = A + B\sin^2\psi + C\sin^4\psi + ... \tag{3.13}$$

degrees per day, where A, B and C are some constants. Different workers have given these constants as (14.37, -2.69, 0) [285], (14.55, -2.84, 0) [182], and (14.05, -1.49, -2.61) [341]. The period of solar rotation varies from 25 days at the equator to more than 30 days at the poles.

The Sun is believed to have rotated much faster in its past phase of early evolution than today. It has shed much of its angular momentum by magnetic braking in the early periods. This has been inferred from *T Tauri* stars having similar masses as our Sun, but being much younger in their evolutionary phase of entering the main-sequence of the *H-R diagram.* They exhibit much larger surface velocities of around 15 km/s as compared to 2 km/s for the present Sun.

3.4 Optical observations of the solar chromosphere

The temperature of the solar atmosphere starts rising from a minimum at an altitude of about 500 km above the photosphere. Under normal conditions, the bright photosphere is directly seen through this rarefied, fainter layer, called *the chromosphere*, which has an average thickness of $\sim$ 2500 km, and is much hotter, T $\gtrsim 10,000°$K.

The physics of the thin outer atmosphere of the Sun is quite different from the physics of the solar interior and the photosphere. As the density drops rapidly above the photosphere, the collisions between particles become less important, hence, the deviations from LTE become important. As a result, ionization and excitation levels cannot be predicted from the Saha and Boltzmann equations any longer. Another consequence is that the magnetic pressure begins dominating over the thermal pressure. As a result, the properties of plasma in this region on a large scale are described by MHD formalism.

The chromosphere is a few thousand times fainter than the photosphere. Therefore it is normally not observed as it is optically transparent in visible wavelengths. It is seen only briefly as a thin red crescent, irregular layer of gases, just before (and after) the totality of solar eclipses, when the bright photosphere gets nearly fully covered by the Moon. The name 'chromosphere' is derived from the bright red colour displays due to its dominant emission in H_α wavelength.

Chromospheric emission lines briefly show up as a flash spectrum during total solar eclipses. Using a suitable chromospheric line, spectrographically it is possible to observe the chromosphere routinely not just around the limb (as during TSEs), but on the entire disk also. For this purpose, narrow passband optical filters centred in H_α 656.28 nm (red line), or singly ionized CaII K 393.4 nm (blue line), are used to block the intense continuum of photospheric light.

Chromospheric H_α is an exceptionally strong absorption line. Contributions to H_α line span from the photosphere (in the line wings) up to the upper chromosphere 1500-200 km (in the line core). Similarly, profiles of the Ca K-line cover the chromosphere from 500-2000 kms. Most of the Fraunhofer absorption lines are weaker, and are formed entirely below the temperature minimum in the photosphere.

Apart from the visible absorption lines, chromosphere can also be observed by radio emissions produced by electrons in the presence of strong magnetic fields. Emissions in shorter wavelength, like UV, also occur in the upper chromosphere, but its observation requires space-based instruments.

FIGURE 3.7: A high resolution H_α full disk filtergram constructed using many small-field images of solar chromosphere shows fine structures, sunspot, bright plages, super granular networks, dark elongated, thread-like filaments, and prominences hanging above the limb (USO/MAST H_α).

3.4.1 Chromospheric features

Chromosphere is highly structured even away from intense magnetic field regions of the sunspots, as the dynamics of hot gases begins to be guided by magnetic fields (Figure 3.7). *Magnetic network structures* are observed with scale sizes comparable to the photospheric supergranules.

Small dark dots, or mottles, are observed embedded within these networks. These are most likely associated with *spicules*, the 'grass blades' like features seen on the limb. In fact, the entire chromosphere is covered with these features having typical diameters of 500 km. They eject material upwards into the hot corona at speeds of 20-30 km/s, reaching $\sim$ 5,000 km above the photosphere, which may last 5-10 min.

Usually bright *plage* (the French word for beach), or chromospheric faculae, are seen distributed around the dark sunspots in the chromosphere.

The sunspot group, as seen in the photosphere, is only the footprint of a more comprehensive three-dimensional entity known as an *active region.* It extends upwards through the relatively irregular and rarefied chromosphere (Figure 3.5), and beyond in the corona. The active regions are usually associated with various forms of solar transients, which encompass a diverse range of phenomena. As we shall see later, the occurrence of most of the solar activity is directly controlled by the interplay of plasma flows, differential rotation between layers at different latitudes, and their interactions with Sun's magnetic fields.

3.4.2 Solar filaments

Filaments are dark ribbon-like elongated features commonly seen in the chromosphere (but not in the photospheric or broad-band images). These are large accumulations of (relatively) cold gas of increased density suspended high in the solar atmosphere by the Sun's magnetic field (Figure 3.8, left frame).

As the filaments rotate on to the limb of the Sun, they appear as bright cloud-like structures against the dark sky background, and hence, traditionally termed the 'prominences' (Figure 3.8, right frame). Large prominences are often visible at times of solar eclipses in the form of small reddish filaments or blobs jutting out in the lower corona. The first photographic image of a solar prominence was captured by Charles A. Young (1834-1908) in the eclipse of 1870.

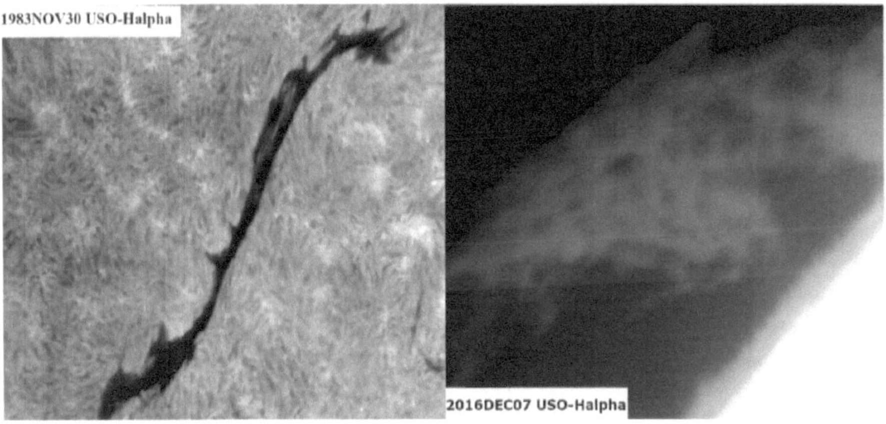

(a) (b)

FIGURE 3.8: Prominence/filaments as imaged in H_α 656.28nm: (a) a dark elongated filament as seen on the disk, and (b) bright cloud-like prominence hanging above the limb. The thickness of fainter chromospheric limb is visible above the photosphere (USO H_α-filtergrams).

The typical thickness, height, and length of filaments are 5000 km, 50,000 km, and 200,000 km, respectively. Their plasma density $\sim 0.5-1.0\times10^{-17}m^{-3}$ is $\sim$ 500 times larger than the ambient coronal density, but they are relatively much cooler (5,000 - 10,000 K). Direct measurements of the prominence magnetic field using the Zeeman and Hanlé effects indicate weak fields of 5-10 G in the prominences.

The prominences protrude far above the average chromosphere, so they are surrounded by 100 times hotter, 500 times lower-density coronal matter at 10^6 K (Figure 3.8). However, they can remain stable for days to weeks. The long life-time of *quiescent prominences* indicates a remarkable stability of their MHD equilibrium. Thus, it is indicated that they are thermally and pressure isolated from the surrounding coronal environment. (For a detailed review, one may refer to [295].)

Prominences are found mainly in two groups: (i) the *polar crown filaments* which are seen in high latitude polar locations, and (ii) filaments located in mid-latitudes. The polar crown filaments appear a few years after a sunspot maximum, and then show a pole-ward migration, reaching the poles around the time of the succeeding maximum.

Some small filaments develop within active regions containing sunspots, but these are usually more active, having shorter life times depending on the state of evolution within the active region. On the other hand, the *quiescent filaments* lie along magnetic polarity inversion lines, separating large areas of weak fields of opposite polarity. They remain stable over several weeks. However, high speed plasma motions of about 5 km/s are observed even in relatively quiescent prominences. They can end their lives through a slow decay process, or sometimes disappear more suddenly as the fields holding them in balance are destabilised.

At the chromosphere they appear to be connected by 'barbs' to the supergranulation cell boundaries. As this connection is gradually lost due to the flux evolution below, or around them, the balance is lost and prominence may eventually erupt. This is called 'disparition brusque' event.

The total mass in a few large prominences is comparable to that in the whole coronal volume. The prominence plasma seems to be draining downward at a rate which may be sufficient to deplete the corona within a few hours. The rapid draining of prominence plasma suggests that a mass supply from the chromosphere below is required.

To observe the chromosphere against the disk, one usually needs a narrow passband filter of 0.05–0.07 nm. But for prominence observations seen against the dark sky background, a 0.3–0.4 nm passband filter is adequate. Another advantage of using a broadband filter is that one can observe faster moving parts of a prominence along the line-of-sight. For observing the limb prominences, another method used is to block most of the strong light of the solar photosphere using an occulting disk, and image only the outer region of the chromosphere. Lyot filters are widely used for routinely observing a variety of chromospheric phenomena.

3.4.3 Possible mechanisms for heating of the chromosphere

Temperature increases in the chromosphere with height above the temperature minimum region of the photosphere, and the rarefied chromospheric gas starts to deviate from local thermal equilibrium (LTE). If some physical processes continuously keep heating the chromosphere, it does not need to find equilibrium with the lower atmospheric levels.

There are the acoustic photospheric fluctuations and the magnetic network which may provide the heating of the chromosphere on a continuous basis. The acoustic fluctuations of the photosphere lead to a power density of the sound waves in the chromosphere (Section 6.1). Its contribution has been estimated to be around 10 Wm^{-2}. This may be sufficient to heat the chromosphere up to 10^4 °K.

The acoustic fluctuations do not reach the coronal altitudes, and thus the mechanism to heat the corona must be related to the magnetic field. In the coronal heights, the magnetic heating processes, through MHD waves or magnetic reconnection (current sheet dissipation) becomes increasingly important. It may be noted that in the absence of the continuous heating process, the chromosphere would cool down within about 20 minutes' time.

3.5 Optical observations of the white light corona

Above the chromosphere, temperature increases very steeply over a very thin layer, called *the transition region*. This must be associated with an equally steep decrease in density for pressure balance requirements. Although, the magnetic pressure is quite substantial, it is not enough to compensate for the steep temperature gradient. The transition layer is very thin, and is observable only from space by the UV and EUV emissions from highly ionized species, produced by the high temperatures (see Section 8.6.1).

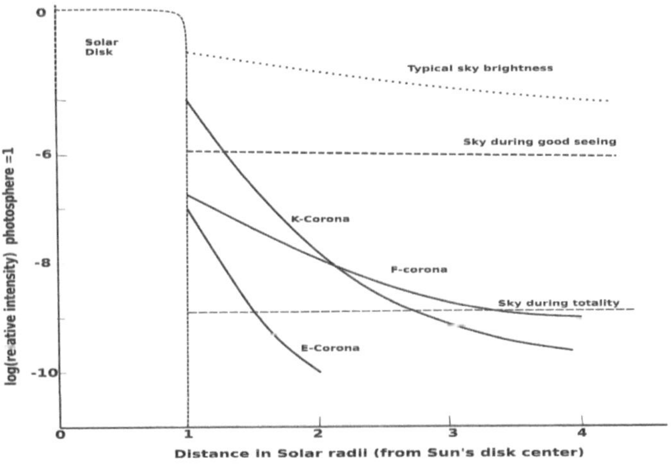

FIGURE 3.9: Typical radial profiles of various components of the solar corona along with sky brightness with the distance from the Sun's disk.

The temperature further increases above the TR with distance up to several million °K, as the density continues to decrease. It is the outermost very tenuous layer of the solar atmosphere, called *the corona*. The corona in white light is produced due to the scattering of photospheric photons by the fast moving electrons present in the corona. However, the density of the corona is about 10^{-12} times less than the photosphere, and the intensity is around 100 million times less than the photospheric intensity. Therefore, it remains normally invisible in optical wavelengths from ground due to the intense glare produced by the scattering of Sun's photospheric light by the dust and molecules present in the Earth's atmosphere. The intensity of this scattered light is far greater than the intensity of the faint corona (Figure 3.9).

The faint glow of white light corona is revealed from ground only when the photosphere is fully covered by the Moon during a total solar eclipse (Figure 2.10a). The Moon cuts off the intense photospheric light falling on the Earth's atmosphere during the short duration of totality. At that phase, only the faint coronal light enters the Earth's atmosphere. This coronal light too gets scattered in the atmosphere; however, its intensity is several times less than the coronal emission. Therefore, historically the total solar eclipses have been eagerly looked forward to, as they provided the rare opportunity when solar corona could be studied.

The first photograph of a solar eclipse was obtained on 28 July 1851 by Berkowski at the Königsberg Observatory (then in Prussia, now Kaliningrad in Russia). Although Warren De Le Rue was not the first to photograph a solar eclipse, his eclipse photos of 1860 are still studied (Figure 1.5a). Eclipse photographic techniques were further improved by using radial gradient filters. These filters were designed to differentially attenuate the innermost, brightest portion of the corona. The resulting photographs allow to discern details of coronal structure out to many solar radii. In modern times, detailed and extended images of solar corona are obtained during eclipses by making composites of several digital images taken at many short to long exposures (Figure 2.10).

3.5.1 The various components of the solar corona

Coronal emission consists of three components: (i) *K-corona*, or 'Kontinuum' (in German for continuum), (ii) *E-corona*, or the emission line corona, and (iii) *F-corona*, or the Fraunhofer corona [47]. The white light K-corona is the main component of the corona extending to several solar radii beyond the solar photospheric disk. It arises due to the scattering of photons from the photosphere by the electrons present in the corona. The large speeds of these coronal electrons smear out the Fraunhofer lines present in the photospheric light.

The K-coronal light is linearly polarized due to the scattering of the photons by the electrons aligned to the coronal magnetic field. Its intensity depends on the electron density in the corona, and is observed to be highly structured. The electron density and the alignment of the coronal magnetic field are determined from the intensity and polarisation measurements carried out during the time of the total solar eclipse [18].

In addition to the white light emission, there are other forms of coronal emissions that are produced in the optically thin coronal atmosphere. These emissions can be observed close to the Sun's limb, and also from space. The E-corona appears due to the emission of radiation by the highly-ionized species present in the million degree hot corona.

The third component of the solar corona, *i.e.*, the Fraunhofer, or F-corona, arises due to the scattering of photospheric light by the slow speed dust particles extending all the way into the interplanetary medium. It is named so because Fraunhofer lines of the photosphere can be seen in the spectrum of the F-corona. It is usually observed as the zodiacal light, a faint emission concentrated along the ecliptic.

Thus, the K- and F-coronae are not truly coronal light, but arise due to reflection of the bright photospheric light by the electrons and fine dust particles, respectively. The plasma frequencies corresponding to the electrons and ions of the corona can be routinely observed in radio wavelengths, and maps of extended solar corona can be obtained without the need for an eclipse.

3.5.2 Discovery of million-degree temperature of the corona

In the coronal spectrum, in addition to emission lines from the chromosphere (such as, H_α, H_β, Helium D3 and others), two emission lines of coronal origin are seen in the red (637.4 nm) and green (530.3 nm) wavelengths (Figure 3.10). These are the strongest coronal emission lines observed in the visible wavelengths, although currently dozens of other coronal emission lines are known. These are known as *forbidden lines* corresponding to the highly ionized Fe XIX and Fe X in the emission spectra taken during total solar eclipses. Their presence indicates very low densities, and high coronal temperatures of over 2×10^6 °K, that is required to ionize iron to such an extent. At these high temperatures, the corona also emits in EUV and X-rays wavelengths from an extended but optically thin envelope that can reach a significant fraction of a solar radius out from the photosphere. These emissions are used in obtaining coronal images, both on- and off-disk, from space.

C.A. Young and W. Harkness first obtained the spectrum of solar corona during a total solar eclipse in North America in 1869, and discovered the coronal emission line at the wavelength of Fe XIV 530.3 nm. The wavelengths of this coronal emission line did not match those from elements known in the 19th century, and its origin remained a mystery. As these bright emission lines at wavelengths did not correspond to any known materials, they were attributed to hitherto unknown '*coronium*'! For an early study on the possible origin of the coronal emission lines, refer to [407].

W. Grotrian had noticed a possible transition whose wavelength matched the red coronal line (637.4 nm). This transition is a forbidden transition, so called because the transition takes a very long time under laboratory conditions of evacuation, where the density of gases is still relatively high. In

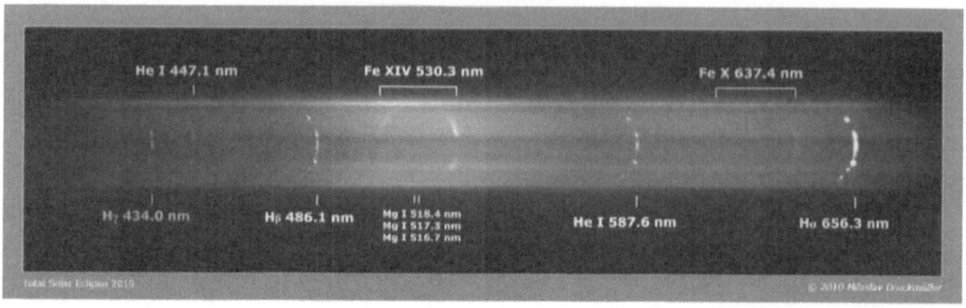

FIGURE 3.10: A slit-less spectrum of the solar corona obtained during TSE 2010: The spectrum of the corona is dominated by two forbidden emission lines of highly ionized iron - Fe XIV line at 530.3 nm and Fe X at 637.4 nm. The spectrum of the upper chromosphere, and prominences consists of hydrogen Balmer series lines at 656.3 nm, 486.1 nm and 434.0 nm, and two helium lines at 587.6 nm and 447.1 nm. Image courtesy of Miloslav Druckmüller.

a tenuous gas like the solar corona, collisions are infrequent, and ions have enough time for such forbidden transitions to take place.

According to the suggestion from Grotrian, Edlén (1943) made a systematic study of UV emissions from highly excited ions in spark discharge experiments. He identified the origin of many coronal lines, and attributed them to forbidden transitions between low-lying fine structure states of heavy and many times ionized atoms [408]. The coronal green line was identified as due to Fe XIV. The best-known lines are the Fe XIV 530.3 nm (green line), the Fe X 637.4 nm (red line), and the Ca XV 569.4 nm (yellow line). In all, he could identify 19 of the 24 coronal emission lines known at that time. Very high temperatures of more than 10^6 °K are needed to produce these highly ionized states.

This fact of high coronal temperature also explains the great extent of the scale height of the corona. If the corona had photospheric temperatures, then its scale height, *i.e.*, the height over which the density decreases by a factor $e(= 2.7)$, is only about 150 km. Therefore, at distances of a solar radius (7×10^5 km) the density would be essentially zero. However, with a temperature of the order of two million°K, the scale height is about 10^5 km, which means its density drops very slowly with radial distance.

3.5.3 The coronal heating problem

What causes the corona to be so hot remains one of the most vexing problems in astrophysics. As the density of the corona is much smaller than that of the photosphere, the thermal energy density of the corona (although it is 200-300 times hotter than the photosphere) is negligibly small compared with the photospheric energy density.

From the second law of thermodynamics, heat cannot flow from the photosphere at lower temperature to the higher temperature corona to raise its temperatures. However, there is no plausible source of energy further out in the corona to heat up the corona; we must assume that some form of energy other than heat is supplied from below the corona to realise its million-degree temperature.

In principle, as the amount of required energy is not so much, it can be taken care of easily by the many solar activities that are continuously operating. The energy flux needed to power the active regions is of the order of $10^4 Wm^{-2}$. On the average only a fraction of this power in electromagnetic radiation would be sufficient to heat the corona. The difficulty, however, is posed by the fact that the corona is optically very thin, so one has to find the suitable mechanism to absorb the electromagnetic radiation. There is enough energy also contained in the solar magnetic field, but how to convert it into heat is a problem.

3.5.4 Possible mechanisms of coronal heating

There are several mechanisms, known from plasma physics, which can help to dissipate the magnetic energy as heat, such as waves, nonlinear instabilities, current sheet dissipation, and magnetic reconnection. Thus, currently there are two broad classes of mechanisms of coronal heating: (i) MHD waves (acoustic and magnetic), and (ii) magnetic reconnections (microflare/nanoflare theory). Both these heating models have their merits and demerits, but the essential issues are nearly pinned down.

Acoustic waves do not reach far into the chromosphere, let alone to the transition region or corona. They steepen into shocks in the rapidly decreasing density regions. However, MHD waves can be excited by the motion of magnetic and acoustic disturbances in or near the photosphere. When these waves propagate outward, they are damped and part of their energy is transformed to heat. Slow mode magneto-acoustic waves propagate and form shocks, thus dissipating in the chromosphere. Fast modes, however, manage to reach transition region between chromosphere and corona, and they are either dissipated (heat) or get reflected back. More recently, there are many evidences of coronal waves detection. They may carry adequate energy to heat the corona.

Large flares can release enormous energy of the order of 10^{25} J in time periods of a few minutes. However, such large flares are too localised, and do not occur very frequently. At most, they can only provide around 1% of heat to sustain the 10^6 °K temperature of the corona. One would rather require small flares occurring uniformly all over the Sun, and on a continuous basis, to explain the coronal heating. As discussed in Chapter 8, the Skylab mission in 1973 revealed the presence of X-ray bright points (XBPs), and their uniform distribution over the whole Sun. They exist during active, as well as in quiet phases of the Sun. They are tiny micro- and nano-flares, and most likely occur due to continuous reconnection processes of the dynamic magnetic field structures, the *magnetic carpet* in the lower atmosphere. Thus, they can supply the substantially large amounts of energy required to explain the heating of the corona (Section 8.8, page 220).

3.6 Optical observations of solar energetic transients

Several types of highly localised, short-lived (transient) explosive phenomena are observed to occur in the outer solar atmosphere, from the chromosphere to the corona. The source of energy for most of these energetic events is believed to be the magnetic energy available in solar active regions. Solar flares, tornados, filament/prominence eruptions, and coronal mass ejections (CMEs) are some of the observed explosive solar phenomena.

The solar transients result in violent bulk mass motion of plasma, accelerated particles, and enhancement of radiation, over the entire electromagnetic spectrum, ranging from γ-rays to radio wavelengths. Some of these aspects are observable in optical wavelengths using specially designed instruments, which we discuss in this chapter. There are other signatures in shorter and larger end of the electromagnetic spectrum, which shall be described in the following chapters.

3.7 Solar flares

Solar flares are sudden, intense brightening in localised areas on the Sun. The catastrophic energy enhancement occurs over the entire electromagnetic spectrum. It is not only the amount of energy released in flares, but also the suddenness of its release which makes the flares so spectacular. For a recent review of solar flare observations, one may refer to [42].

Normally, flares are not observable in the photosphere, except in some exceptionally energetic and rare class of events, the so-called, *white light flares* [187]. However, they can be routinely observed in the chromospheric H_α wavelength from ground-based observatories. They usually appear as elongated bright patches of emission on either side of dark filaments tracing magnetic field polarity inversion lines (Fig. 3.11). As flare emissions progress at both sides of the inversion line, a post-flare "loop" structure evolves. The bright ribbons separate out and move away from the inversion line as the flare progresses, indicating propagation of energy release site and a reconnection process of associated magnetic field structures occurring high above in the corona (Chapter 8).

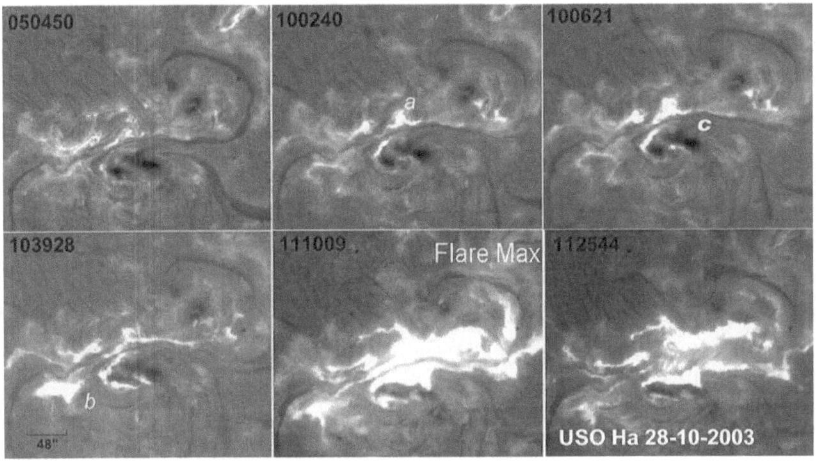

FIGURE 3.11: A large two-ribbon flare in the active region NOAA 10486 on 28 October 2003 observed in H_α 656.3 nm wavelength. Elongated bright patches of emission are seen separating away with time (in HHMMSS UT) on either side of a dark filament along the magnetic field polarity inversion line (*USO* H_α*-filtergrams*).

Flare energy, time and size have a wide range: $10^{17} - 10^{26}$ Joules, released in nano-flares to major flares over a time period of a few minutes to several hours. The energy released in a typical large flare may be several times 10^{25} J, the equivalent of several billion hydrogen bombs. For a comparison of such energies, one may remember that the luminosity of the Sun is 4×10^{26} Js^{-1}.

Apart from the enhancement of electromagnetic radiation, flares are usually associated with accelerated particles, hydrodynamic plasma flow, blast and shock waves, erupting sprays, interplanetary disturbances, *etc.*

Figure 3.12 shows an example of flare phenomena occurring at the limb, showing the vertical structure and extent of a typical flare. Flare brightening, along with associated material ejection, or *surges* of ascending, and then descending motions of the chromospheric material is evident, with upward velocities of 50–200 km/s, reaching heights of about 100,000 km. Larger speed *sprays*, more explosive mass eruption events, are also observed where fragments of material are ejected out at velocities of up to 2000 km/s, *i.e.*, far greater than the escape velocity on the Sun.

Flares were erroneously considered as only a chromospheric phenomenon for many decades after their discovery. However, now from the mounting evidences obtained from space-borne instruments operating in invisible wavelengths, it is clear that flares are associated with all the layers of the Sun's atmosphere. With the development of observational techniques in other

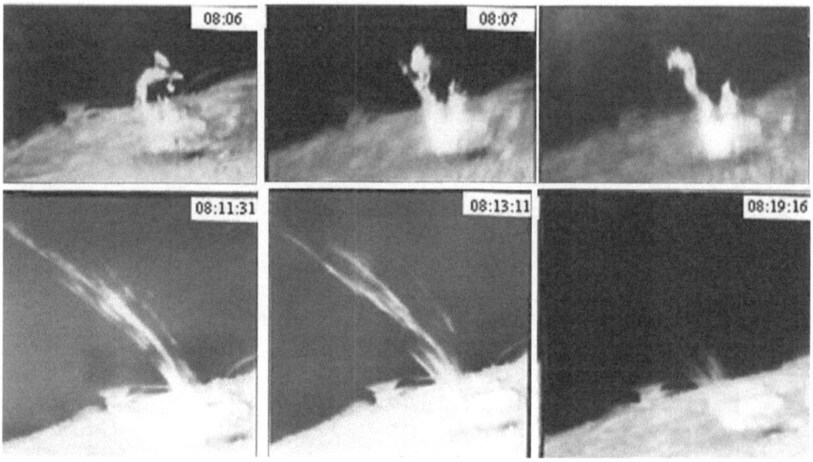

FIGURE 3.12: An eruptive chromospheric mass ejection observed on 1984 March 15 in H_α (USO H_α filtergrams adapted from [10]).

wavelengths, it is now clear that the chromospheric flare is just one facet of a much more complex 3-dimensional process of energy release.

3.7.1 Source of flare energy

Flares are believed to derive their energy from the stressed magnetic structures of the active region. This inference is based on the observational fact that most flares occur in active regions in close proximity to sunspots, which are seats of strong magnetic fields. Also, no other form of energy, *viz.*, gravitational, thermal, or nuclear fusion, can explain the amount of flare energy released, considering the physical conditions existing in the solar atmosphere.

The thermal and gravitational energies are not adequate to account for the large flare energy release as $E_{thermal} = 3 \times 10^{21}$ J, and $E_{grav} = 3 \times 10^{21}$ J, as deduced from appropriate parameters for solar atmosphere. Furthermore, the nuclear fusion process is not viable in the outer layers, except perhaps at the time of very energetic flares when temperature may increase to exceedingly high levels. This leaves only the magnetic energy E_{mag}, which can be expected to explain the large energy release of the order of 10^{25} J in major flares.

There are several processes operating during the evolution of solar active regions that can lead to the storage of magnetic energy. Only a part of this stored energy, called the *free energy*, is suddenly released by some trigger mechanism. The various observed manifestations seem to be only the secondary responses to the original energy release process, converting magnetic energy into radiation enhancement, particle energy, and heat. The diversity of flares, flare-associated phenomena, and the ongoing search for the catastrophic trigger mechanism make flare research an important challenge in solar physics. For more details on the MHD processes operating in flares, one may refer to [335].

3.7.2 Conditions for flare onset

There have been many attempts to find the necessary and sufficient conditions for *flare onset*, using various observational parameters. Most of these efforts pertain to the magnetic field parameters, such as sunspot area, sunspot rotation, their magnetic field strengths, steep magnetic gradients, shear (stress) of magnetic field configuration, or helicity, large flux motions, and flux emergence/cancelations, *etc.* [376]. The potentially flaring locations are usually associated with magnetically complex δ-configuration defined as umbrae of opposite magnetic polarity.

Large flares may also occur in association with *eruptive filaments* (prominences) in locations of low magnetic fields. Smaller flares can occur in quiet regions on the boundaries of the magnetic network, and even in the network interiors. A huge research literature is available on these topics spread over the last several decades. However, a complete understanding of the physics of flares remains an outstanding problem.

As revealed by maps of photospheric vector magnetograms, ARs are locations of strong magnetic fields and twists, where large flares preferably occur [11]. The build-up process of the unstable configuration for large flares is often well observable in photospheric magnetic fields. As the flares derive their energy from the magnetic field, it is expected that the extent of magnetic non-potentiality should relax toward a lower energy state after the excess or *free energy* available in the active region is released. There are some direct or 'tracer' observations which indicate such a relaxation of the non-potential configuration after a large flare, for example, evolution of H_α arcades observed during a large duration flare [93]. The restructuring of magnetic fields with large flares is also reflected by the dramatic formation and evolution of post-flare loops observed in chromospheric images.

The search for magnetic field changes associated with flares has continued since the 1930s [138]. But, most of the earlier results were unreliable due to poor sensitivity and the spatial resolutions of magnetographs used. It is suggested that any observed magnetic field changes may arise from flare-induced line profile variations, and thus may not reveal real change in the magnetic field parameters. However, recent high-quality, high-cadence observations have provided mounting evidence of rapid, permanent changes in the longitudinal and transverse magnetic fields during solar flares ([11], [382], [356]).

3.8 Stability of prominences/filaments

The issue of prominence stability, *i.e.*, the dense prominence material's support against gravity was addressed originally by Kippenhahn and Schlüter (K–S), according to which a horizontal magnetic field is bowed downward by the prominence mass [210]. A Lorentz force acting to balance gravity is generated by a current running transverse to the field lines, along the filament axis. The magnetic field associated with this current alters the original magnetic field, and the filament is supported by the tension in the bowed field lines, whose foot-points are anchored into the photosphere. A prediction of this model is that the magnetic vector should thread the filament in the direction joining the two magnetic polarities observed in either side of the filament. The lower latitude prominences having heights below 3×10^4 km, which occur in active latitudes, are generally consistent with the *K-S* model.

Another model was later put forward by *Kuperus and Raadu (K-R)* [226], associating the filaments with material condensing within a current sheet. The support against gravity of the dense filament plasma is provided by the vertical gradient in magnetic field-line tension. As seen in this model, the closed field-lines thread the prominence both in the direction expected from straight connections between polarities observed to either side of the prominence, and also in the opposite direction. The higher quiescent prominences, such as *polar crown filaments*, generally exhibit the magnetic field structure consistent with the K–R model. More detailed models explain the formation and support of prominence by thermal instability at the tops of several loops in an arcade structure, into which material from the chromosphere accumulates (see articles in [307]).

An objection to these models is that although the quiescent prominences are observed lying along the polarity reversal lines, the field-lines do not run perpendicular to the length of the prominence, but almost parallel to it. In fact the magnetic geometry could be in general helical, as seen in many prominences, particularly well seen during their eruption [346]. It is clear that the theoretical models do not match with the observations, and one major problem remains in the accurate measurements of magnetic fields in and around the prominences.

3.8.1 Eruption of filaments/prominences

As the connection of the prominence (filament) is gradually lost due to the evolution at lower layers, or emergence/submergences of magnetic fluxes in the neighbourhood, the magneto-static balance of the prominence is disturbed. When the static balance is lost, it turns into explosive, *eruptive prominence.* There are also cases observed when filaments once erupted are reformed within a few hours to days.

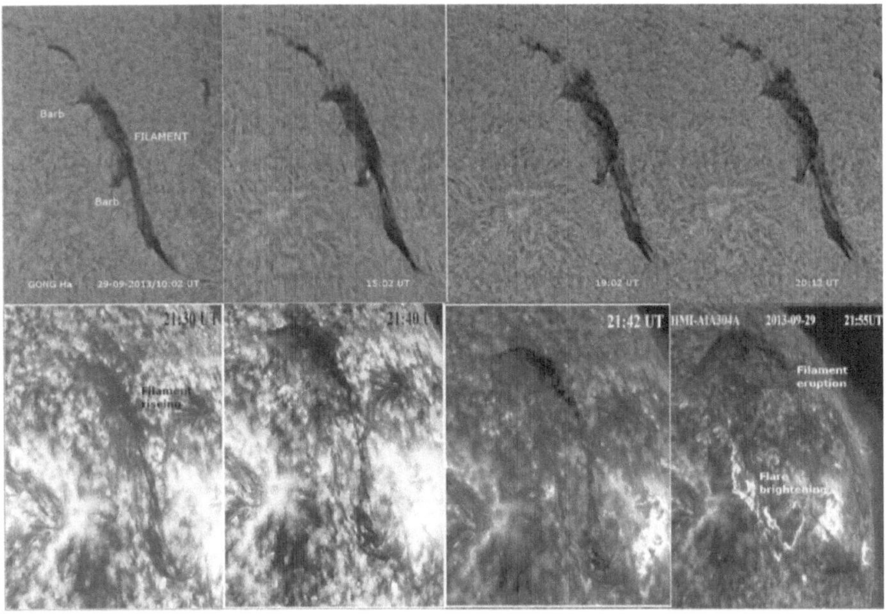

FIGURE 3.13: A filament eruption leading to a flare and a CME observed on 2013 September 29 observed in H_α (GONG filtergrams) and SDO-AIA303Å images (Courtesy of GONG and SDO-AIA Science Teams).

In active regions filaments can erupt in association with solar flares and/or CMEs. In such events, filament oscillation motions are observed followed by a *disparition brusque* event, triggered by a thermal or dynamical instability. While erupting, they rise off the Sun over the course of a few minutes, or even hours (Figure 3.13). The violent phenomenon of prominence eruption implies an explosive rearrangement of the magnetic structure, and its ejection into the extended corona. *Partial or failed eruptions* are also observed. The eruption of filaments is accompanied by other signatures, for example, a flare brightening of the underlying chromosphere. Many of the large-scale coronal mass ejections (CMEs) detected by space-borne instruments appear to be outward moving remnants of erupting prominences.

In many cases, prominence eruptions appear to be spontaneous, as no other observable events seem to have occurred in the neighbourhood, such as a flare, to have acted as a catalyst for eruption. In some events, filament eruption may be related to a large flare at a remote location that triggered a shock wave, called a *Moreton wave*. Magnetic flux emergence in close proximity, or local and large-scale photospheric motions are other triggers leading to filament eruptions.

Viewed at the limb, quiescent prominences mostly appear as curtains of vertical thread-like structures, and occasionally they look like tornados with rotation along the axis of their structures. Solar tornados are dynamical, helical magnetic structures mainly observed as a prominence activity. They apparently do not erupt, but continue to rotate for several hours before subsiding [294].

Helicity is a commonly observed feature associated with eruptive prominences, both on microscopic as well as on macroscopic scales. In some cases, two or more tubes of matter are seen helically intertwined in a rope-like structure [346]. The outward eruptions occur at speeds of the order of a few hundred km/s, and eventually they disappear. Multi-wavelength observations contribute significantly to the understanding of erupting filaments as in the case of flares.

3.9 Coronal mass ejections (CMEs)

In the white light (optical) structure of the rarefied solar corona, seen during total eclipses, large-scale streamers extending to several solar radii are observed. Rapid motions and brightness changes in the white light corona have been noticed during the short periods of TSEs, and also using some ground-based and space-borne coronagraphs (Figure 2.10, page 49). There are occasionally most spectacular and eruptive transients, known as the coronal mass ejections (CMEs), observed being ejected out through the solar corona.

The visual drawings made during the historically most extensively studied total solar eclipse of 18 July 1860 depicted a very peculiar feature in the SW-portion of the corona. These perhaps represented the first visual record of a CME in progress [110]. However, CMEs were first discovered from space in 1973 by OSO-7 (Orbiting Solar Observatory-7), and then by the Skylab mission (1973-74) using an occulting disk based white-light coronagraph. More than 24 such coronal transients were observed using Skylab coronagraph's continuous photographic coverage in the first 4 months [253]. These observations helped in determining the structure, motions, and life times of these transient events.

CMEs are important solar transient phenomena, like the flares, as they influence the physical conditions of the interplanetary medium. Some 10^{13} kg of mass, angular sizes of 40-70°, are ejected in CMEs with speeds reaching up to $2000 km/s$ (Fig. 3.14). Thus, their kinetic energy may be as large as $\sim 10^{25}$ J, *i.e.*, comparable to or even more energetic than large flares. Most of the matter in CMEs originates from the lower atmosphere, and thus is not coronal mass as such. But, the word 'coronal' refers to the fact that CMEs are observed in the corona.

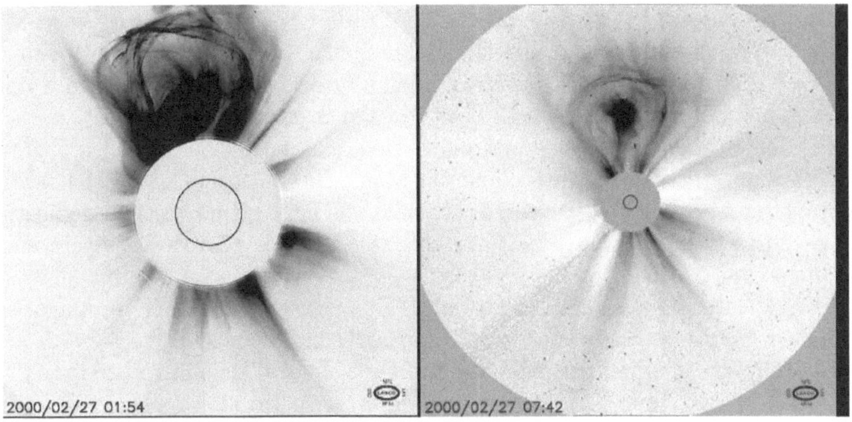

FIGURE 3.14: Large Angle Spectrometric COronagraph (LASCO) on-board the SOlar and Heliospheric Observatory (SOHO) image of a white light CME of 27 February 2000 at (left) 01:54 UT, and (right) 07:42 UT (inverted colour images). The black circles denote the edge of the photosphere. Courtesy of SOHO/LASCO Science team.

Current knowledge of CME kinematics indicates that the ejection starts with an initial pre-acceleration phase characterised by a slow rising motion, followed by a period of rapid acceleration away from the Sun until a near-constant velocity is reached.

The speeds of ejection vary widely, with a maximum of over 1500 km/s for the most energetic events, but down to about a few tens of km/s for CMEs with no associated activity, generally seen near a solar activity minimum period. Observations have established that CMEs are a very common phenomenon occurring both during the minimum and maximum phases of solar activity; the rate of occurrence varying from $\sim$ 1 in two days during solar minimum, to $\sim$ a few per day during solar maximum.

3.9.1 Structure of CMEs

A typical CME usually has a three component structure: a cavity of low electron density, a dense core (the prominence, which appears on coronagraph images as a bright region embedded in this cavity), and a bright leading edge (Fig. 3.14). But not all these may be present in all the CMEs. The temperature of the core is around 10^4 K, and $\sim 2 \times 10^6$ K at the frontal structure. CMEs are identified by attributes, such as its width, position angle, speed and acceleration.

A sub-class of CMEs, known as, the *halo CMEs*, originate on the disk. They propagate directly towards the Earth, and hit the terrestrial surroundings within 2-3 days of their origin on the Sun. Halo CMEs may be geo-effective if they are 'front' halo (Earth-directed). There are 'back' (anti Earth-directed) halo CMEs too, which are generally difficult to distinguish from the 'front' ones, when observed using space-craft located in near-Earth orbits.

CMEs in the solar wind were detected as magnetic clouds or ejecta, from the data obtained by the Interplanetary Magnetic Platform (IMP), Helios and Voyager spacecrafts [67]. Several space missions have been launched dedicated to extensively monitoring the Sun's corona and its explosive activities, such as flares and CMEs.

Using interplanetary scintillation (IPS) observations, CME propagation has been monitored over the entire Sun-Earth distance [254], [255]. Solar origins of intense geo-magnetic storms and the role of CMEs have been studied by [347]. CMEs represent one of the more energetic and geoeffective manifestations of solar activity, and play an important role in impacting space weather and the terrestrial environment in a number of ways (Section 8.7.6, page 219).

3.9.2 CME mechanism

The possibility that flare-associated CMEs are propelled by a pressure pulse due to the flare is unlikely in view of the fact that many CMEs appear to precede the flare. Because CMEs are initiated in the solar corona, which is dominated by magnetic energy, and are as energetic as the flares, it is plausible that their energy source must be magnetic.

The CMEs are essentially large-scale magnetic structures expelled from the Sun due to MHD processes involving interaction between plasma and magnetic field in closed magnetic field regions, such as active regions and filaments/prominences. A CME is believed to happen because of a large-scale departure from equilibrium when the magnetic and gravitational forces no longer balance the coronal gas expansion, and lifts off with constant speed, either slow or fast. For further details, one may refer to [143], and the articles in [145].

It is still rather unclear as to how the CMEs are initiated although the recent space-craft observations have provided extensive information about the source regions and the initial phase of CMEs. Flux emergence near such pre-existing closed field structures has been considered as a possible trigger of CMEs and filaments [119], [386].

There are CMEs associated with flares, and also those without any perceptible chromospheric association. However, a large flare is invariably associated with a CME. Also, a large number of CMEs are associated with eruptive prominences, which occur throughout the activity cycle, and not just during solar maximum phase. Formation and eruption of prominences are the central issues of CME initiation, and there are reports that consider the eruption of prominences as the cause of CMEs [402]. However, there is a complex relationship between the filament, coronal cavity, and the frontal structure before and during the eruption. It is clear that multi-wavelength, multi-instrument data are required to construct a complete picture of the flare and/or CME phenomenon.

Accordingly, the CME may not occur as a result of a flare or erupting prominence, but all three may occur due to the loss of equilibrium. Coronal mass ejections (CMEs), prominence/filament eruptions, and flares are thus three different types of large-scale eruptive transient phenomena that occur in the solar atmosphere. It is believed that they are closely related, different manifestations of a single physical process.

3.10 Summary

A large variety of structures, activities and explosive transients occur on the optically visible layers of the Sun, *i.e.*, the photosphere and chromosphere. These observations are routinely carried out from a large number of ground-based observatories located around the world. The observations have made rapid progress, by increasing spatial and temporal resolutions, and have contributed immensely towards our understanding of several aspects of these phenomena. However, the optical observations only provide a limited idea as one would require additional simultaneous coverage of the events over the higher layers of the Sun in order to construct a full 3-dimensional scenario (Chapters 7, and 8). It is evident that coronal dynamics plays a complex role in the magnetic energy build-up process. Some of the flare associated activity is also indicated in the higher corona that is accessible only from space in short wavelengths. There are also radio emissions from the corona during the quiet, active, and transient phenomena that provide a lot of information on the various physical processes. These are important issues in solar physics and shall be discussed in the following chapters.

Chapter 4

Solar Magnetic Field and Activity Cycles

4.1 Solar magnetism

Magnetic fields play an important role in various cosmic objects at varying scales, including in our Sun [297]. A detailed understanding of solar magnetic field is, therefore, essential for understanding the variety of structures, and their evolution observed in the photosphere and other layers of outer solar atmosphere. Of particular interest is the study of the storage of magnetic energy and their sudden outbreak in the form of explosive transient events that cause various types of disturbance in space weather, and terrestrial environment. Thus, the task of measuring solar magnetic fields is of central importance in solar physics.

George Ellery Hale in 1908 made the pioneering efforts to measure magnetic fields in sunspots [155]. This was one of the most fundamental discoveries in solar physics in the early twentieth century. It triggered new fields of spectroscopy and polarimetry in solar and stellar physics that are combined for exploiting the information embedded in the light coming from distant heavenly objects. The inter-relationship between the general magnetic field of the Sun and the localised sunspot magnetic field became clear with the development of the magnetograph by H. Babcock in 1953 [23]. With the increasing improvement in sensitivity and accuracy of measurements, it has become possible to measure relatively weak magnetic fields away from sunspots. The advancements in instrumentation and observations further led to the discovery that the polar field of the Sun reverses during the 22-year cycle. These findings led Babcock to propose a model, which is still the most accepted model, for the near cyclic behaviour of solar magnetic fields [24].

The early magnetographs measured only the line-of-sight (or, longitudinal) $B_{||}$ component of the solar photospheric magnetic field. Such measurements provided the magnetic flux polarities, and their distributions over the Sun. The distribution of magnetic flux at the solar surface spans all scales, ranging from global dipolar structure seen during solar eclipses, down to the limit of achievable angular resolutions of present-day telescopes. E. N. Parker pointed out that the convective motions of foot-points (sunspots), or magnetic fluxes, lead to deform or stress the overall magnetic field structures [298]. For studying the evolution of these stresses, one needs full vector magnetic field measurements. Such studies are useful in understanding the energy storage process required in eruptive phenomena. For a complete measurement of the magnetic field, thus, it is not enough to measure only the longitudinal $B_{||}$ component, but also the more challenging, transverse (or horizontal) component $B_{\perp}$of the magnetic field.

There are several proxies in the solar atmosphere which are often used to indirectly infer the associated magnetic fields, such as g-band bright points, faculae, and sunspots (in the photosphere), CaII H and K plages, fibrils, filaments in chromospheric H_α line, coronal loop structures in EUV and X-rays, *etc.* However, it has been a rather difficult observational task to measure magnetic fields quantitatively in the upper layers of the Sun due to the lack of suitable spectral lines. The situation is improving, of late, due to new developments in radio and IR observations. Also, gyroresonance produces polarized radio spectra, that is being used to estimate coronal magnetic fields.

Due to very low plasma β (ratio of gas pressure to magnetic pressure) in the upper chromosphere and corona, the magnetic field in these layers can be considered to be nearly force-free. As another recourse, extrapolation techniques are employed, which use the photospheric vector magnetic fields as the boundary conditions to compute magnetic fields at the higher layers.

4.2 Magnetic fields in the solar photosphere

From the middle of 19th century, astronomical spectroscopy experienced a great development along side the laboratory discoveries. In 1866, Sir Joseph Norman Lockyer observed the spectrum of a sunspot, which revealed a conspicuous widening as compared with the normal solar photospheric spectral lines. Spectral lines formed in the presence of a magnetic field are split into different polarized components. This phenomenon, however, is known as the Zeeman effect, after the Dutch physicist, Pieter Zeeman. He found in 1896 a conspicuous widening of the sodium D lines, in a laboratory experiment, when an electromagnet was switched on.

The Zeeman effect causes the splitting of atomic energy levels in atoms when formed in strong magnetic fields (Figure 4.1a). Thus, a line that radiates at a wavelength λ_0 in the absence of a magnetic field is split into two or three components. The resulting components of the spectral line are polarized, depending on the strength and direction of the magnetic field.

In the magnetic field B a level with total angular moment J splits into $2J + 1$ sub-levels with different M_J:

$$E_{J,M} = E_J + \mu_0 g M_J B = E_J + \mu B. \tag{4.1}$$

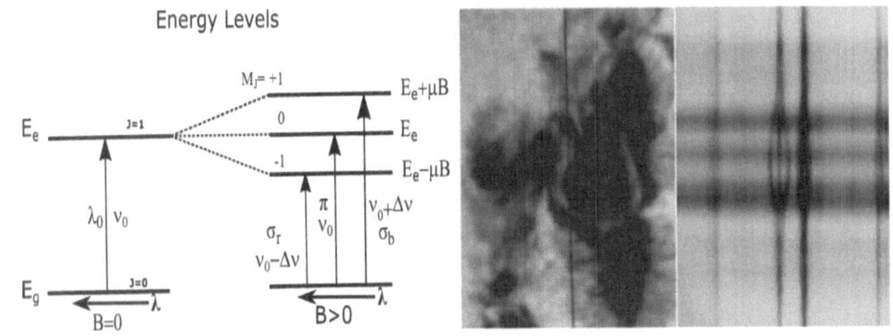

(a) (b)

FIGURE 4.1: (a) A magnetic sensitive spectral line in B=0 and B≠0. (b) Observed splitting of the absorption line FeI 525.0 nm over a sunspot (Courtesy: William Livingston).

Transitions are allowed between levels with $\Delta J = 0, \pm 1$, and $\Delta M = 0(\pi), \pm 1(\sigma_b, \sigma_r)$, and splitting is determined by the Landé factor $g(J, L, S) = 1 + [J(J+1) + S(S+1) - L(L+1)]/[2J(J+1)]$.

Zeeman splitting of a magnetically sensitive spectral line is given by:

$$\Delta\lambda_B = \lambda - \lambda_0 = \frac{e}{4\pi c m_e} g^* \lambda^2 B \approx 4.67 \times 10^{-5} g^* \lambda^2 B. \tag{4.2}$$

where $g^* = gM - g'M' = -1, 0, +1$, and g, g', M, M' are the Landé factors and magnetic quantum numbers of the lower and upper states of the transition, and λ is measured in cm and B in Gauss.

The number of observed components depends on the orientation of the magnetic field with respect to the observer. When the line-of-sight is in the direction of the magnetic field, the observer sees two shifted σ components having circular polarization of opposite sense. When the magnetic field is perpendicular to the line-of-sight, the observer sees an unshifted π component, linearly polarized perpendicular to the magnetic field, and the two σ components, which are now linearly polarized along the magnetic field. Thus the observation of the degree of polarization is important in the determination of the magnetic field on the Sun.

Even for a strong magnetic field, such as in sunspots, the Zeeman effect is quite weak. Also, the Doppler and thermal (collisional) broadening make observing any clear physical Zeeman separation even more difficult. For example, only in sunspots with relatively strong magnetic field $B = 0.30T$ is the shift $\Delta\lambda_B/\lambda_0$ in optical wavelengths of the order of 10^{-6}. In regions of weaker fields, away from sunspots, the splittings are extremely small and often, unresolvable. Therefore, one takes recourse to measuring the state of polarization of the spectral line.

The Zeeman effect of spectral lines is widely used to measure the line-of-sight strength and direction, and the total strength of the Sun's magnetic field. The main requirement is selecting suitable lines corresponding to atoms with large magnetic moments, and regions that are dense enough to make the signal observable. From Equation 4.2, the splitting is proportional to λ^2. Therefore, spectral lines with longer wavelengths, such as in the IR region, provide better results, and are more promising. Also, high spatial resolution observations are needed for identifying features of magnetic fluxes at the smallest scales, which may get cancelled out in coarser resolution images if opposite polarity fluxes are present.

The Zeeman effect can produce distinctly separated multiple lines, when observed at very strong fields of sunspots, where the physically measurable splitting directly gives the total intensity of the magnetic field (Figure 4.1b). This method is used for the absolute calibration of magnetic fields, while applying over the polarization measurements of weaker field areas away from the sunspots.

In the following sections, we discuss the subject of polarization of light, and spectro-polarimetric methods used in solar polarimetry and magnetic field measurement.

4.3 Polarization of radiation and magnetism

Information about the physical conditions of astrophysical objects like the Sun and other stars is encoded in their radiation, particularly in the spectral lines contained in their spectra. This is contained not only in the intensity, but also in the state of polarization of the radiation, as a function of frequency. Polarization of radiation is essentially produced when the spatial symmetry is broken in the physical process that generated the observed radiation [352].

The breaking in symmetry can be caused due to various physical processes, such as (i) macroscopic magnetic (Zeeman and Hanle effects), and electric (Stark effect) fields, (ii) oblique reflection (for example, non-axisymmetric optical components in a telescope), or (iii) anisotropic excitation process (radiative or collisional).

Collisional excitation by directed particle beams may lead to *impact polarization* as in solar flares. Scattering of photons at free electrons (*Compton* or *Thomson scattering*) makes the radiation linearly polarized as in the solar white-light corona. In the same manner, the scattering of radiation by atoms and molecules also leads to linearly polarized light (*Rayleigh or resonance scattering*). For details of various physical processes, which can give rise to polarized radiation, one may refer to [325], [228].

Polarimetric accuracy is one of the most important goals of modern astronomy. But, along with polarimetric accuracy, high spectral, spatial, and temporal resolutions, with the wide spatial and spectral coverage, are other important aspects in any astronomical observations.

4.3.1 The Zeeman polarimetry

The full state of Zeeman polarization of radiation, coming from the solar atmosphere in the presence of magnetic fields, can be specified in terms of the so-called four *Stokes parameters* (I, Q, U, V) [73]. The final state of the polarization of light after passing through the solar atmosphere is obtained by solving the equation of radiative transfer [229]:

$$\mu \frac{dI}{d\tau} = (1 + \eta_0 A)(I - S) \tag{4.3}$$

where, Stokes vectors I=[I(τ), Q(τ), U(τ), V(τ)], and the source function S(τ) are the function of continuum optical depth τ. Here, Stokes Q and U correspond to linear polarization, V corresponds to circular polarization, and I is the total continuum intensity.

The matrix **A** in Equation 4.3 is given by

$$A = \begin{pmatrix} \eta_I & \eta_Q & \eta_U & \eta_V \\ \eta_Q & \eta_I & \rho_V & -\rho_U \\ \eta_U & -\rho_V & \eta_I & -\rho_Q \\ \eta_V & \rho_U & -\rho_Q & \eta_V \end{pmatrix} \tag{4.4}$$

where the elements of the matrix **A** in Equation 4.4 are given by

$$\begin{aligned} \eta_I &= 1 + \frac{\eta_0}{2}\{\phi_p \sin^2\theta + \frac{1}{2}[\phi_b + \phi_r](1 + \cos^2\theta)\} \\ \eta_Q &= \frac{\eta_0}{2}\{\phi_p - \frac{1}{2}[\phi_b + \phi_r]\} \sin^2\theta \cos 2\varphi \\ \eta_U &= \frac{\eta_0}{2}\{\phi_p - \frac{1}{2}[\phi_b + \phi_r]\} \sin^2\theta \sin 2\varphi \\ \eta_V &= \frac{\eta_0}{2}[\phi_r - \phi_b] \cos\theta \end{aligned}$$

and

$$\begin{aligned} \rho_Q &= \frac{\eta_0}{2}\{\psi_p - \frac{1}{2}[\psi_b + \psi_r]\} \sin^2\theta \sin^2 \cos 2\varphi \\ \rho_U &= \frac{\eta_0}{2}\{\psi_p - \frac{1}{2}[\psi_b + \psi_r]\} \sin^2\theta \sin 2\varphi \\ \rho_V &= \frac{\eta_0}{2}[\psi_r - \psi_b] \cos\theta. \end{aligned}$$

Here, θ and ϕ are interpreted as the *inclination angle* of the magnetic field vector with respect to the propagation direction, and the *azimuth angle* of magnetic field B with respect to Stokes Q positive direction, respectively.

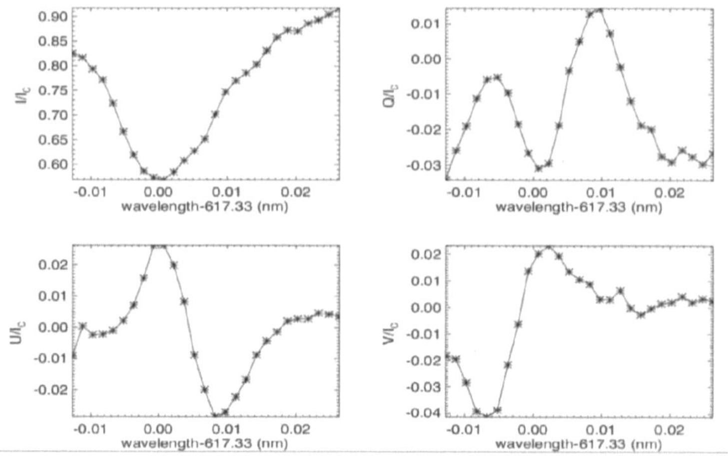

FIGURE 4.2: Profiles of Stokes parameters I, Q, U, and V at a point in the penumbra (at the star mark) in Figure 4.3. With permission from [363].

The interpretation of observed polarization, *i.e.*, the Stokes profiles in terms of magnetic field vector is done by fitting them with theoretical profiles under *Milne-Eddington model* atmosphere assumptions. Solutions of radiative transfer equations in this model are known as *Unno-Rachkovsky solutions* [371] [309].

An example of the Stokes profiles I, Q, U, and V is shown in Figure 4.2 for Fe I line centred at wavelength 617.33 nm at a selected position in the penumbra of NOAA AR 12740. For more details about the theoretical formalism and effects of various atmospheric parameters, one may refer to [94].

4.3.2 Solar magnetic field using Zeeman polarimetry

Spectro-polarimetry is the basic tool to record the polarization spectrum of the Sun (or stars). From the observed spectra, we can then extract the required information, and deduce the physical properties of the remote location from where the observed radiation originated.

K.O. Kiepenheuer and H. W. Babcock independently developed the first solar magnetograph in 1953 for measuring circular polarization, V, at the wings of the spectral line FeI 525 nm [209], [23]. A solar polarimeter is essentially a spectrograph having additional optical elements to analyse the polarization properties of solar light (Fig. 4.4).

The aim of solar polarimetry is to measure the Stokes vector $\boldsymbol{I}$ (the four Stokes parameters I, Q, U and V) with the highest possible polarimetric accuracy, spectral, spatial, and time resolution, and spectral and spatial coverage. The Stokes I is a sum of left and right circular polarization, and the Stokes

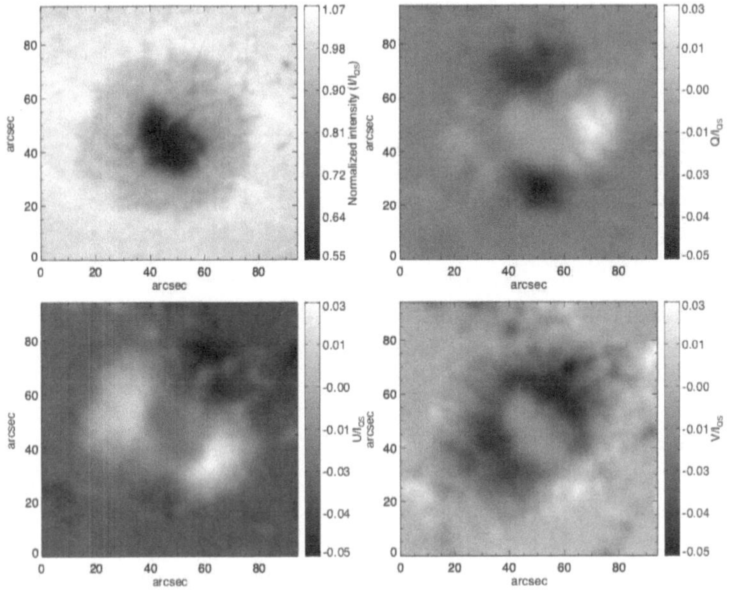

FIGURE 4.3: Stokes images of NOAA AR 12470 (top panel) I and Q, (bottom panel) U and V obtained using vector mode of operation for NOAA AR 12470 at 617.3nm. With permission from [363].

V is the difference between these. The Stokes parameters can be directly measured observationally as sums and differences of measured intensities by observing polarized signals at different polarizer orientations:

- $I = I_{lin}(0°) + I_{lin}(90°) = I_{lin}(45°) + I_{lin}(135°) = I_{circ}(right) + I_{circ}(left)$,
- $Q = I_{lin}(0°) - I_{lin}(90°)$,
- $U = I_{lin}(45°) - I_{lin}(135°)$,
- $V = I_{circ}(right) - I_{circ}(left)$,

where, V gives the longitudinal component B_L of solar magnetic field B. For constructing a vector magnetogram, one needs to measure all the 4 Stokes parameters, for which a technically more complex polarization analyser is required ([363], Figure 4.3). The transverse component of magnetic field B_T is based on the measurement of linear polarizations U and Q. These are much weaker in strength as compared to V, therefore, larger telescope apertures, and longer integrations are required to be carried out very rapidly to adequately improve signal-to-noise ratios. Thus, magnetic maps (or, magnetograms) of sunspots and their surroundings are obtained.

4.3.3 Degradations of polarimetric measurements

Solar polarimetry has several factors that limit the final accuracy of the measurements. These can be categorised as:

- *Environmental polarization*: It includes the polarization of radiation arising due to scattering in the atmosphere. It is also called *seeing-induced polarization*. This can be avoided only if the observations are carried out on-board a spacecraft, located outside the degrading effects of the Earth's atmosphere.

- *Instrumental polarization*: It is related to the polarization analyser system, or the polarimeter itself. The image-forming optics generally introduces instrumental polarization, and thus corrupts the solar polarization signal. This effect can be avoided if the polarimetric analysis is performed before the light encounters any beam-steering optics. A solution is to build *polarization-free* telescopes, which are rotationally symmetric with respect to the optical axis, until the beam reaches the secondary focus, where the polarization analysis is carried out. Even then, residual polarization effects are unavoidable, *e.g.*, due to the stresses in the entrance window, or other optical imperfections in the telescope.

Thus, a careful calibration of the environment, and the polarization analyser system is essential before one can correctly interpret the results. For details, one may refer to [94].

4.3.4 The observational technique

The task of the optics of the polarization analyser is to encode the incoming polarization information into intensity differences, since the detector systems used are only sensitive to photon numbers (intensity), and not to polarization. After the optical encoding of the polarization has taken place, and the beam leaves the optical analysis package, any oblique reflections or instrumental polarization may occur without being of any consequence for the polarization analysis.

For the polarization analysis, a linear polarizer is needed for measuring the components of linear polarization (Stokes Q and U). For measuring circular polarization (Stokes V), a retarder and a linear polarizer are employed. The linear polarizer may be replaced by a polarizing beam splitter, which gives two output beams of orthogonal polarization states.

Modulation of the polarization state may be done (i) mechanically, *e.g.*, with a rotating wave plate, which is a slower process, or (ii) electro-optically, without involving any moving parts; a much faster method. The electro-optical modulator as a variable retarder is advantageous for modulating at frequencies in the kHz range, much above the seeing frequencies [257], [258].

To record all four Stokes parameters simultaneously, two such non-moving variable retarders having their optical axes at different position angles are

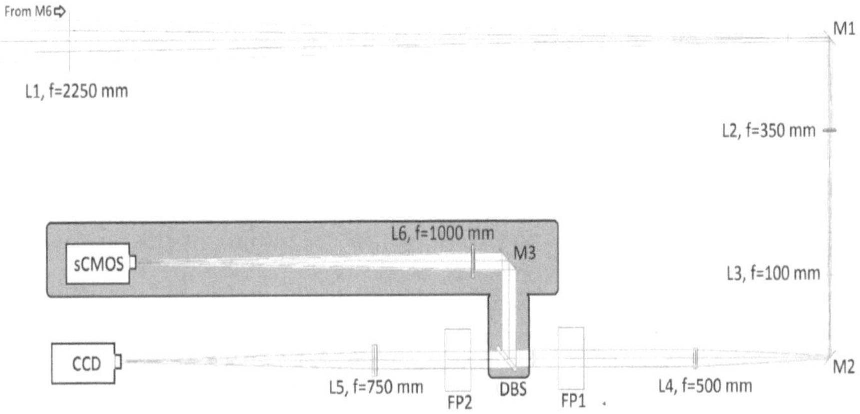

FIGURE 4.4: Schematic of the optical layout of USO's MAST Vector Magnetograph (USO photo). A f/12 beam from the telescope is collimated using the lens L1, which then passes through the narrow band imager consisting of two FP etalons (FP1 and FP2) and a pre-filter. The polarimeter consists of two LCVRs and a polarizer kept between the CCD and imaging lens (see Figure 14 [363]).

needed. Wavelength selection can be made with a spectrograph, a narrow-band filter, or a *Fourier Transform Spectrometer*(FTS). The 2-D image at the end focus of a spectrograph represents (x; λ)-space, *i.e.*, one spatial dimension (x) and the wavelength dimension (λ) are covered for each given value of the other spatial coordinate (y). With a narrow-band filter one covers (x; y)-space for a given value of λ. The detector system can be placed in the exit focus of any wavelength-selecting system, be it a spectrograph, narrow-band filter, or an imaging FTS.

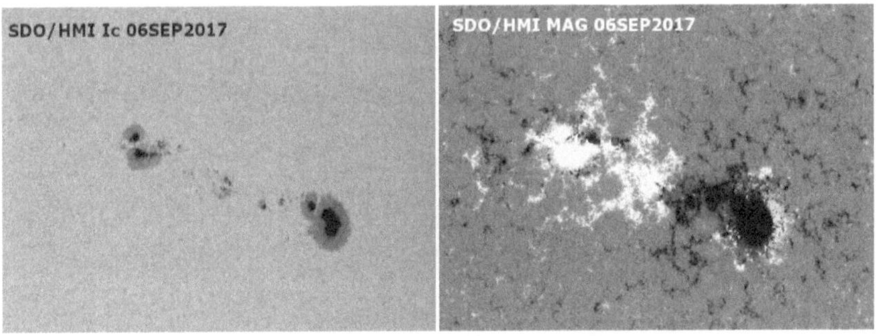

FIGURE 4.5: A solar photosphere image of the sunspot group AR 2674 of September 6, 2017, along with its corresponding magnetic field map showing the magnetic polarity or flux distribution. Courtesy of SDO/HMI Science Team.

The optical layout of an imaging spectro-polarimeter, developed as one of the back-end instruments of USO's Multi-Applications Solar telescope (MAST), shows the observing system of a solar magnetograph (Figure 4.4). It is used for studies involving the dynamics of solar magnetic and velocity fields. This system consists of a narrow-band filter and a polarimeter. The polarimeter includes a linear polarizer, and two sets of liquid crystal variable retarders (LCVRs). The instrument is intended for simultaneous observations in the spectral lines 617.3 nm and 854.2 nm, which are formed in the photosphere and chromosphere, respectively [363].

Similar magnetographs are being operated in many ground-based observatories, and also in space-borne instruments, which routinely provide photospheric magnetic maps, or, magnetograms, of solar active regions (Figure 4.5), or the entire solar disk (Figure 4.8). Such maps show the distribution of the magnetic field and provide a good tool to study the birth and evolution of active regions.

Sunspots at the photosphere are only the 2D foot-prints (cross-section) of an extended 3D-flux tube system extending from the interior to the overlying atmospheric layers. In order to obtain their magnetic field structures with height, one needs to measure magnetic fields at the chromosphere and above too. However, there are several difficulties in such measurements, such as the lack of suitable spectral lines, rapidly decreasing magnetic field strengths, and increasing thermal broadening of the lines, *etc.*

4.3.5 Emergence of solar flux tubes

The magnetic field emerging from the solar interior in active regions is responsible for many of the structures seen on the Sun in different wavelengths. In the photosphere, the structures are mostly in the form of sunspots. The magnetic flux tubes emerging from one sunspot are often observed to connect to another spot of opposite polarity.

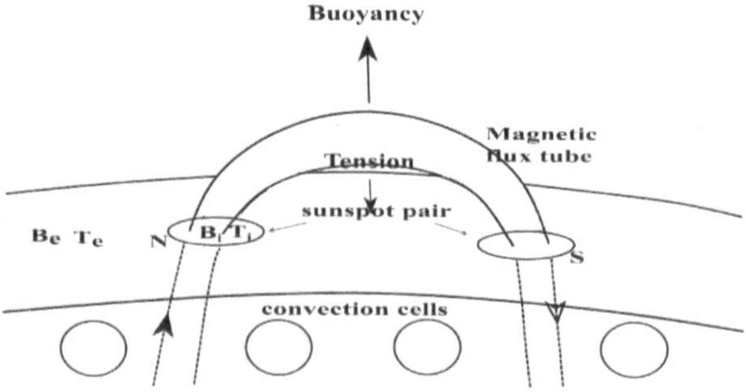

FIGURE 4.6: Emergence of a flux tube from sub-photospheric internal layers.

In order to understand the emergence of sunspots from the solar interior, let us consider a flux tube of higher magnetic field strength B_i embedded in an ambient medium of weaker magnetic fields $B_e << B_i$ (Figure 4.6). Considering the balance of kinetic (thermal) and magnetic pressures, one has [299]:

$$\frac{k_B T_e \rho_e}{m} \approx \frac{k_B T_i \rho_i}{m} + \frac{B_i^2}{2\mu_0}, \tag{4.5}$$

where the exterior and interior of the flux tube are denoted by 'e' and 'i', respectively. Clearly, the flux tube will rise due to buoyancy if $\rho_e > \rho_i$. If this buoyancy is greater than the magnetic tension that tries to pull the flux tube back, one has

$$(\rho_e - \rho_i)g > \frac{B_i^2}{\mu_0 L}, \tag{4.6}$$

where L is the distance between the foot-points of the flux tube. The condition for minimum separation of the foot-points, is therefore given by

$$L > \frac{2k_B T}{m} = 2h_p, \tag{4.7}$$

where h_p is the pressure scale length. Accordingly, if L is too small, *i.e.*, the foot-points are too close, the tension dominates over the buoyancy, and the flux tube is pulled back.

A sunspot group, as seen in the solar photosphere, is surrounded in an extended three-dimensional magnetized region, called an *Active Region.* This region appears brighter (hence, hotter) than its surroundings, as seen in various wavelengths, such as, H_α, UV, EUV and X-rays. The magnetograms usually start showing the signature of flux emergence, associated with the birth of sunspots, well before they actually emerge out. The sunspots are related to convective motions below the photosphere. However, it is not clear as to how deep are they magnetically connected within the convective zone.

4.3.6 Nomenclature of active regions

Previously sunspot groups were named or numbered by individual observatories, such as Mt. Wilson Number. National Oceanic and Atmospheric Administration, USA (NOAA), started naming the active regions (which generally consist of sunspots) with NOAA numbers, such as NOAA AR0001, observed on January 5, 1972. Active regions are being assigned numbers consecutively according to their appearance on the visible side of the disk.

As the Sun rotates around its axis, some long-lasting active regions may reappear at the E-limb of the Sun after transiting across the far-side of the Sun. However, they are given new numbers when they reemerge.

On June 14, 2002, the active region numbering NOAA AR 10000 appeared. Thereafter, only the last 4 digits are usually retained to refer to the active

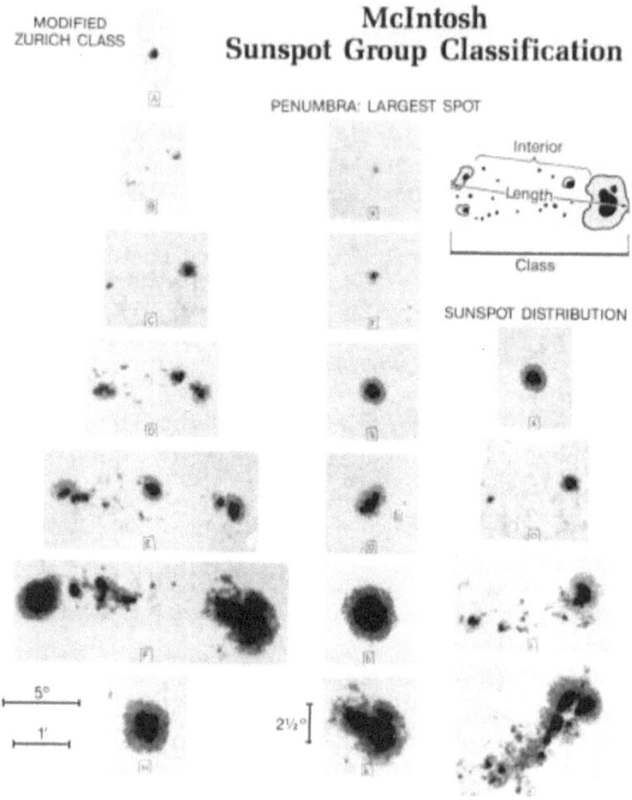

FIGURE 4.7: The 3-Component McIntosh classification of sunspots (Fig 1 page 253 with permission from [268])

region for convenience. For example, after AR9998, 9999, 10000, ... AR 10484, 10485 are usually referred to as NOAA AR0484, 0485, *etc.*

4.3.7 Classification of sunspots/ARs

Various types of sunspots are observed. They are classified based on their morphological structure and magnetic complexities. McIntosh classification is based on a three-component definition of sunspot groups, having the general form Zpc, where Z is the modified Zurich class, p is the type of the main sunspot (primarily related to penumbra), and c is the degree of compactness or spot distribution (Figure 4.7). This classification requires only white light (continuum) observations. It is based on observational features such as unipolar or bipolar, presence or absence of penumbra, length of sunspot group, presence of spots between the leading and trailing sunspots, *etc.* [268].

Another classification of active regions, which is more popular and in common use, is based on their magnetic complexity. Magnetic classification provides a simple way to describe the configuration and complexity of the magnetic flux, and sunspots in a solar active region (AR). The Mount Wilson (or Hale) magnetic classification system was originally put forward by Hale in 1919 [156]. According to this classification scheme, one has:

- α: a single dominant spot, often linked with a plage of opposite magnetic polarity,
- β: a pair of dominant spots of opposite polarity (bipolar, *i.e.*, having a leader and a follower polarity),
- γ: complex groups with irregular distribution of polarities,
- $\beta\gamma$: bipolar groups having more than one clear polarity inversion line, and
- δ: when umbrae of opposite polarity are present within a single penumbra.

More than half of the observed sunspot groups usually fall in α or β classification [194]. Larger sunspot groups are often more complex, of $\beta\gamma$ or $\beta\gamma\delta$ class, that can be very active, and generally produce the most energetic solar flares.

4.4 Measuring the weak magnetic fields: the Hanle effect

Another diagnostic used to determine weaker magnetic fields is called *Hanle effect* diagnostics, which is still relatively unknown among stellar astronomers. Scattering of unpolarized incident radiation is known to produce linear polarization. A magnetic field modifies this polarization via the Hanle effect by depolarizing it and rotating its direction. It provides an important extension of the diagnostic for some fundamental problems in solar physics which are not adequately dealt with by the Zeeman effect alone.

Experiments to describe the polarization from resonance line scattering date back to the early 1900s (*e.g.*, see [274]). The influence of a magnetic field on the line polarization was explained first by Hanle for his dissertation [158]. He described the effect of the magnetic field in semi-classical terms as arising from the precession of an atomic damped harmonic oscillator. Quantum mechanically, the effect is understood in terms of an interference effect that occurs when the degeneracy of the magnetic sub-levels in the excited level is partially lifted, and has applications to many topics in atomic physics (*e.g.*, see the review [277]).

Although the Hanle effect was discovered in 1924, its potential for magnetic-field diagnostics has been realised only recently. Astrophysically, the Hanle effect has been applied almost exclusively in the solar context, especially to measuring magnetic fields in the solar chromosphere and corona. In the weak magnetic field regime, the Zeeman effect does not contribute significantly to the linear polarizations. In the circular polarization measurement, however, both Zeeman effect and Hanle effect may contribute.

A line-of-sight magnetic field can also generate circular polarization by the longitudinal Zeeman effect, which occurs both when the excitation is by radiation (as in scattering) and by collisions. While the field-strength sensitivity to the Hanle effect (in the linear polarization) depends on the ratio between the Zeeman splitting, and the natural width of the excited state, the longitudinal Zeeman effect (the circular polarization) depends on the ratio between the Zeeman splitting and the Doppler width. The Hanle effect is therefore sensitive to much weaker magnetic fields, of just a few Gauss up to around 300 Gauss, than the usual Zeeman effect. Zeeman and Hanle effects are in a sense complementary to each other as the Hanle effect is sensitive to weak fields and the fields of mixed polarity, as compared to the Zeeman effect.

In laboratory applications, the observed polarization phenomena in a known magnetic field are used to obtain information on the atomic structure. In astrophysical applications, however, the problem is the other way round, as the magnetic field is the unknown quantity. Here, the magnetic field needs to be derived from the observed polarization of the radiation from the celestial object, using the knowledge about the atomic physics involved [350]. For a detailed description of the physics of the Hanle effect and polarized radiation transport with application to solar studies, one may refer to [351].

The relative importance of the Zeeman effect polarization in the Stokes V observation depends on the efficiency of the *He I* ionization process. The fact that filaments appear as dark absorption features against the solar disk suggests that coherence scattering should make an important contribution to the polarization signals of the filaments and prominences [242]. The shape of the linear polarization line profiles also shows strong signatures of scattering polarization. Therefore, the Hanle effect should have strong contribution to the observed polarization signals in filaments. Diagnostics of Hanle effect has been used for measuring solar prominence magnetic fields by many researchers ([52], [227], [238], [50]), and also for turbulent photospheric magnetic fields [349]. Various applications of the Hanle effect in astrophysics have been reviewed in [237] and [206].

4.5 Solar Dopplergraph/Magnetograph instruments

The concept of Doppler-shift of spectral lines present in the solar spectrum has been exploited for detecting photospheric motions since 1960 [234]. A solar Dopplergraph instrument is essentially a spectrograph which analyses the intensities of a velocity sensitive spectral line profile at its blue and red wings, and the difference gives a measure of the radial, or L-O-S, component of velocity [60].

These instruments work with monochromatic sunlight selected around a suitable solar spectral line, such as FeI 525.0 nm or NaD 1589.6 nm. The light is then rendered linearly polarized and fed to an electro-optical modulator. This modulator is usually a special crystal, such as $KH_2PO_4 - KDP$, which is uniaxial in the absence of an electric field. By applying electric voltage parallel to the optical axis, these crystals can be made biaxial. The light, polarized parallel to the slow axis, is retarded by $\lambda/4$ by applying a high voltage, usually of the order of a few kV, so that the combined light emerging out is rendered circularly polarized. Repeating this process by reversing the sign of the applied voltage, the fast and slow axes change their roles, and circularly polarized light of opposite sense emerges. The difference between the two signals then provides the measure of line-of-sight (Doppler) velocity.

Usually it is required to carry out the integration of the above two operations, as in the case of magnetograms, in order to improve the S/N-ratio, and obtain a temporally averaged solar Dopplergram. However, the integration must be carried out faster than the time over which the atmospheric seeing changes occur. Such filter-based Dopplergraph instruments have been in use for producing two-dimensional solar velocity maps, the solar Dopplergrams, since the early 1970s.

Dopplergrams of chromospheric prominences/filaments have been used to study their activation and eruption. Similarly, sunspot Dopplergrams readily show the signature of the Evershed flow in their umbrae/penumbrae. Continuous and long time-series of photospheric Dopplergrams obtained over several days have been very useful in precise measurement of the oscillation modes of the Sun [59], which we shall discuss in the next chapter.

4.5.1 The GONG Doppler instrument

Global Oscillations Network Group (GONG) is an international network of ground-based observatories which is devoted to observing the solar photospheric magnetic and Doppler velocity fields using the Ni I 676.8 nm line. The heart of the GONG Dopplergraph instrument is a polarizing Michelson interferometer having a path difference of about 30,000 waves. This is constructed to have a wide angular field and to be thermally stable [162]. The cosine-squared transmission pattern produced by the interferometer is scanned across the filtered spectrum of the Sun by a rotating wave plate.

Modulation is produced by the presence of the Fraunhofer line. Three images are obtained for each cycle of modulation to provide the phase, amplitude, and brightness of each pixel of a full disk image of the Sun. Such images are integrated over a duration of 60 s to obtain solar Dopplergrams. The phase of the modulation is a good measure of Doppler shift (`https://gong.nso.edu/instrument/`). The GONG instrument also provides full disk magnetograms, and H_α filtergrams.

The continuous observations of solar Doppler images from the ground-based networks, such as, GONG, and from space-borne instruments, such as SOHO-MDI, and more recently, SDO-HMI, have made important contributions to the studies of solar internal structure and dynamics involving the tool of helioseismology (Chapter 6).

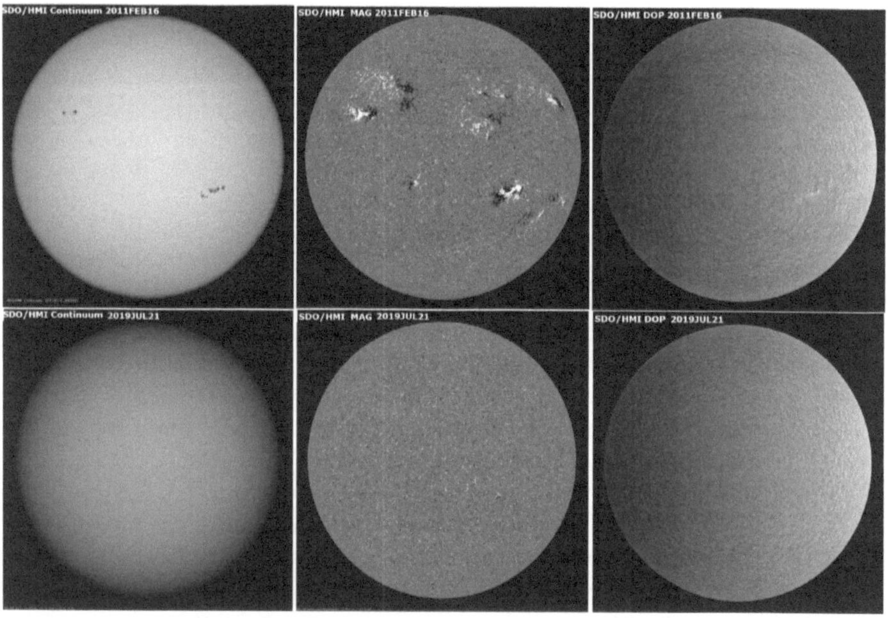

FIGURE 4.8: SDO/HMI continuum image, magnetogram and Dopplergram obtained on (top row) February 16, 2011, of maximum solar activity phase, and (ii) July 21, 2019, of minimum activity phase, of the current solar cycle. Courtesy of NASA/SDO and the HMI science teams.

4.5.2 The Helioseismic and Magnetic Imager (HMI)

The Helioseismic and Magnetic Image (HMI) onboard Solar Dynamics Observatory (SDO) is a space-borne instrument which measures photospheric Doppler velocity and magnetic fields. It samples Fe I-line at six wavelength positions symmetrically around the line centre at rest. It consists of a combination of a Lyot filter and two Michelson interferometers, with a set of three retarders: $\lambda/2$, $\lambda/4$ and $\lambda/2$ between them. There is a beam splitter after these, dividing the light between two CCD cameras, acquiring full-disk images of the Sun. HMI provides full-disk Doppler, intensity, and magnetic images at 1 arcsec resolution (4096×4096-pixel images) of the solar photosphere. Figure 4.8 shows these images taken by SDO/HMI near the maximum activity phase on February 16, 2011, and the corresponding images at the low activity phase on July 21, 2019. (`http://jsoc.stanford.edu/ajax/lookdata.html`) [89]. The HMI also obtains all three components of the photospheric magnetic field for constructing vector magnetograms.

4.6 Chromospheric and coronal magnetic fields

Magnetic fields dominate the plasma dynamics in the upper atmosphere, particularly in the chromosphere, transition region and the corona. Although the magnetic field strengths are much weaker in these layers, as compared to the photosphere, they play a central role in the structures, dynamics, and heating mechanisms. However, observational measurement of magnetic fields in these upper layers is not easy or straightforward.

Photospheric magnetic field is routinely measured by using the Zeeman effect and properties of associated polarized radiation in suitable magnetic sensitive spectral lines. However, this technique is not so easily applicable to the chromosphere, transition region and corona, because of poor Zeeman sensitivity to weaker magnetic fields. In addition, there is large thermal broadening of chromospheric lines, as compared to the Zeeman splitting. Furthermore, the Zeeman effect is observationally limited to regions of relatively strong magnetic fields, *i.e.*, in active regions. Infrared lines are preferable due to the advantage of λ^2 in Zeeman splitting, and the relatively large widths of coronal lines [240]. We shall discuss those observations in Chapter 7.

The Hanle effect offers a strong diagnostic of regions of weak magnetic fields such as the solar corona, where a number of spectral lines with different but complementary sensitivity ranges are present. The Hanle effect of spectral lines provides an attractive alternative, as it is sensitive to weaker magnetic fields than the Zeeman effect, and is unaffected by thermal broadening.

Spectroscopic diagnostics in UV and EUV wavelengths, observable only from space-based instruments, provide measurements of plasma densities, temperatures, and velocity, but not the magnetic field in the upper atmosphere of the Sun. The coronal observations of the Sun in EUV and X-ray wavelengths show bright magnetic loops, X-ray bright points, sigmoid structures and coronal holes in "unipolar" field regions. These discoveries showed the importance of direct measurement of magnetic fields in the chromosphere and the transition region.

There are many UV and EUV emission lines that are produced in the chromosphere and transition region, where the Lyman-α 121.6 nm line is the brightest. It is predicted to be sensitive to magnetic fields of 100 Gauss (as in chromospheric active regions), and even to fields as low as $\sim$ 10 Gauss of quiet chromosphere. There are plans for making precise measurement of the polarization profile of the Lyman-α line, using a rocket-borne sounding experiment that aims to make the first ever measurement of the Hanle effect polarization. The Hanle effect in the UV and IR spectral lines is a largely unexplored diagnostic. Recent advances in understanding and modelling of the Hanle effect show promise for retrieving magnetic field information from Hanle polarization measurements [313].

MHD simulations are being carried out to study the expected magnitude of the signal for typical coronal magnetic fields for selected spectral lines in the UV (H I Lyman α) and IR (neutral helium line He I 1083 nm) wavelength ranges. These may provide reliable diagnosis of coronal magnetic fields [51], [313].

Other methods for coronal magnetic field measurements are based on radio emissions. These provide strengths of fields only, but may help in constraining the magnetic field measurements in the corona ([390], [136]). Further, in decimetric and longer radio wavelengths supra-thermal electrons accelerated in active regions provide a bright emission through collective processes (noise storms) that exceed the emission of hot thermal plasma by few orders of magnitude. At distances of $2 - 3R_{\odot}$ plasma structures in white light show most prominent coronal streamers. Their structure suggests closed field structure at low heights and open field structure at the top due to streaming slow solar wind.

In the absence or limited availability of direct measurements, coronal magnetic fields are usually approximated through MHD modelling and extrapolation of photospheric measurements (*e.g.*, [392]). However, these have their own limitations as extrapolation models are generally based on the assumption of force-free magnetic fields at the lower boundary, or the photosphere, which is actually not correct.

4.7 The solar activity cycle

In the late 19th century it was realised that occurrence of sunspots varies with an 11-year cycle. Christian Horrebow had already mentioned about a possible periodic variation way back in 1776. However, the sunspot cycle was not truly discovered until 1844, when Heinrich Schwabe reported that his observations over the previous 18 years indicated the presence of a cycle with a period of about 10 years [332]. The epoch-making work of George Ellery Hale and collaborators further confirmed that sunspots appeared and disappeared over a quasi-periodic pattern of about 11 years [156].

4.7.1 The indices of solar activity

Solar activity indices quantify different aspects of the level of solar activity and their effects by measurable values. The most commonly used index of solar activity is based on sunspot number. The weighted number of individual sunspots and/or sunspot groups is calculated from visual solar observations, using a technique initially developed by Rudolf Wolf (1816-1893) of the Zürich observatory in 1848. His goal was to overcome the difficult task of comparing sunspot observations from different astronomers, using different instruments and observing techniques. Accordingly, the Zürich or Wolf sunspot number (WSN) R_z was defined as:

$$R_z = k(10G + N), \tag{4.8}$$

where, G is the number of sunspot groups, N is the number of individual sunspots in all groups visible on the solar disc, and k is a correction factor to compensate for differences in observing conditions, observational techniques, and instruments used by different observers.

Since Wolf, several workers have attempted to further improve, and standardise the sunspot data. These are the standard international sunspot number (R_i), and the Group Sunspot Number (GSN). Another indicator of the level of solar activity is the flux of radio emission from the Sun at a wavelength of 10.7 cm (2.8 GHz frequency). Flux $F10.7$ has been shown to follow the sunspot number quite closely. This flux has been measured daily since 1947. It is an important indicator of solar activity, because it tends to follow the changes in the solar UV that influences the Earth's upper atmosphere, and ionosphere.

Solar activity is also indicated from the frequency of occurrence of solar transient and explosive phenomena, such as flares and CMEs. In 1852, Edward Sabine announced that the sunspot cycle period was identical to that of *geomagnetic activity*, for which reliable data had been accumulated since the mid-1830s. This marked the beginning of solar-terrestrial interaction studies [211].

4.7.2 The 11-year sunspot cycles

Wolf succeeded in reliably reconstructing the variations in sunspot number as far back as the 1755-1766 cycle, which is conventionally known as 'Cycle 1'. All subsequent cycles thereafter are numbered consecutively. The numbering of cycles starts as zero from 1749, and the cycle beginning in September 1996 is numbered as 23. The ascending phase of the current cycle 24 began around 2009-10, which started descending in 2015-16, reaching its minimum level in early 2019. It is expected that the next Cycle 25 would commence in late 2019.

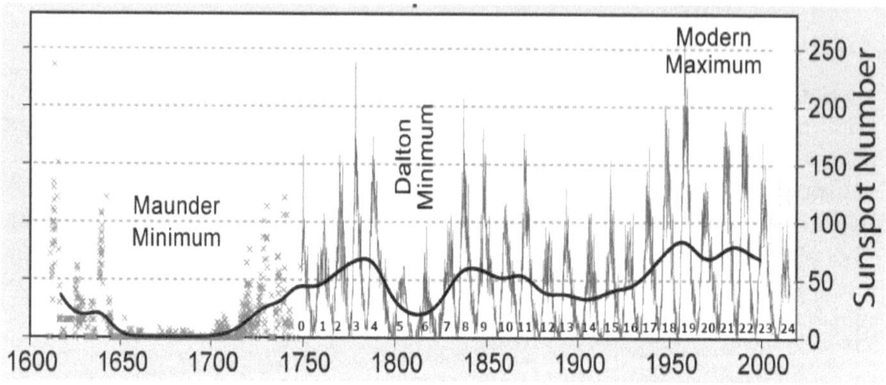

FIGURE 4.9: The 11-year solar activity cycle indicated by monthly averaged sunspot numbers with time over a 400-year period of sunspot observations. (Figure by Robert A. Rohde prepared as part of the Global Warming Art project.)

The sunspot numbers (or, sunspot area) represent the level of activity on the Sun, and its variation with time. The number of sunspots, and other structures, on the Sun go through tremendous changes with time. There are periods when no sunspots are observed. When plotted over several years, these parameters show a quasi-periodic behaviour (Fig. 4.9). The amplitude, shape and temporal width, or span, clearly differ from one cycle to another. The time span of cycles between maxima varies in individual cases, between 8 and 15 years, with an average length of ~11.1 years. On average, spottedness rises to a maximum in around 4 years and then drops to a minimum in around 7 years (for a recent review, see [169]).

The amplitude, and the shape of the peaks vary from one cycle to another. The strongest recorded recent solar maximum occurred in 1957 during Cycle 19. That year was designated as the International Geophysical Year (*IGY*). There was an increasing trend of the peak sunspot numbers, with the exception of cycle 20, during the last century. Cycles 23 and 24 were weaker than the previous two, *i.e.*, cycles 21 and 22. It may be expected that the coming maxima of Cycle 25 would be even smaller than the recent ones. This trend may be due to the effect of a superposition of longer periods on the 11-year cycle, such as the *Gleissberg cycle* of about 80 years.

4.7.3 The long period solar activities

There are some interesting signatures of prolonged periods of very low sunspot activity, such as the Maunder Minimum (1645-1715), and Spörer's minimum (1450-1550), observed in more recently recorded sunspot number data. During the Maunder Minimum period, that lasted an incredible 70 years, the 11-year solar cycle seemed to have broken down completely. This long quiet period of low sunspot numbers coincided with a *Little Ice Age*, observed in Earth's northern hemisphere. Low solar activity was believed to have played a role in that 17th century period of abnormal cooling.

For some unknown reasons, the sunspot cycle revived itself in the early 18th century, and has carried on ever since with the familiar 11-year period. What triggered the Maunder Minimum or exactly how it influenced Earth's climate? Are there signs that it might happen again? These are some perplexing questions, which still remain to be answered.

From the pattern of sunspot activity, the amplitude of cycles is observed to vary non-randomly; a series of larger maxima appears to follow a series of lower ones. This indicates that long period sunspot cycles are also operative. Several reconstructions of solar activity on millennial time-scales have been performed using tree rings, deep lake sediments and ice cores.

The solar cycle modulation of cosmic rays leaves its imprint in the concentration of the radioisotopes ^{14}C in tree rings and ^{10}Be in ice cores [113]. These isotopes are produced in Earth's stratosphere by the impact of galactic cosmic rays on ^{14}N and ^{16}O. The ^{14}C gets oxidised to form CO_2, which is taken up by plants, in general, and trees, in particular, where it becomes fixed in annual growth rings. The ^{10}Be gets oxidised and becomes attached to aerosols that can precipitate in snow, where it then becomes fixed in annual layers of ice.

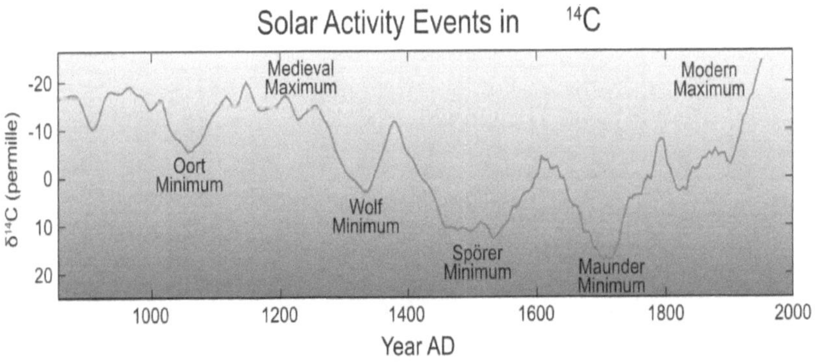

FIGURE 4.10: Long period solar activity events as inferred from the ^{14}C dating records for the period of 1050-1900. Figure by Leland McInnes `https://commons.wikimedia.org/w/index.php?curid=6397030`

The solar cycle modulation of the cosmic ray flux can then lead to solar cycle related variations in the atmospheric abundances of ^{14}C [354] and ^{10}Be [41]. The connection between solar activity and radioisotope concentrations is complicated by the transport and storage of these radioisotopes. Nonetheless, estimates of solar activity levels over time scales much longer than the 400-year sunspot record are obtained (Figure 4.10)(detailed review in [372]).

These reconstructions of solar activity reveal *grand minima* like the Maunder Minimum as well as *grand maxima* similar to the last half of the 20th century. In fact, the late Precambrian annual deposits (varves) laid down around 680 million years ago show that 11-year cycles were still operative! Long period activities have been inferred, such as the 80-year Gleissberg Cycle, and anomalous periods of solar activity, *i.e.*, Grand maximum (1100-1250 AD), Spörer minimum (1450-1550) (Fig. 4.10).

4.7.4 The butterfly diagram of sunspots

When latitudes of sunspots in the visible solar hemisphere are plotted against the time axis, an interesting butterfly (time-latitude) diagram results (Fig. 4.11). Here, each Carrington rotation is averaged in longitude at each value of latitude, and the result is placed in one column of the map.

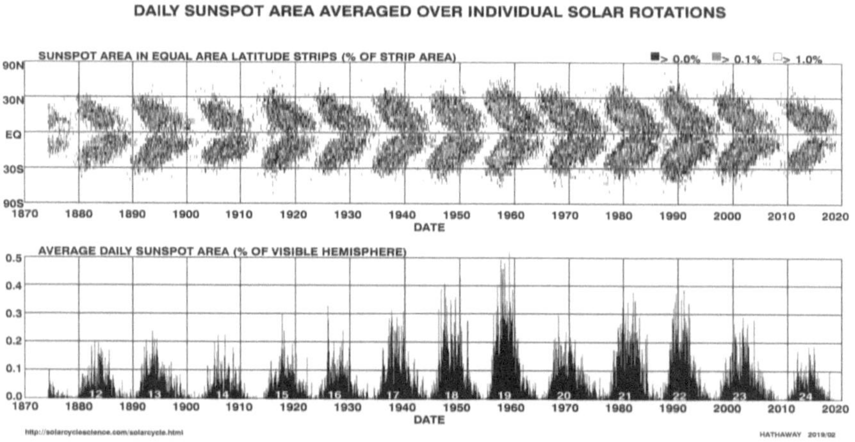

FIGURE 4.11: Butterfly (time-latitude) diagram of sunspot occurrence (credit: D. H. Hathaway/Stanford University).

It shows that sunspots occur in certain patterns in two latitude zones, called the *activity zones*, located nearly symmetrically around the solar equator. With the progress of a solar cycle, a zone migration of sunspots occurs towards lower latitudes (Spörer's law). Shortly before the sunspot minimum phase, first sunspot groups corresponding to the next cycle may appear at high latitudes. Interestingly, at times sunspots belonging to two consecutive cycles may be present as a new cycle starts.

4.7.5 The Waldmeier effect

Waldmeier (1935) first showed that an inverse correlation exists between the length of the ascending phase of a cycle (rise time) and the peak sunspot number of that cycle [381]. The sunspot cycles are asymmetric, *i.e.*, the time to rise to maximum is less than the time to fall to the minimum (Figure 4.12). Thus, larger amplitude cycles take less time to reach the maximum of the cycle than do the small cycles. The best fit through the Sunspot Number data gives the relationship, Rise Time (in months) $\approx 35 + 1800/\text{Amplitude}$ (in Sunspot Number).

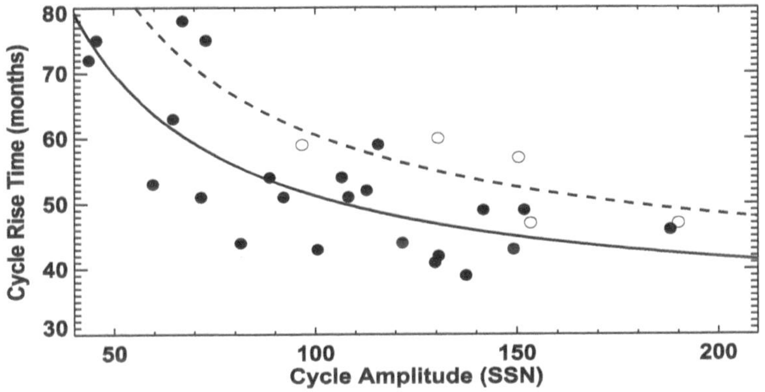

FIGURE 4.12: Waldmeier effect: rise time and amplitude relationship. (credit: D. H. Hathaway/Stanford University)

In a more recent analysis of the *Waldmeier effect*, using Zürich sunspot number data, it was shown that the correlation is present when all cycles are accounted for. However, there are significant deviations from the average pattern for individual cycles [202]. Results are also sensitive to different indices of sunspots that are used [172].

The Waldmeier effect is an important aspect of sunspot cycles for its possible use in predicting the peak, or amplitude of a solar cycle, just after it commences. It may be important to use the sunspot area data, instead of sunspot numbers, and look for the Waldmeier effect. This is because the sunspot area is better related to surface magnetic fluxes associated with the sunspots, an important ingredient in dynamo-based predictions of solar cycles [101].

There are other relationships found between different characteristics of cycles, for example, between a cycle period and the amplitude of the next

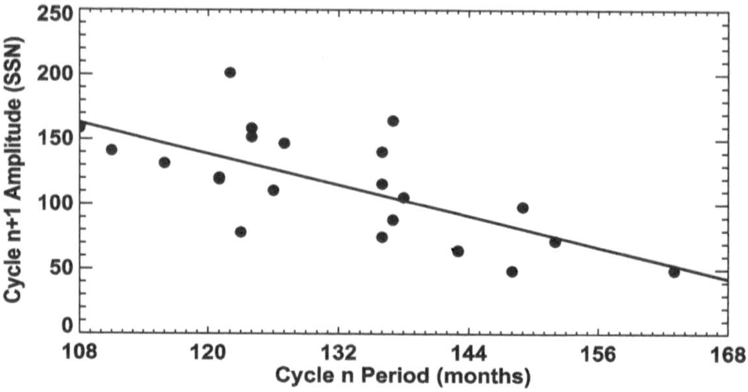

FIGURE 4.13: The amplitude and period relationship - Spörer's law (credit: D. H. Hathaway/Stanford University)

cycle ([170], [344]). The correlation is quite strong and significant at the 99% level as shown in Figure (4.13).

4.7.6 Joy's law of active region tilts

Sunspots represent magnetic activity on the surface of the Sun and frequently appear as bipolar groups. The tilt angle of bipolar sunspot groups is defined as the angle of the line connecting the centroids of each polarity with the E-W line along constant latitude. Axis of sunspot groups (or active regions) is observed to have a tendency to be tilted relative to the solar equator, such that the leading spot is usually closer to the equator than the following spots. Tilt is typically reported as positive, when the leading sunspot is closer to the equator in either hemisphere, than the following spot (Figure 4.14a).

A.H. Joy studied drawings made by Carrington and Spörer in the 1800s, and discovered that the tilt angle changed with the latitudinal location of the sunspot pairs, regardless of the hemisphere. Statistical analysis of the relationship between the tilt angle and latitude revealed a linear dependence. In low latitudes, the sunspot pair axes are nearly parallel to the Sun's equator. But with increasing latitude, the mean inclination increased to a maximum of about 12° (*c.f.*, Figure 5 in [156]). Figure (4.14b) shows the result derived from a more recent statistical study of the sunspot drawings of Carrington and Spörer ([184], [340]).

While Joy's law is observed statistically over sufficiently large samples, the tilt angles of individual sunspot regions are noisy over their lifetimes, and often identified as the *anti-Hale effect.* The overall trends in tilt angle are difficult to observe for individual active regions. However, its evolution in a particular

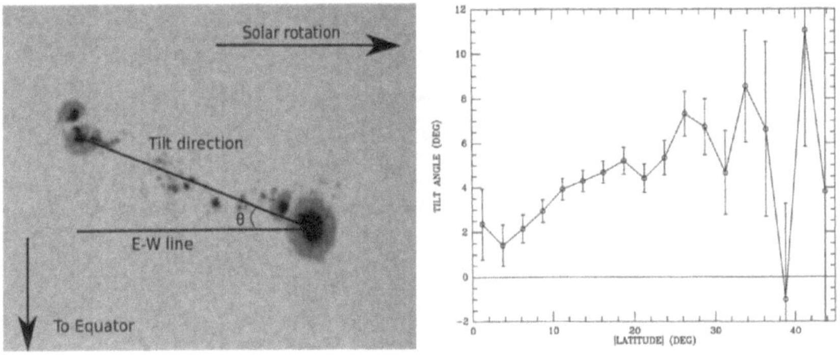

(a) (b)

FIGURE 4.14: (a) The tilt-angle between the line connecting two magnetically opposite polarity sunspots, and the E-W line. (b) Tilt angle variation with latitude according to the Joy's law (by permission from [184]).

active region, from the period of its emergence to decay, provides some insight about the associated formation and distribution of magnetic fields.

4.8 Solar magnetic cycles

4.8.1 Hale's polarity law of sunspot groups

Once Hale discovered the existence of magnetic fields in sunspots, other patterns were soon observed within a solar cycle [156]. Sunspots often appeared in groups with a pairing of opposing magnetic polarities. If a leading sunspot in a pair was of positive polarity in the northern hemisphere, then almost all leading sunspots in the northern hemisphere were generally of the same polarity for that cycle.

It was observed to be consistently the opposite in the southern hemisphere. The behaviour is reversed at the end of each 11-year activity cycle. This trend of sunspot polarity led to the belief that the magnetic field in the Sun was systematically organised on a global scale. Figure (4.15) illustrates the way magnetic polarity typically behaves in a sunspot group. This is 'Hale's polarity Law'.

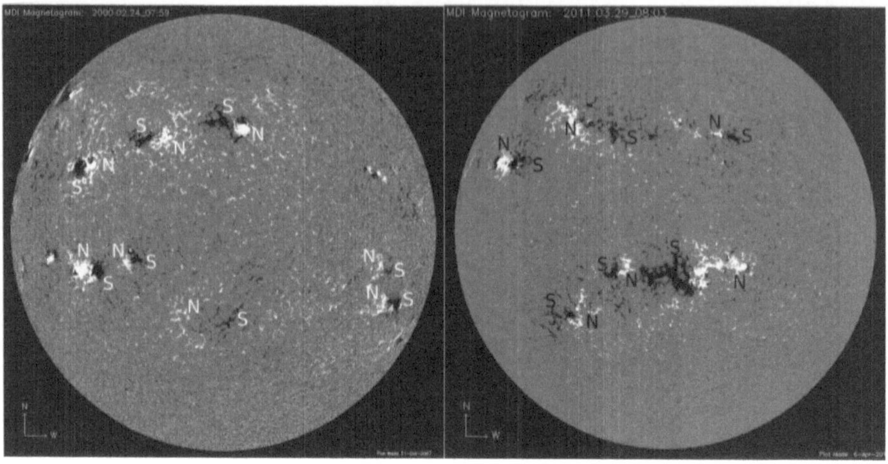

(a) (b)

FIGURE 4.15: Full-disk line-of-sight magnetograms from SoHO/MDI instrument (a) Cycle 23/February 24, 2000, and (b) Cycle 24/March 29, 2011. Bright (dark) areas represent magnetic field directed towards (away) from us. North is at top, and east is to the left. (Credit: SOHO is a project of International Cooperation between ESA and NASA).

4.8.2 The magnetic butterfly diagram

A *magnetic butterfly diagram* results when magnetic flux latitudes are plotted with time. Full disk magnetograms have been available since 1975. The global magnetic field of the Sun is observed to undergo an interesting 22-year cyclic behaviour, as seen in the magnetic butterfly diagram (Figure 4.16). This diagram is made by stacking, side-by-side, the line-of-sight magnetic flux averaged over all longitudes.

From these diagrams, the emergence of magnetic flux in active regions is observed to migrate to lower latitudes with solar cycle. The combined effects of convection and meridional circulation lead to the magnetic flux transport to high latitudes, and thus cause reversals of the polar magnetic fields in phase with the activity cycle [384].

The reversal of the Sun's general magnetic polarity takes place at the poles around the maximum phase of sunspot activity. Thus, the period of a solar magnetic cycle is $\sim$ 22 years, or two sunspot activity cycles, in average duration, to complete the polarity reversals of both the polar and equatorial (sunspot) fields [169].

The noticeable features of magnetic butterfly diagram are: (i) opposite polarities in opposite hemispheres, (ii) gradual migration of active region belts towards the equator, (iii) poleward migration of the following polarity flux, (iv) polar magnetic fields, (v) reversal of polarities in active region belts, and polar fields from one cycle to another, after 11 years.

FIGURE 4.16: Butterfly (time-latitude) diagram of longitudinally averaged line-of-sight magnetic flux. (credit: D. H. Hathaway/Stanford University).

At a glance, it would be easy to assume that the largest magnetic events on the solar surface are the foremost contributors to the global magnetic landscape. However, small-scale magnetic activity contributes greatly to the global flux balance, where the flux emergence rate decreases as the scale size of bipolar active regions increases [166].

The magnetic butterfly diagram further strengthened the argument for a global structuring of solar magnetic field. It is also fundamental to dynamo theory, as any model must describe not only the orientation of polarities in the hemispheres, but the switching of polarities between consecutive solar cycles in the course of 22 years.

4.9 Total solar irradiance variation

Total Solar Irradiance (T.S.I.) is the energy flux at a given distance (1 A.U. for the Earth) per unit area and time, observed at the top of Earth's atmosphere. It provides a measure of solar luminosity, $L_\odot = 3.844 \times 10^{26} W$. The first attempts at its measurement were carried out independently, at around the same time, by a French astronomer, Claude Pouillet (1790-1868), and a British astronomer John Herschel (1792-1871). Their inferred value was about half of the accepted modern value of $1367 \pm 4\ W/m^2$, because they did not account for the absorption of the solar radiation by the Earth's atmosphere.

The *total solar irradiance* was often termed the *solar constant*, which is misleading, as we know that the luminosity of Sun has varied from 0.72 $L_\odot$

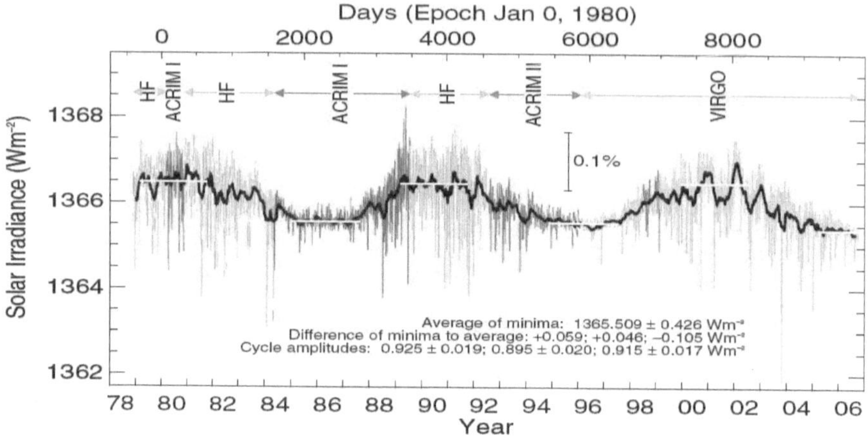

FIGURE 4.17: Two and a half solar cycles of Total Solar Irradiance (TSI). Fluctuations in irradiance correspond well with the cycling of sunspots. (*Courtesy of SOHO/Virgo consortium. SOHO is a project of international cooperation between ESA and NASA.*)

to the present 1.0 $L_\odot$ during the Sun's 4.6×10^9 years of life, according to the stellar evolution theory. In addition, the recent results from space-borne instruments also show irradiance-variation over shorter periods, in phase with the 11-year activity cycle with relative amplitude of 0.04 (Fig. 4.17) ([122], [393]).

The total solar irradiance variation is generally explained in terms of a combination of decreases due to blocking by dark sunspots and increases by bright faculae and plages, both of which are positively correlated with solar magnetic activity. These two effects tend to nearly cancel each other. The net positive correlation of the total irradiance is explained as a slight excess of facular brightening over sunspot darkening. Other sources of luminosity variation that are not related in this simple way to solar activity cannot be ruled out (for a review, see [343]).

4.10 Solar dynamo models

It is generally accepted that the solar magnetic cycle is produced by a magnetohydrodynamic dynamo operating in the solar convection zone and the solar differential rotation. The general dynamo theory began in the 1930s, motivated mainly by the existence of the Earth's magnetic field, thought to be generated by a dynamo operating in its molten interior.

Major goals for solar cycle simulation can be carried out within *kinematic* dynamo theory, in which the principal equation is the MHD induction equation, where the velocity field v is assumed to be given, and not affected by the evolution of the magnetic field. This is a reasonable starting point in a hydrostatic object, like the Sun, where pressure and gravitation balance each other, and the Lorentz force $J \times B$ is negligible, *i.e.*, the field can be assumed to be force-free. The induction equation is linear in such a situation, which helps to make the analysis easier.

In reality, however, the magnetic force may not be negligible and v becomes a function of B. The induction equation is then nonlinear, and the analysis requires a simultaneous solution of the momentum equation. This essentially implies that one has to solve a *dynamic* (magnetohydrodynamic) problem. In the case of the Sun, it involves a combined solution of the convective motion and magnetic field generation, requiring extensive and complex computer simulations.

A simple model describing the process of magnetic field regeneration in the Sun was proposed by Babcock (1961) [24], and later amended by Leighton (1969) [233]. The magnetic field of the Sun is described in relation to the rotational axis geometry of the Sun. At solar minimum, the dipole magnetic poles of the Sun are in close proximity to the rotational poles of the Sun. Magnetic field lines are primarily N-S in direction, or *poloidal,* stretching from one pole to the other. These lines follow the classical model of a magnetic dipole. There are two basic processes that are involved in this and most other dynamo models:

- The ω-effect of shearing motions due to the differential rotation of the Sun. It strengthens the magnetic field, stretches and aligns it with the flow, such that the poloidal field lines are pulled in the direction of the equatorial rotation and are eventually wrapped in the E-W direction around the Sun. Magnetic field lines that are primarily in the E-W direction are called *toroidal*. This conversion of poloidal magnetic field lines to a toroidal orientation by way of differential rotation is called the ω-effect (Figure 4.18A).

- The α-effect of lifting and twisting the magnetic fields, introduced or furthered by Parker (1955) [299], is thought to convert portions of the toroidal magnetic fields back to poloidal fields. Flux tubes rise through the solar surface due to magnetic buoyancy (Figure 4.18B). As the middle span of these tubes rises faster than the 'anchored' ends of the tube, a loop forms. Sunspots are formed when the loop, tilted with respect to a line of constant latitude by the Coriolis force, breaks through the surface of the Sun. The magnetic field 'frozen' in the tube has now been tilted, and redirected away from the Sun in a poloidal direction (Figure 4.18C).

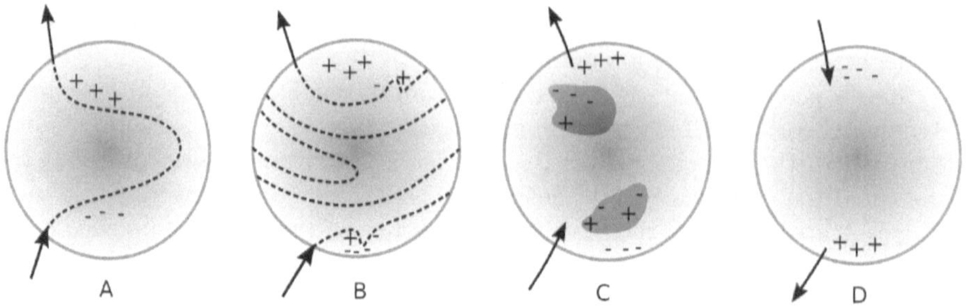

FIGURE 4.18: Schematic of Babcock-Leighton model: (A) The ω-effect transforms a poloidal field into a toroidal field. (B-C) Toroidal magnetic field lines are twisted into loops, and rise toward the solar surface due to the magnetic buoyancy. (D) The α-effect transforms a toroidal field into a poloidal field.

The $\alpha-\omega$ model describes the regeneration of toroidal magnetic fields into poloidal fields (Figure 4.18D). Although a simple kinematic $\alpha - \omega$ dynamo is unlikely to give a correct representation of the solar dynamo, it is possible to adjust the parameters so that an oscillatory behaviour corresponding to the solar cycle can be reproduced.

Joy's law provides observational support for the operation of the so-called "α-effect" in solar dynamo action in relation to the solar cycle. Tilts and slopes are reported to vary with cycle and hemisphere, and there are periods when Joy's law is nearly absent. From the study of Cycles 17-21 data, there is a suggestion that stronger cycles are associated with smaller tilts. Tilt has been attributed to Coriolis forces acting on plasma draining from the apex of rising flux tubes, as suggested by Schmidt (1968) [329]. Models of rising flux tubes are able to simulate observed tilt angle behaviour if the magnetic field is restricted within 60-160 kG [104].

More recent reports show no tilt angle dependency on field strength as would be necessary if the tilt were caused by Coriolis forces [218]. Mt. Wilson Observatory data from 1917-1985 were used to analyse the active-region tilt angle as a function of solar cycle, hemisphere, and longitude, in addition to the more common dependence on latitude [265]. The slope of Joy's law is found to increase more steeply for tilt angle measurements obtained from Mount Wilson magnetograms, as compared to those obtained from Debrecen Photohelographic Data ([387], [31]). This discrepancy could be due to the contribution of plage areas that tend to have greater tilt inclinations than adjacent sunspot groups.

4.11 Prediction of solar cycles

Solar dynamo models have made considerable advancement, but the quest remains to build a dynamo model that simulates the most prominent features of a solar cycle and allows prediction of the amplitude and other properties of cycles to follow. It is evident that the Sun does not maintain a clockwork of regular activity or magnetic cycles, and an accurate prediction has been rather difficult and controversial, with conflicting predictions made by various researchers. For example, there were predictions for Cycle 24 using the flux transport model and the solar dynamo model that gave widely varying amplitudes ([100], [80]).

Predicting the behaviour of a sunspot cycle, however, is fairly reliable, once the cycle is well under way, about 3 years after the minimum in sunspot number occurs [170]. Relationships have been found between the size of the next cycle maximum and the length of the previous cycle, the level of activity at sunspot minimum, and the size of the previous cycle. A number of techniques based on widely varying aspects of polar flux, geomagnetic indices [385], *etc.*, are used to predict the amplitude of a cycle during the time near and before sunspot minimum. A synthesis of solar cycle prediction techniques is described in [171]. For more details, one is referred to a review [304].

4.12 Summary

Solar magnetic fields from the smallest observable scales to the global scales play important roles in determining the level of activity from the active regions to solar activity cycles. Most of the energetic solar transients derive their energy from the stored magnetic energy in active regions and filaments. Therefore, the measurement of magnetic fields from photosphere to the upper layers of solar atmosphere is one of the most important tasks in solar physics. It is also essential to measure the horizontal and vertical (Doppler) velocities, which interact with magnetic fields. A variety of diagnostic tools based on the Zeeman and Hanle polarizations have been developed with improved accuracy over the year, helping in our understanding of the observed phenomena.

The large-scale solar magnetic fields, and their interaction with solar differential rotation, clearly drive the solar activity cycles. It is very important to understand the Sun's activity cycles as they have a direct role in governing the heliospheric structure, and space weather, and in driving many of the solar-terrestrial effects. The reconstructions of solar activity suggest that the Sun spends about 1/6th of its current life in a grand minimum phase, and about 1/10th in a grand maximum. The existence of long cycles shows that the Sun does remember how active it was in the previous cycles, and there is a link between the 11-year cycles. A successful theory of solar dynamo should explain why the solar cycle is so irregular and how it regenerates itself after having it turned off ([24], [77]). There is still much to learn about the way the Sun operates its activity cycles.

Chapter 5

The Invisible Solar Interior: Neutrinos as the Probe

5.1 Introduction

Most of what we know of the physics of stars is heavily based on our knowledge of the Sun. The structure and dynamics of the distant stars can not be resolved, because they appear as point-like objects even with the largest telescopes. On the other hand, the complex and varying face of the Sun's outer layers can be observed at the finest temporal and spatial scales. Granulations and sunspots, *etc.*, are only some of the surface signatures of the vigorous physical processes taking place in the solar interior.

Our knowledge of the Sun is not complete unless we can also infer its internal structure and dynamics. The interior of the Sun, however, is hidden underneath the optically thick outer 'skin', the photosphere of the Sun. Therefore, it is not possible to *observe* the Sun's interior by any instrument directly.

In the absence of a *direct* view, our knowledge of the Sun's interior has traditionally relied on mathematical formulations of the physical processes operating within a star. While modelling the solar interior, we have the advantage of a lot more knowledge about the Sun than for any other star. But, it also makes the modelling that much more challenging, because any useful solar model must yield all its observed properties.

Thankfully, there are some other methods of inferring about the internal structure of the Sun. These are based on the radiated energy flux, the neutrinos emitted from its central core, and the observed photospheric oscillation modes. Of these, the energy flux is a strongly averaged source of information, while the neutrinos are rather difficult to detect and interpret. This chapter is devoted to theoretical formalism and to the solar neutrinos for testing our understanding of the Sun's interior.

During the last few decades, precise observation of solar surface oscillations has made rapid progress. It has become an important tool for inferring the structure of the solar interior. More importantly, as we shall see, it has also helped in resolving the perplexing problem of solar neutrino deficiency. Its major success has been not only in making some new discoveries about solar interior, but also in contributing to the standard theory of particle physics. We shall discuss some of these issues in the following sections. Solar oscillations will be discussed in Chapter 6.

5.2 Modelling the Sun

A solar model is the solution to a set of mathematical equations describing the physical processes occurring within the Sun. These equations are often divided into two subsets: *the structure equations*, and *the chemical evolution equations.* In the very simplest cases, these equations are coupled, nonlinear first-order *ordinary differential equations.* But more realistic models involve nonlinear, coupled first-order partial differential equations. In either case, the equations cannot be solved analytically and must be solved numerically. Therefore, a solar model is usually comprised of a set of tables describing the internal conditions, such as the chemical composition, density, opacity, luminosity, mass, pressure, temperature, *etc.*, at different depths in a star, here the Sun [153].

5.2.1 Standard solar model (SSM)

The Sun is believed to have begun its life by contraction of a gaseous *proto-sun* composed mainly of hydrogen and helium. The initial composition was altered by a steady increase of helium due to nuclear burning as the gravitational collapse of the proto-sun enhanced the density and temperature at the central core.

A standard solar model is the one where the physical inputs are not changed to bring the model in to better agreement with the Sun. The input physics (nuclear reaction rates, diffusion coefficients, opacity tables, equation of state) are input as they are. The agreement or otherwise between the Sun and the model is an indication of how good the input parameters are.

The two basic forces at play in a star are: (i) *gravity* that causes it to collapse, and (ii) *pressure* that causes it to expand. The interplay between these forces occurs throughout a star's life from its birth to death. During the evolution of a star, there are stages when gravity wins, and the star begins to collapse (*i.e.*, at the birth from a molecular cloud → proto-star → to the present epoch). There are also stages, when pressure wins, and the star begins to expand (the end phase → red giant → planetary nebulae).

For the most part though, gravity and pressure balance each other, and the star maintains a hydrostatic equilibrium *i.e.*, as in the main sequence stage of the *H-R diagram*(Figure 1.8, page 21). The Sun is currently in this stage of its life in the hydrostatic equilibrium.

The major assumption for constructing a standard solar model is that the Sun is considered to be a spherically symmetric ball of gas in hydrostatic balance. All physical parameters are functions of only the radial distance, r, from its center. The small effects of non-radial distortions due to rotation, magnetic field, and tidal forces are assumed to be negligible in this approximation.

There are several other ingredients and assumptions that are required in constructing the standard solar models, such as the standard nuclear and neutrino physics, a uniform initial chemical composition, no significant mass loss or accretion, no mixing of nuclear reaction products outside the convection zone, no transport of energy or momentum by waves, *etc.* Auxiliary input physics needed are the opacity, microphysics involving nuclear energy generation in the core, transport of radiation (or energy), the equation of state, and turbulent convection. The macro-physics involves diffusion of helium and heavy elements, and change of elemental abundances by various processes.

5.2.2 Basic equations of solar structure

As the stars are assumed to be spherically symmetric in standard solar models, we can express their properties using a set of *1-D* equations. The main theoretical formalism of a solar model include the equations corresponding to the conservation laws, as follows.

Conservation of mass: In absence of flows, the mass of a volume element must be accounted by its density. At any point r, the volume of a small part of a spherical shell is given by $dv = r^2 sin\theta d\theta d\phi dr$ in spherical coordinates (r, θ, ϕ) (Figure 5.1b).

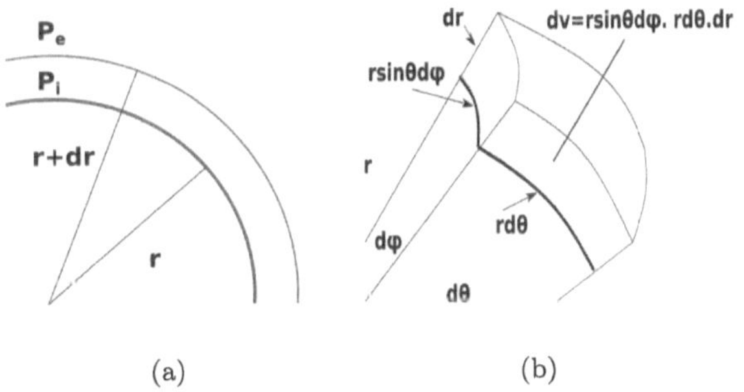

FIGURE 5.1: (a) Geometry of a spherical shell. (b) Volume of an infinitesimally thin section of the shell.

In the spherically symmetric case, the volume of a spherical shell of thickness dr can be obtained by integrating over $\theta = (0, \pi)$, and $\phi = (0, 2\pi)$, which provides $dV = 4\pi r^2 dr$. If one assumes a constant density ρ over this infinitesimally thin shell, the mass dm of this shell is given by

$$dm = \rho dV = 4\pi r^2 \rho dr. \tag{5.1}$$

The radius of a star undergoes a change in the course of its evolution. However, the mass of stars, particularly of a low mass star like the Sun, does not change significantly. Therefore, it is customary to use mass, and not radius, as the independent variable, and one writes:

$$\frac{dr}{dm} = \frac{1}{4\pi\rho r^2}. \tag{5.2}$$

It is assumed here that the Sun is in a steady state. But, in general, $r = r(m;t)$, and hence, one must use the partial derivative form:

$$\frac{\partial r}{\partial m} = \frac{1}{4\pi\rho r^2}. \tag{5.3}$$

Conservation of momentum: Let us now consider the pressure acting over the thin shell of thickness dr. The pressure (force/area) acting outwards at the inner (r), and outer ($r + dr$) boundaries are P_i and $P_e = P_i + (dP/dr)dr$ (Figure 5.1a). The difference in pressure on the shell is thus $f_p = P_i - P_e = -(dP/dr)dr$.

Similarly, one can obtain the force due to gravity acting inward on the shell as mass $\times$ acceleration $= -(4\pi\rho r^2)dr \times g$, and hence, force per unit area due to gravity, $f_g = -(4\pi\rho r^2)dr \times g/(4\pi r^2) = -g\rho dr$. As the Sun is assumed to be in hydrostatic equilibrium, one has $f_p + f_g = 0$. Therefore, one can write

$$\frac{dP}{dr} = -g\rho. \tag{5.4}$$

Using conservation of mass ($dm = 4\pi r^2 \rho dr$), and $g = -Gm/r^2$, one obtains

$$\frac{dP}{dm} = -\frac{Gm}{4\pi r^4}. \tag{5.5}$$

or, in the partial derivative form,

$$\frac{\partial P}{\partial m} = -\frac{Gm}{4\pi r^4}. \tag{5.6}$$

It is to note that this equation is valid only for the static state, therefore, it does not describe collapse or expansion of the proto-star.

What if the forces acting on the shell do not balance?

In this case, the shell will experience acceleration, described by the equation of motion:

$$\frac{d^2r}{dt^2} = -4\pi r^2 \frac{dP}{dm} - \frac{Gm}{r^2} \tag{5.7}$$

Case (i) : The hydrostatic equilibrium, $(d^2r/dt^2) = 0 \Rightarrow dr/dt = 0$.

Case (ii): When pressure vanishes: free fall or collapse occurs, as using order of magnitudes, one has $(d^2r/dt^2) \approx (R/\tau_{ff}^2) = g = (Gm/r^2)$. The characteristic time for free-fall $\tau_{ff} = (R_\odot/g_\odot)^{1/2} = 27$ minutes, for the Sun.

Case (iii): When gravity vanishes: expansion of the star occurs, as one has $(d^2r/dt^2) = -4\pi r^2\ dP/dm = -(1/\rho)dP/dr$. This gives, in the order of magnitudes, $R/\tau_{exp}^2 \approx P/(\rho R)$. Hence, the characteristic time of expansion $\tau_{exp} = R_\odot(\rho/P)^{1/2} \approx R_\odot/c_s$, turns out again 27 minutes for the Sun, where, the sound travel time $c_s = (\Gamma_1 P/\rho)^{1/2}$, and in general, $\Gamma_1 = (\frac{\partial lnP}{\partial ln\rho})_{ad}$.

Conservation of Energy: Let us define dl the luminosity of a shell, such that *i.e.*, $\int_0^m (dl/dm)dm = L$. An equation of continuity must be satisfied by radiation, where dE/dt is the rate of energy production per unit thickness of a shell of radius r, and L is the net flux of energy passing through the shell. Then,

$$\frac{dE}{dt} + \frac{dL}{dr} = 0 \Rightarrow \frac{dL}{dr} = -\frac{dE}{dt} = 4\pi r^2 \rho\varepsilon \Rightarrow \frac{dL}{dm} = \varepsilon. \tag{5.8}$$

Here, ε $(kg^{-1}s^{-1})$ is the energy generation rate by thermonuclear reactions in the gravitationally stabilised central core. It is assumed that the energy produced by nuclear reactions equals the photon luminosity of the Sun.

The mean energy generation rate for the Sun can thus be inferred as $\bar{\mathrm{E}}=L_\odot/M_\odot$. It is to note that during the main sequence evolution, the interior of the Sun is very close to thermal equilibrium, and the heating/cooling effect $\partial S/\partial t$ is small (S being the entropy) and may be neglected.

The temperature and pressure estimates for Sun's central core

From the momentum conservation equation, $dP/dm = -(Gm)/(4\pi r^4)$, which on integration gives

$$-\int_0^M dP = -\int_0^M \frac{Gm}{4\pi r^4} dm \rightarrow P_c = P_s + \frac{GM^2}{8\pi r^4} \tag{5.9}$$

where, P_c and P_s are the pressures at the core, and surface, respectively. Now, assuming $P_s << P_c$ and setting $r = R_\odot, M = M_\odot$, the estimates of pressure and temperature at the solar core can be obtained from

$$P_c \sim \frac{GM_\odot^2}{8\pi R_\odot^4}, T_c \sim \frac{GM_\odot m_H}{k_B R_\odot} \tag{5.10}$$

where, we have used the ideal gas law at the core, as $P_c = nk_BT_c = (\rho k_B T_c)/m_H$, and k_B = Boltzmann's constant, n = number density (atoms/cc), and density $\rho = M_\odot/(4\pi R_\odot^3)$. This gives the estimated value $T_c \sim 2.7 \times 10^7$K; close to the accepted value, $\sim 1.5 \times 10^7$K.

The equation of state: The pressure P arises from momentum transfer by particles and photons. In the case of the Sun, it has two components: the gas pressure P_G and the radiation pressure P_R. Radiation pressure is important only in the outer atmosphere and the solar wind. For the present solar model one can thus apply the perfect gas law, given by

$$P = \frac{\rho RT}{\mu}, \tag{5.11}$$

where R is the gas constant, and μ is the mean molecular weight (in *a.m.u.*). The constituents of the gas are usually denoted by X (the fractional abundance of H), Y (the fractional abundance of He), and Z (the fractional abundance of all the other heavier elements). The ionization of elements adds particles to the gas and reduces μ, *e.g.*, the mean molecular weight of fully ionized electron-proton plasma is half of neutral hydrogen gas.

Equations of Chemical Evolution: The equations of chemical evolution depend solely upon the nuclear processes that are operating in the interior of the star. For example, in the Sun, the *pp*-chain of fusion reaction is considered to be most important.

In a simple model, one may usually keep track of five elements: H, two species of He, C, and N. With certain simplifying assumptions, one can construct equations of chemical evolution with reaction rates between the species, their abundances, and diffusion rates. The total abundance is taken as unity, *i.e.*, $X + Y + Z = 1$, where X and Y are varying with depth and time in the star, while Z remains constant.

5.2.3 Procedure for calculating a solar model

A star is considered to have homogeneous composition at the zero age (the proto-star phase). It has begun to derive most of its luminosity from nuclear reactions. Thus, the period early in its contraction from a cloud of gas and dust is neglected. The Standard Solar Model (S.S.M.) considers evolving a one solar mass ($M_\odot$) stellar model at zero age to the present age of the Sun.

The abundance of elements in the zero age solar model is generally estimated from primordial meteorites. Along with this abundance information, a reasonable guess at the zero-age luminosity is then converted by an iterative procedure into the correct value for the model. The set of basic governing differential equations are integrated numerically.

The star is considered to be made up of spherically symmetric shells, with conditions slowly changing from shell to shell. The laws of physics relate one shell to the other that allow the physical quantities to be numerically determined in each layer. Calculating a solar model means the determination of pressure, temperature, and the chemical composition as a function of mass or radius throughout the Sun.

Numerical integration is carried out in finite steps making use of the equations of state, giving relationships for the pressure, the opacity and the energy generation rate in terms of the density, temperature and composition. In principle, an integration might be carried outward from the center $r = 0$, where natural boundary conditions on mass and luminosity are $M_\odot = 0$, and $L_\odot = 0$. The central pressure and temperature, P_c, and T_c are actually unknown, but the estimated values are obtained from the hydrostatic equations as described in Section 5.2.2.

The surface boundary conditions for the solar structure equations are taken as $R = R_\odot$, $M = M_\odot$, $L = L_\odot$, $P = 0$, and $T = 0$. Any discrepancy from the measured values of the Sun's luminosity, surface abundances, *etc.*, can then be used to refine the model.

It is necessary that these models agree with the known equilibrium properties of the Sun: the radius [8], the observed luminosity [393], the mass, the composition of the photosphere (inferred from meteoritic abundances and spectroscopic measurements) $Z/X = 0.0245 \pm 0.0015$ [151], and the age of the Sun $4.6 \pm 0.004 Gy$. The computations involve additional hypothesis, and an appropriate theory (*e.g., mixing-length*) (Section 5.4.4) for the treatment of solar convection to simplify its theoretical description. The profiles of temperature, density, pressure, and sound speed show rapid decrease from the core through the interior as r increases, except some departures near the surface. There are several standard models constructed using different assumptions for the solar internal structure, such as the one given by [66]. These models obtained for the interior are then matched to the outer solar atmospheric model (such as in [377]).

We shall now discuss the physics of energy generation in the core and its propagation through the various layers of solar interior as inferred from solar models.

5.3 Energy production in the central core of the Sun

The luminosity of the Sun and other stars is due to the powerhouse operating in the solar interior, its central engine. The Sun reached this main sequence stage after going through the course of its evolutionary phases lasting over ~50 million years from the proto-sun stage. The central core of the Sun is very dense, containing a mass $\sim 0.5M_{\odot}$ within the radius $\sim 0.25R_{\odot}$, implying an extremely high density $\sim 1.53 \times 10^5 kg.m^{-3}$. The temperature and pressure of the core are $T_{core} \sim 1.57 \times 10^7\ ^{\circ}K$, and $P_{core} \sim 250 \times 10^9$ Earth-atmospheres, respectively (Figure 5.2). These extreme conditions of density, temperature and the confinement of hot plasma in the core by gravity lead to the thermo-nuclear fusion process, the basic source of energy generation in stars.

The likely nuclear fusion reactions in the Sun's core depend on the abundance of species, and the probability of reactions under the prevailing conditions, such as the temperature and pressure there. Hydrogen is the most abundant element, followed by helium, carbon, nitrogen, oxygen, *etc.*, in the Sun, and Sun-like stars. Therefore, at the high temperature, pressure in the core, and the lowest nuclear charge of H nuclei (protons), the most likely nuclear process to occur is the *pp*-chain reaction. As the Coulomb repulsion increases between charged nuclei as the product of their nuclear charges, the force of repulsion involved between the protons is the lowest.

The conversion process by the *pp*-chain reaction involves several steps of different nuclear reactions, given schematically as follows:

Step 1: The fusion of two protons forms deuterium H^2, a positron and an electron neutrino. This reaction proceeds extremely slowly and essentially controls the solar nuclear burning process.

$$_1H^1 + {}_1H^1 \rightarrow {}_1H^2 + e^+ + \nu_e \ (Q = 1.44MeV). \tag{5.12}$$

Step 2: The deuterium, produced in Step 1, fuses with another proton within a very short time of around 10 seconds as follows:

$$_1H^2 + {}_1H^1 \rightarrow {}_2He^3 + \gamma \ (Q = 5.49MeV). \tag{5.13}$$

Step 3: As the nucleus Li^4 is unstable, a further proton capture is unlikely, and in the next step, He^3 nucleus, formed in Step 2, fuses with another He^3 (formed earlier in another $p-p$ reaction) and gives He^4:

$$_2He^3 + _2He^3 \rightarrow \ _2He^4 + 2\ _1H^1 + \gamma\ (Q = 12.86MeV). \tag{5.14}$$

For each occurrence of the 3rd reaction, first two must have occurred twice, so the net result is:

$$4\ _1H^1 \rightarrow \ _2He^4 + 2e^+ + 2\nu_e + 2\gamma + 26.7MeV. \tag{5.15}$$

Here, it may be noted that the four protons (or, 4_1H^1 nuclei) taking part in the *p-p* reaction involve more mass (4.0331 a.m.u.), than the resulting total mass of the $_2He^4$ nucleus (4.0026 a.m.u.), two positive electrons and two electron neutrinos, where 1 atomic mass unit (a.m.u.) = 1.66×10^{-27} kg. This loss of mass is released as energy of the Sun as the reaction occurs continuously, involving a large amount of hydrogen mass in the core. The Sun loses $\sim 6 \times 10^9$ kg/s corresponding to its luminosity of 3.8×10^{26} W.

The energy generation rate can be described as a simple power law to more complicated expressions involving nuclear reaction rates. This energy appears as production of γ-ray photons, or as kinetic energy of the nuclei. About 2% of the released energy is also carried out by the neutrinos. However, this does not contribute to solar photon luminosity, as the neutrinos do not interact with the material, while propagating outwards through the solar interior.

There are possibilities of other branches of nuclear reactions proceeding to form *Be* and *Li* nuclei by the fusion of He^3 and He^4 nuclei produced in Steps 2 and 3, which are energetically less important. However, they are the ones that are responsible for producing the higher-energy neutrinos first detected in neutrino experiments from ground-based neutrino observatories. These reactions proceed as follows:

$$_2He^3 + _2He^4 \rightarrow \ _4Be^7 + \gamma\ (Q = 1.59MeV). \tag{5.16}$$

The Be^7, thus produced, can react with an electron, and then with a proton, to produce He^4 as follows:

$$_4Be^7 + e \rightarrow \ _3Li^7 + \nu_e\ (Q = 0.86MeV), \tag{5.17}$$

$$_3Li^7 + _1H^1 \rightarrow \ 2\ _2He^4\ (Q = 17.35MeV). \tag{5.18}$$

There may be another alternative to the above reaction, where the Be^7 nucleus may first react with a proton to form boron, thereafter eventually producing He^4 through the following sequence:

$$_4Be^7 + _1H^1 \rightarrow \ _5B^8 + \gamma\ (Q = 0.13MeV), \tag{5.19}$$

$$_5B^8 \rightarrow \ _4Be^{8*} + e^+ + \nu\ (Q = 15.08MeV), \tag{5.20}$$

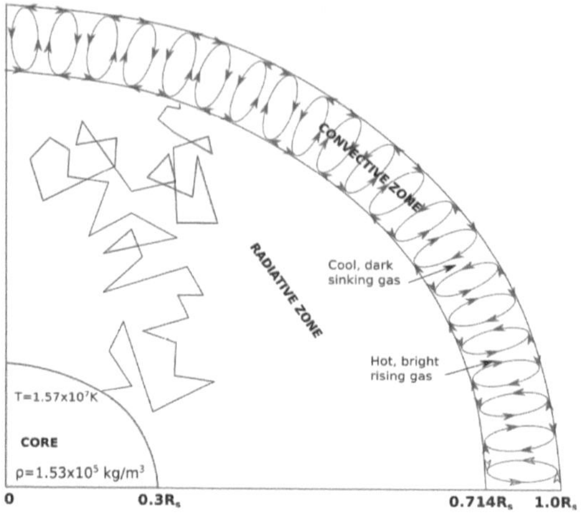

FIGURE 5.2: A schematic diagram showing the internal structure of the Sun: Core, Radiative Zone and Convective Zone.

$$_4Be^{8*} \rightarrow 2\ _2He^4\ (Q = 3.00MeV). \tag{5.21}$$

In these reactions, (*) represents excited nucleus.

There is a second mechanism, called the *CNO-cycle*. However, around 99% of the Sun's energy comes from the *pp*-chain, and only about 1% from the *CNO*-cycle [43]. But, it may be an important source of energy generation in massive stars. The net result of the *CNO*-cycle, as is the case with the dominant *pp*-reactions, is to form *He* from four protons, while, *C* and *N* isotopes serve only as catalysts leading to the formation of *He* nucleus as a by product. Some reactions involving about 0.1% N^{14} produce O^{16}, F^{17}, *etc.*, and contribute heavy isotopes back to the *CNO*-cycle.

The energy generation rate by thermonuclear reactions in the gravitationally stabilised solar fusion reactor ε $(kg^{-1}s^{-1})$ is determined by the reaction rate per unit volume, the energy generated per reaction, and the density, and is given by

$$\varepsilon = \varepsilon_0 X_1 X_2 \rho T^n f_0. \tag{5.22}$$

For *pp*-reactions, $X_1 = X_2 = X$ (hydrogen abundance fraction), and for *CNO*-cycle, $X_1 = X$, and $X_2 = X(C + N)$, *i.e.*, the carbon and nitrogen mass fraction. The temperature dependence of $\varepsilon_{pp} \sim T^{4.5}$, but $\varepsilon_{CNO} \sim T^{20}$ at around $T \sim 1.5 \times 10^7$K.

5.4 Energy transport through the solar interior

What happens to the photons and neutrinos that are produced in the *pp*-nuclear fusion occurring in the central core of the Sun? While the neutrinos pass through the overlying layers unhindered, because they are very weakly interacting particles, the energy transport is not so smooth for the photons (γ-rays) released from the core. They cannot reach us directly, as they need to make their way out through the dense interior.

There are three basic mechanisms for energy transfer in a medium: (i) *conduction*, where the energy is transmitted by electrons moving in a medium, (ii) *radiation*, where the energy is transmitted by photons, and (iii) *convection*, where the energy is transmitted by bulk motion of a gas or liquid. In the Sun, the conduction process is unimportant, because the density of the Sun is too low for energy to be transferred rapidly by this mechanism. However, in stars with extreme densities, such as in white dwarfs, conduction is important.

The relative importance of radiation, or convection, in solar interior depends on several factors. It varies with position within the Sun from deep interior outwards to the surface. Accordingly, the solar interior, above the top of the core to the outer visible layer of the Sun, can be divided in to two zones. This division is essentially based on the variation of the opacity of the intervening medium from low to high, and the efficiency of the energy transfer process.

5.4.1 The radiative zone

Due to the still very high density of material in the region above the core, the photons produced in the core get continuously absorbed, and re-emitted as they travel outwards by the hot plasma medium. The most efficient process for energy transport is radiation in this region, therefore, it is termed the *radiative zone*.

In this very hot region, spanning from around 0.25 $R_\odot$ to around 0.7 $R_\odot$, the material's ability to block light, *i.e.*, the opacity, is low. The temperature falls from $\sim 8 \times 10^{6\circ}$K at the outer core to $\sim 2 \times 10^{6\circ}$K in this zone. However, the gas density is still very high, decreasing from around 20,000 $kg\ m^{-3}$ near the lower boundary to around 200 $kg\ m^{-3}$ at the upper boundary. Therefore, photons continuously get absorbed by the ions in the gas, and are re-emitted in random directions (Figure 5.2). Thus, a photon executes a random walk from the core as it traverses outwards. The *mean free path* of photons here is of the order of 1-2 cm. Due to this random-walk process, photons are trapped in the radiation zone for a long time, taking more than ~10,000 years before actually reaching the edge of the radiative zone, as compared to its direct flight time within the solar interior of about 2 sec!

According to Wien's displacement law of black-body radiation the radiative zone at an average temperature of 8MK is dominated by radiation peaked at $\lambda_{max} \sim 0.25$ nm, *i.e.*, it is full of X-rays.

5.4.2 The convective zone

The radiative zone is surrounded by a relatively cooler and more opaque envelope beyond $\sim 0.7R_\odot$ to the outer surface. The temperature falls in this envelope from the high of about $2\times10^{6\circ}$K at the top of the radiative zone to much cooler of about 6000°K at Sun's visible surface. Also, the density ρ decreases from around 200 $kg\text{-}m^{-3}$ to $\sim 0.2kg\text{-}m^{-3}$.

Photons continuously get absorbed in the lower layers of this region, as the temperature gets sufficiently low here to allow atoms to form, thereby increasing the opacity of the medium. This leads to the development of a temperature gradient between the lower and the upper layers. As the temperature decreases further outwards, more and more ions recombine. Partially ionized and neutral atoms absorb radiation much more efficiently than the fully ionized plasma.

Convection occurs when the temperature gradient gets larger than the adiabatic gradient (the rate at which the temperature would fall if a volume of material were moved higher without adding heat). Where this occurs, a volume of material moved upward will be warmer than its surroundings, and will continue to rise further. Therefore it becomes less efficient for energy to move by radiation in this zone. Heat energy starts to build up, and then begins to move upward by convection, in the form of huge cells of circulating gas (Figure 5.2).

Convection is a much faster way of energy transport than radiation in an opaque medium. It takes only about 10 – 30 days for the heated gas cells to climb through the convection zone. In this process the gas also cools rapidly.

Finally, convective cells reach the photosphere breaking into cells of smaller scale sizes in a turbulent cascading process. Its signature is seen as the granular patterns on the photosphere, or the *granules* (Section 3.3.2, and Figure 3.5 on page 58). The thin photosphere absorbs the energy coming from the convection zone, and radiates it out as a black body at the effective temperature of $5778^\circ K$.

5.4.3 Onset of convection: the Schwarzschild criterion

The condition for the onset of convective instability can be described by the change in vertical temperature gradient dT/dr. Let us consider a vertically stratified gas in hydrostatic equilibrium, in which the temperature, T', of a parcel of gas of density, ρ', is increased slightly by ΔT from the temperature, T, of the surrounding medium of density ρ. This will increase its pressure, P', while decreasing the density, such that, $\rho' < \rho$ (Figure 5.3a).

The parcel will thus experience a force due to buoyancy, causing it to rise upwards to a distance r where its density drops to the same value as the ambient medium. The temperature of the ambient medium at this location is,

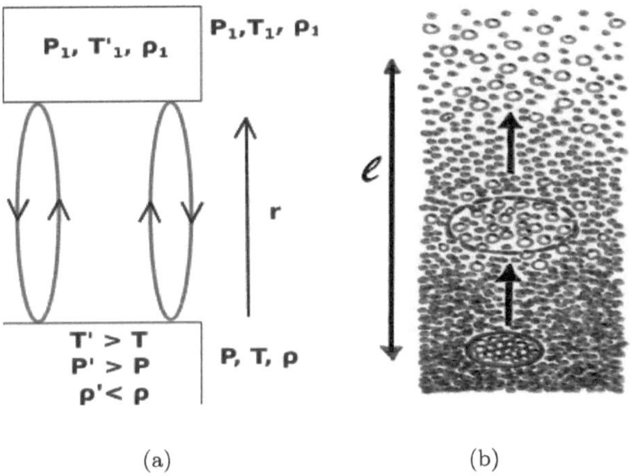

FIGURE 5.3: (a) Schematic diagram illustrating the Schwarzschild criterion for onset of convection. (b) Illustration for mixing length of heat transport by turbulence.

$$T_1 = T + (dT/dr)_{rad} \times r. \tag{5.23}$$

The temperature decreases with height r as $(dT/dr)_{rad} < 0$. In the optically thick solar interior, the time scale of the energy (heat) exchange is long compared to the sound travel time across the parcel. Therefore, when assuming the pressure balance, the cooling of the parcel can be considered adiabatic. Thus, the instability condition can be expressed as

$$T_1{}' = T' + (dT/dr)_{ad} \times r = T + \Delta T + (dT/dr)_{ad} \times r. \tag{5.24}$$

Let us consider the following cases:

- If $T_1' > T_1$, or $(dT/dr)_{rad} > (dT/dr)_{ad}$, the bubble will continue to rise. This is the *Schwarzschild's convective instability criterion* according to which convection sets in when $(dT/dr)_{rad}$ begins to dominate over $(dT/dr)_{ad}$.
- If $T_1' < T_1$, the bubble begins to contract and becomes heavier than the ambient medium. Therefore, it sinks downwards to its original position.

5.4.4 Mixing length and convective overshoot

A theory of heat transport by turbulent convection was first given by L. Prandtl, and was applied to the Sun by Bierman and others in the 1930s. In this, called the *mixing length theory*, it is assumed that heat is transported by motions of fluid parcels driven by buoyancy. The size of these fluid parcels, termed the mixing length l, is comparable to the local pressure scale height, $h_p = -[d(lnP)/dr]^{-1}$, where l determines the distance travelled by the fluid parcel before mixing with the ambient medium, transferring its excess heat (Figure 5.3b).

The mixing length is described by a simple relation $l = \alpha h_p$, where α is the mixing length parameter. By changing the value of α, the efficiency of convective heat transport can be varied, typically taken between 1-3 to reproduce the correct diameter of the Sun. It gives the size scale of a convection cell (height or depth of cell given by the mixing length), since the horizontal extent of the cell cannot be much larger than the mixing length due to mass conservation.

Due to their inertia, the packets of gas reaching the boundary of the convection zone pass into the convectively stable layers, where they are braked and finally stopped. This is called the *overshooting convection.* The typical width of the overshoot layer is of the order of $h_p \approx 150$km in the photosphere. This phenomenon occurs both at the bottom as well as the top boundaries of the convection zone, and is considered to be important. At the top boundary, granulations are the signatures of the convective overshooting. At the bottom boundary, the overshoot layer allows magnetic field to be stored, and it is considered to be the seat of the solar dynamo.

The velocity of a buoyant parcel can be calculated from this theory based on some basic assumptions, as $v = [gl(\Delta T/T)]^{-1}$. Here, $\Delta T = [(dT/dr) - (dT/dr)_{ad}]l = (v^2T)/(gl)$ is the excess of the temperature of the fluid parcel from the surroundings. These are in good agreement with observed size l, velocities v, the life-time ($l/v \sim 10$ min), and the temperature (brightness) contrast of the observed photospheric granules.

5.5 Are the standard solar models reliable?

Solar standard models, as described by the mathematical formalism in the previous sections, had been the primary and only way that allowed to make inference about the solar interior. But, the question is, how do we test their correctness, or otherwise [26]? In the late 1960s, detection and flux measurement of the neutrinos that are produced inside the Sun were considered to provide a test of the mathematical models.

Around the same period, another new tool began developing that was based on solar oscillations observed on the photosphere. This technique had its bearing on geoseismology that has been used to infer the inner structure of the Earth. Therefore, solar seismology, or *helioseismology*, promised to provide a more direct probe of the solar interior [82]. It is essentially based on time series analysis of solar surface (photospheric) oscillations.

These two methods serve as complementary probes, which enable us to peep inside the Sun, and deduce thermal and chemical composition profiles, as well as the rotation and magnetic fields prevailing in the solar interior. In fact, as we shall see, helioseismology not only provided 'direct' inferences about the structure and dynamics of the solar interior, but also helped resolve a problem related to the fundamental property of neutrinos.

5.6 Solar neutrinos: messengers from the central core of the Sun

Neutrinos were predicted by Wolfgang Pauli in 1930, and first observed much later by Cowan and Reines in 1956 in a reactor experiment [315]. According to the standard model of particle physics, neutrinos are charge-less, mass-less, and very weakly interacting fundamental particles.

As discussed in Section 5.3, page 122, solar electron neutrinos are created in the core of the Sun at every step of the *pp*-reactions that power the Sun. These weakly-interacting neutrinos escape easily from the Sun into space isotropically. Neutrino fluxes are sensitive to the temperature and composition of the core, or on the rate of energy generation (Equation 5.22). It was, therefore, hoped that the temperature dependence of nuclear reaction rates in the production of neutrinos would give an independent method for the determination of the Sun's central temperature, and thus provide a check of the solar models.

With this motivation, solar neutrino detection experiments were conceived (a) to test the physics of the nuclear reactions operating in the solar core, (b) to confirm the nature of thermonuclear energy generation in the Sun: *pp*-chain, or CNO cycle, and also (c) to study the basic properties of the neutrinos.

Experiments to detect solar neutrinos began in the 1960s. In 1964, Raymond Davis, Jr., and John N. Bahcall proposed an experiment to test whether converting hydrogen nuclei to helium nuclei in the Sun is indeed the source of the Sun's energy. Before the first neutrino observations in 1967 the solar models had evolved. One could predict with confidence the neutrino production rate of 2×10^{38} neutrinos per second from nuclear reactions; 2 ν per reaction. Therefore, at the distance of the Earth (1 A.U.), the number of neutrinos crossing per square meter per second was estimated to be $\sim 6 \times 10^{14}$.

With the enormous flux of neutrinos released from the Sun, it is estimated that about 100 billion neutrinos per second are passing through our body. However, they interact so rarely, and so weakly, with matter that one does not feel them passing through! Due to their very nature, their detection requires huge detectors in volume. The detection of the weakly interacting neutrino flux is measured in Solar Neutrino Units (SNU), where 1 $SNU = 10^{-36}$ captures per target atom per second.

Neutrino observatories are often built deep underground in order to reduce effects of cosmic rays and other background radiations. Several neutrino observatories based on various techniques were built in Japan, Europe, and Canada, designed to detect solar neutrinos through their rare interactions with large reservoirs containing water, heavy water or even cleaning fluids.

There are three basic types of solar neutrino detectors: the ^{71}Ga experiment has the lowest energy threshold (233 keV), the threshold of the ^{37}Cl is 834 keV, and the large water detectors have the highest threshold of about 5 MeV.

5.6.1 Neutrino detectors: the Homestake experiment

The first solar neutrino experiment, Homestake gold mine, located 1480 metres deep underground, in South Dakota began in 1967. It used 615 tons of ordinary cleaning fluid, tetrachloroethene, C_2Cl_4, whose chlorine is converted to argon according to the reaction:

$$^{37}Cl + \nu_e \longrightarrow {}^{37}Ar + e^-(\geqslant 0.834MeV).$$

Here, the neutrino energy threshold is given in the bracket. The capture rate is dominated by the ^8B neutrinos contributing 5.9 SNU, with the ^7Be neutrinos making a contribution of 1.1 SNU.

Unfortunately, the Homestake experiment reported solar neutrino counting rate of 2.56 ± 0.23 and NOT 7.6 ± 1.25 SNU, as predicted by the solar standard models for the chlorine experiment. This puzzling deficit in the neutrino counting rate by nearly a factor of 3 over the prediction came to be known as the *solar neutrino problem*, first observed in the mid-1960s [91], and summarised in an important paper [87].

5.6.2 Neutrino detectors: Kamiokande experiment

The Kamiokande water experiment started operating in 1987 with a 680 ton water tank located 1 km underground in the Kamioka mine in Japan [128]. Water detectors detect neutrinos through Cérenkov light from elastic ν-e$^-$ scattering (*ES reaction*), if the recoil energy of the electron is at least 5 MeV.

$$e^- + \nu_x \longrightarrow e^- + \nu'_x(\geqslant 5MeV).$$

Here, ν_x corresponds to all three flavours of neutrino $(\nu_e, \nu_\mu, \nu_\tau)$. The upgraded Super Kamiokande (SK) that became operational in 1996 also used the water Cérenkov effect [127]. Due to the emitted radiation, it is possible to determine their arrival direction, and ascertain if they really came from the direction of the Sun. These experiments are sensitive only to the high-energy ^{8}B neutrinos released by the *pp*-chain, high temperature dependent reaction rate. The measured flux from the SK experiment too was found to be deficient by about 50% over the flux predicted by SSM.

5.6.3 Neutrino detectors: GALEX, SAGE and GNO

There are three other radiochemical experiments, GALLEX, SAGE and GNO, that used a gallium detector [157].

$$^{71}Ga + \nu_e \longrightarrow {}^{71}Ge + e^- \ (\geqslant 0.233MeV).$$

These are capable of detecting the low-energy *pp*-neutrinos, while the chlorine experiments detected intermediate, and the water-based (ES) experiments were sensitive to high-energy neutrinos.

These detectors reported the solar neutrino counting rate of 74 ± 5.0 SNU, while the SSM prediction of the neutrino capture rate for gallium experiments is 128 ± 8 SNU. Thus, again a deficit in observed neutrinos resulted, only confirming the discrepancy between the measured, and more refined calculated theoretical models.

Thus, it turns out that all the initial results on detection of neutrinos met with an unexpected problem. These experiments reported far too few neutrinos by detectors on Earth-based neutrino observatories, as compared to the predictions from the standard solar models (Table 5.1). Each of the experiments faced the *neutrino deficiency problem*, which posed a huge question mark on the correctness of standard solar models.

Another possibility, if the models were to be correct, could be related to the neutrinos themselves! Perhaps it was not well understood how the neutrinos, produced in the solar core, behaved as they travel large distances in reaching the Earth-based detectors. This possibility would give rise to a new physics of the fundamental particles having implications not just for solar physics, but also for such diverse aspects as the early universe, the explosion of stars, and laboratory experiments on the Earth.

TABLE 5.1: Predicted and measured fluxes of neutrinos

Experiment	Chlorine	Gallium	SK	SNO
Threshold (MeV)	0.834	0.233	5	5
Predicted Flux (SNU)	7.7 ± 1.1	130 ± 7	1.0 ± 0.15	
R	0.33 ± 0.3	0.55 ± 0.03	0.465 ± 0.015	0.36 ± 0.015

where, R = (measured neutrino flux)/(Predicted neutrino flux), and 1 SNU = capture of one neutrino per second per 10^{36} target atoms.

5.6.4 Non-Standard solar models

Several so-called *non-standard solar models* were suggested to address the possibility that the solar standard models (and hence, that of all stellar interiors) are incorrect in some significant way. It was suggested that a 5-10% lower core temperature might give neutrino flux consistent with Homestake detector. But, it soon turned out that reduction in the core's temperature would lead to even larger suppression of high-energy ^8B neutrino flux to which the Super Kamiokande experiment was exclusively sensitive.

Paradoxically, the Homestake experiment that detected the intermediate, as well as high-energy neutrinos would also then show an even larger reduction in the neutrino counting rate. Thus, it was realised that by reducing the core temperature, it is not possible to get a solar model, which simultaneously matched both with the Homestake and Super Kamiokande experiments.

Attempts were then made to reduce the relative abundances of heavy elements Z. It turned out that it would reduce the opacity κ, and thus the temperature gradient. Also, it did not seem to be plausible for solar-type stars. Moreover, this suggestion was not supported by the new inferences emerging from helioseismology.

The standard solar models neglect rotation and internal magnetic fields. Attempts were made to relax these assumptions by having a rapidly rotating solar core that would help in lowering the thermal pressure. Similarly, a strong internal magnetic field would help lower the central temperature. But this should be very intense to have a significant contribution to the pressure of the core. All such proposals essentially led to a slight reduction in the central temperature, resulting in a further lowering of the flux of high-energy neutrinos.

New evidence from helioseismology showed that the solar model predictions were quite reliable [26]. Precise measurements were made of the sound speed throughout the solar interior using helioseismic observations. The measured sound speeds agreed to a precision of 0.1% with the sound speeds calculated for theoretical models of the Sun.

These measurements suggested that the theoretical solar models were so accurate that they must also correctly predict the flux of solar neutrinos. Therefore, it became clear that rather than expecting non-standard solar models to resolve the neutrino problem, it was required to focus on the neutrinos themselves. This was soon to be revealed by the new neutrino detection experiments, such as the *Sudbury Neutrino Observatory.*

5.6.5 The Sudbury Neutrino Observatory (SNO)

SNO used a detector of 1000 tons of heavy water (D_2O), located at a depth of over 6000 metres in a nickel mine in Sudbury, Canada. For their first measurements, the SNO used the detector in a mode sensitive only to the solar electron neutrinos ν_e by Cérenkov radiation (sensitive to both charge, or *CC*, and neutral current, or *NC* reaction).

The deuterium isotope present in heavy water is dissociated by a neutrino. It was able to study the same higher-energy solar neutrinos, in a different way, that were investigated previously by Kamiokande and Super-Kamiokande ordinary-water detectors. The SNO is capable of measuring the ^8B neutrinos (> 5 MeV) through the following reactions:

$$\nu_e + d \longrightarrow p + p + e^- \text{ (CC reaction)},$$
$$\nu_x + e^- \longrightarrow \nu'_x + e^- \text{ (elastic scattering)},$$
$$\nu_x + d \longrightarrow \nu'_x + p + n \text{ (NC reaction)},$$

where $x = (e, \mu, \tau)$ corresponding to the three known types of neutrinos. Nuclear fusion in the Sun produces only neutrinos that are associated with electrons, ν_e. The two other types of neutrinos are muon neutrinos ν_μ and tau neutrinos ν_τ, which are produced along with ν_e, for example, in laboratory accelerators or in exploding stars.

SNO was able to distinguish between the total neutrino flux, and the ν_e flux (results reported in `http://www.sno.phy.queensu.ca/`). It reported that neutrinos were not missing, but the ν_e produced in the Sun's core converted in ν_τ and ν_μ. Thus, SNO found that the neutrino problem actually lies with neutrino physics and not with SSM.

5.7 Resolution of the solar neutrino problem

Several experiments found results that suggested the disappearance of electron-neutrinos, ν_e, produced in the solar core, while propagating within the Sun, or between the Sun, and the Earth. The standard model of particle physics assumes that neutrinos are massless. If neutrinos have non-zero mass, neutrino "flavour" oscillations arise as already suggested by Gribov and Pontecorvo back in 1969 [152], and later by Bahcall [25]. A physical derivation of the mechanism was subsequently presented by which matter (electrons) could modify the vacuum neutrino oscillations [395]. It was already being realised that the resolution of neutrino deficiency lies in particle physics.

Many theoretical physicists, however, were not in favour of accepting solar neutrino oscillations until the first results of the SNO experiment were published. This is because, according to them, the standard model of physics was too beautiful and too successful to abandon. But it was being gradually recognised that new particle physics was required to explain what happened to the uncounted neutrinos if the Homestake and Kamiokande solar neutrino experiments were to be correct.

The solution to the inference that new physics was indeed required came in 2001 with the announcement of the Solar Neutrino Observatory (SNO). If the standard model of particle physics was right, the fraction measured by SNO and the fraction measured by Super Kamiokande should be the same. All the neutrinos should be electron neutrinos, but the fractions were different. Combining the SNO and the Super Kamiokande measurements, the SNO collaboration determined the total number of solar neutrinos of all three types (ν_e, ν_τ, ν_μ,) as well as the number of just ν_e.

Electron neutrinos constitute about a third of the total number of neutrinos. The total number of neutrinos of all types agreed with the number predicted by solar standard models. Thus, the neutrino problem was essentially with the difference between the total number of neutrinos and the number of only the electron neutrinos. The missing neutrinos were actually present, but in the form of the more difficult to detect muon and tau neutrinos.

Solar neutrinos have a multiple personality disorder. They are created as ν_e in the Sun, but on the way to the Earth they change their type. For neutrinos, the origin of this personality disorder is a quantum mechanical process called *neutrino oscillations*. Thus, as suggested in 1969 by Gribov and Pontecorvo [152], the lower-energy solar neutrinos switch from electron neutrino to other types as they travel from the Sun to the Earth.

5.8 Summary

Our knowledge of the Sun's interior, hidden under the opaque photosphere, has been traditionally derived from theoretical models based on the global properties of the Sun. Solar standard models have been constructed based on theoretical formulations and numerical modelling, energy production in the central core, and its transport through the radiative and convective zone. Detection of solar neutrinos produced in the hot and dense central core provides an important tool to infer about the *solar interior*. They were used as a probe and tool for testing the reliability of solar standard models. There were early set-backs due to observational deficiency as compared to the predicted neutrino flux from standard models, the so-called, neutrino problem. From a series of experiments, however, the problem was eventually resolved, and standard solar models were vindicated. The Nobel Prize in Physics 2015 was awarded jointly to T. Kajita (Super-Kamiokande Collaboration University of Tokyo, Japan) and A. B. McDonald (Sudbury Neutrino Observatory Collaboration Queen's University, Canada) "*for the discovery of neutrino oscillations, which shows that neutrinos have mass*" and the strange ability of neutrinos to transform among the three flavours.

Chapter 6

Probing the Invisible Solar Interior: Helioseismology

6.1 Introduction

New neutrino detectors were being built in the 1960s to resolve the mystery of missing neutrinos from the solar interior, and their possible consequence for the solar standard models. Around the same time, another new tool was emerging that would have tremendous potential in probing the structure of the 'invisible' solar interior. In 1960, Leighton and his coworkers, discovered that the surface of the Sun undergoes oscillations with a period centred at 5 minutes (3 mHz) [234]. These oscillatory motions were observed, using temporal sequence of spectroscopic observations, as Doppler shifts of photospheric spectral lines. Simultaneously measuring the relative intensities in the red and blue wings, and taking the difference in the two observations indicated the motion of parcels of gases toward or away from the observer. Figure 6.1 shows velocity profiles at various positions along one of the slits with time. The distance between the curves for adjacent points is equal to a velocity of 400 m/s [280].

The possibility of oscillatory motions in the solar atmosphere was, in fact, already suggested theoretically by Whitney (1958) [391]. After the first observational discovery, several others also detected similar periodicity using different instruments ([117], [287], [337], [290], [141]). Various theoretical interpretations were put forward to interpret these oscillations. There were suggestions that they may be compressions of gas caused by the sound waves in the solar atmosphere [271]. Some attributed it to the atmosphere bobbing up and down like a buoyant cork on water ([328], [369]).

After nearly a decade of such observations, Ulrich (1970) [370], and independently, Leibacher and Stein (1971) [232] came up with their radical ideas to explain these observations. They interpreted the oscillations as trapped acoustic waves (or p-modes) in resonant cavities between the Sun's surface and an inner turning point. Once the nature of these solar oscillation modes was established, it was soon realised that these can be used for probing the 'invisible' solar interior. The study of the solar interior using oscillations has been termed *helioseismology* ([99], [149]) in analogy with *geoseismology* that

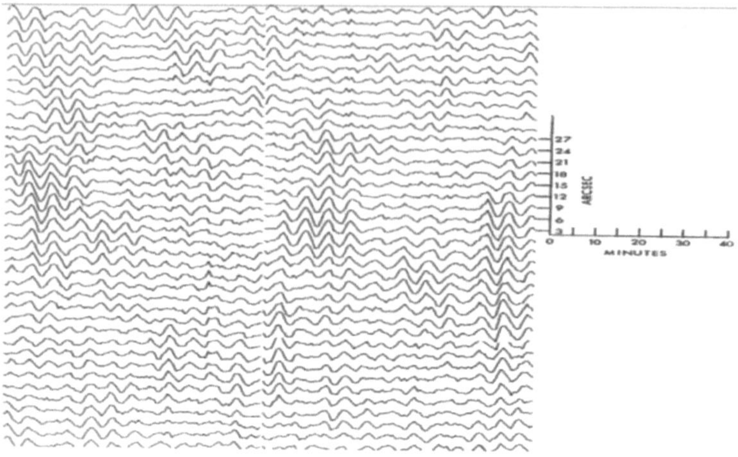

FIGURE 6.1: Doppler motions of gas parcels along a spectrographic slit show a wave-like pattern of troughs and crests at 5-minute periods. Figure with permission from [280].

helps to learn about Earth's interior by monitoring seismic waves produced by earthquakes.

Helioseismology has helped in the resolution of the neutrino problem, infer the location of the base of the convection zone in the outer envelope of the Sun, and about the core of the Sun. It turns out that accurately measured oscillation frequencies provide stringent constraints on the admissible solar standard models. Helioseismology provides a method, along with the SSM, to infer Sun's age independently of the method from meteorites.

6.2 Solar oscillations - Helioseismology

The observations by Deubner (1975) [98], and independently by Rhodes, Ulrich and Simon (1977) [316] confirmed the theoretical hypothesis about the modal nature of solar oscillations. These observations found the power of oscillations to be concentrated in a series of ridges in the $k-\omega$ diagnostic diagram, showing a spectrum of discrete frequencies. Claverie et al. (1979) [86] studied the oscillations in Doppler velocity observations integrated over the entire solar surface, and detected different modes of oscillations for low degree. Similar results were obtained by Grec et al. (1980) [150] using observations from the south pole.

The Sun oscillates in over 10^7 of acoustic modes with extremely low velocity amplitudes which are of the order of a few cm/s [161]. The mode spectrum extends from 0.6mHz to 5.5mHz; having maximum velocity amplitude of around 15 cm/s. Solar global modes of oscillations propagate deep in the Sun. Each oscillation mode is sensitive to the solar interior's physical conditions where the amplitudes of the mode are the greatest, therefore, precise identification and accurate measurement of oscillation modes is the mainstay of *helioseismology*. It is generally believed that these modes may be either intrinsically overstable, or stochastically excited by turbulence in the near-surface convection zone immediately beneath the photosphere [147]. They are maintained by the restoring effect of the acoustic (thermal) pressure.

6.2.1 Standing waves in acoustic cavities: organ pipes

Sound waves are parcels of compressed and rarefied gas. Gas pressure acts as a restoring force to bring the gas back to an undisturbed state. Sound waves are reflected from a boundary as in an organ pipe.

A standing wave arises from two waves travelling in opposite directions, reflected from boundaries, forming an acoustic cavity in which waves travel repeatedly. At the two boundaries, nodes (with zero displacement) are established (Figure 6.2b). Condition for a mode is that its wavelength must be twice the distance between the boundaries (the fundamental modes), or equal to submultiples of the fundamental wavelength (overtones or harmonics).

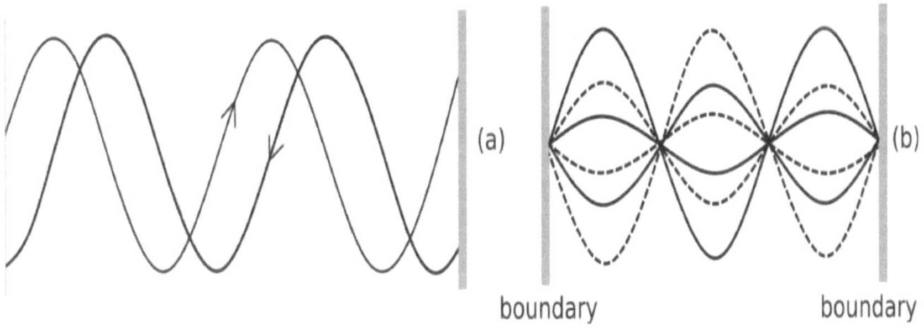

FIGURE 6.2: Wave motion illustrating the formation of oscillations: (a) wave reflected from a single boundary, (b) a standing wave formed by a wave reflected at two boundaries located at the nodes.

6.2.2 The Sun as a giant organ pipe

Why does the Sun act as a resonant cavity? Convection excites stationary sound waves, which are reflected from two boundaries, and establish a variety of modes. Acoustic cavities exist inside the Sun's interior due to density and temperature gradients that reflect, or refract the sound waves. Each solar mode is trapped in cavities of varying depths, depending on horizontal wavelength - affected by internal temperature, composition, flows.

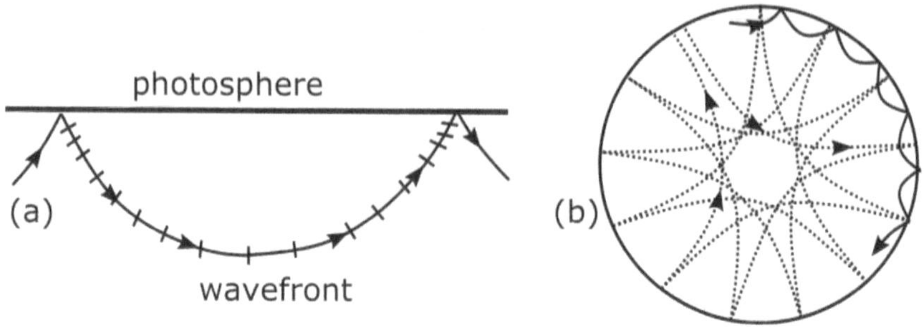

FIGURE 6.3: (a) Wave trapped in cavities inside the Sun's surface, (b) waves of different wavelengths penetrate to different depths inside the Sun.

TABLE 6.1: Intensity variation at the solar surface

Convection (granulation)	Sunspots	Solar oscillations
15 %	factor of 10	10^{-6}

A wave propagating towards the Sun's centre experiences increasing temperature, and hence, the sound speed as $c_s \propto T^{1/2}$. Therefore, while propagating towards the interior from the surface at an angle, the deeper edge of the wave-front (plane perpendicular to the direction of the wave) travels at a faster speed than its upper edge. As a result, the wave is gradually refracted at the lower boundary of its cavity, and turns upwards, travelling in an arc. While travelling upwards, the wave undergoes a total internal refraction at the Sun's surface due to the abrupt decrease of density, and thus a standing wave is created (Figure 6.3a,b).

6.3 Physics of trapped solar global modes

The solar photosphere shows the signature of the modes with different spatial patterns and temporal frequencies. Figure 6.4 shows the power spectrum of these surface oscillations which represents the superposition of all the modes. In fact, such oscillations are expected to be present in all stars having convective outer layers. The spectrum is dominated by the acoustic, or $p-$*mode* band, up to the cut-off frequency (~5.3 mHz). Above this frequency a mode-like structure appears, corresponding to the so-called *High Interference Peaks* [HIPs], which are due to partial wave reflection in the solar atmosphere.

Although the solar oscillations are best observed using Doppler velocities, the intensity observations too reveal similar variations. However, they have been used to a lesser extent for identifying the modes of oscillations. For

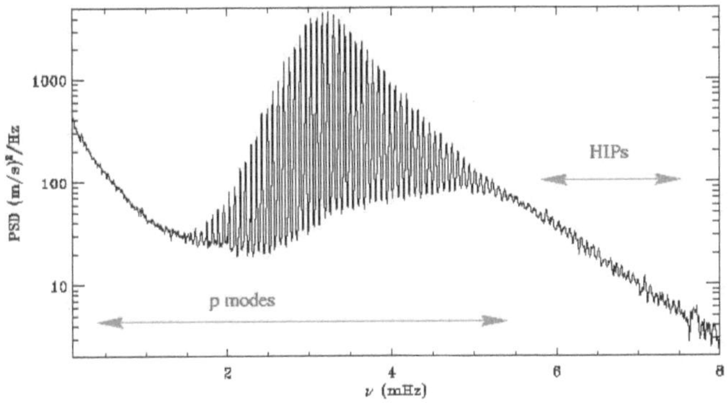

FIGURE 6.4: Power spectrum of global Doppler oscillations, derived from 800 days of measurements by GOLF. (Courtesy of SOHO/[GOLF] consortium. SOHO is a project of international cooperation between ESA and NASA.)

TABLE 6.2: Velocity fields (line of sight component) at the solar surface

Earth's rotation and orbital velocity	Solar rotation	Convective motions	Solar oscillations
500 m s^{-1}	2000 m s^{-1}	1000 m s^{-1}	< 1 m s^{-1}

comparison, the variations in velocities and intensities associated with solar oscillations and some other parameters/features are provided in Tables 6.1 and 6.2.

6.3.1 Characterisation of oscillation modes

The simplest oscillation the Sun, or a star, can undergo is a radial one, in which the star expands and contracts radially and spherical symmetry is preserved during the oscillation cycle. From a mathematical point of view, the differential equation describing the radial displacement is of the Sturm-Liouville type, and thus allows eigen-solutions that correspond to an infinite number of eigenfrequencies. The smallest frequency corresponds to the fundamental *radial oscillation* mode, with the period that is inversely proportional to the square root of the mean density of the star. Radial oscillations are characterised by the radial wavenumber n : the number of nodes of the eigenfunction between the centre and the surface of the star.

If transverse motions occur in addition to radial motions, one uses the term *non-radial oscillations*. The oscillation modes are then not only characterised by a radial wavenumber n, but also by non-radial wavenumbers l and m.

The analysis of oscillation modes involves decomposition of time series of full-disk Doppler (or, intensity) images data. The oscillations can be best

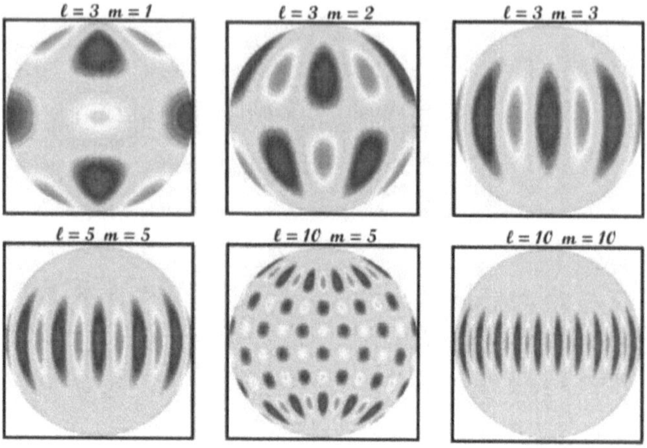

FIGURE 6.5: The surface structures of modes with various ℓ and m values are shown, where the whiter regions between darker (moving radially inward) and lighter (moving radially outward) areas are the nodal boundaries.

described by spherical harmonic functions for an observational parameter, such as velocity or intensity fluctuations at the surface:

$$\tilde{\sigma}(r,\theta,\varphi,t) = \sum_{nlm} \sigma_{nl}(r) Y_l^m(\theta,\varphi) \exp(i\omega t),$$

where $\sigma_{nl}(r)$, and Y_l^m (spherical harmonic functions of latitude θ, longitude φ) describe the radial and angular structures of modes with given degree ℓ, azimuthal number m and radial order n. The temporal part is written as $e^{i\omega t}$.

Thus, a mode with frequency $\nu_{nlm} = \omega_{nlm}/2\pi = 1/P$ is characterised by three integers: The degree $\ell \geq 0$ represents the number of surface nodal lines, the azimuthal number $|m| \leq \ell$ denotes the number of such lines that pass through the rotation axis of the Sun, and the radial order n ($-\infty < n < \infty$) gives the number of nodes in the interior. For each ℓ, there are $(2\ell + 1)$ degenerate m modes, due to the neglect of rotation in standard solar models. The degree ℓ is a measure of the horizontal component of the wavenumber, $k_h(r) = \sqrt{l(l+1)}/r$ at radius r. The surface patterns of some non-radial oscillations are graphically depicted in Figure 6.5.

The modes with $m = 0$ are standing waves, also called the axisymmetric or zonal modes. Non-zonal modes are running waves. We take the following convention with respect to the rotation: positive m - *values* denote modes that move opposite to the rotation (retrograde modes), while negative m - values are associated to motion in the direction of the rotation (prograde modes). The modes with $\ell = |m|$ are called sectoral, and those with $0 \neq |m| < \ell$ are *tesseral.*

6.3.2 Solar (stellar) oscillations: the basic formalism

The Sun is a gravitating, gaseous sphere in hydrostatic equilibrium, primarily due to its gravity and pressure acting against each other (Section 5.2.2, page 117). Stability against perturbations and the modes of oscillations of a non-viscous body are described by the following equations:

(1) The mass conservation equation:

$$\frac{\partial \rho}{\partial t} + \nabla.(\rho \vec{v}) = 0. \tag{6.1}$$

(2) The momentum conservation equation (or equation of motion):

$$\rho \frac{\partial \vec{v}}{\partial t} + \vec{v}.\nabla \vec{v} = -\nabla p + \rho \vec{g}, \tag{6.2}$$

where $\vec{g} = -\nabla \Phi$, and $\nabla^2 \Phi = 4\pi G \rho$ is the Poisson's equation relating the self-gravity Φ with density ρ.

(3) The energy conservation equation:

$$\rho \frac{dQ}{dt} = \rho \varepsilon - \nabla.\vec{F}. \tag{6.3}$$

For small perturbations, these equations can be linearised to obtain corresponding dispersion relation describing the waves and oscillations in a given standard solar model [84], as follows:

$$\rho_1 + \nabla.(\rho_0 \vec{\xi}) = 0, \tag{6.4}$$

$$\rho_0 \frac{\partial^2 \vec{\xi}}{\partial t^2} + \nabla.p_1 - \frac{\rho_1}{\rho_0} \nabla p_0 + \rho_0 \nabla \Phi_1 = 0, \tag{6.5}$$

$$\nabla^2 \Phi_1 = 4\pi G \rho_1. \tag{6.6}$$

Thus, we have six equations for six unknowns: ρ_1, p_1, Φ_1, and three components of $\vec{\xi} = \left[\xi_r(r), \xi_h(r)\frac{\partial}{\partial \theta}, \frac{\xi_h(r)}{\sin\theta}\frac{\partial}{\partial \phi}\right] Y_l^m(\theta,\phi) e^{i\omega t}$ which is a vectorial displacement of a gas parcel from its equilibrium position. The Eulerian perturbations (at a fixed position in the Sun) of various parameters are denoted by subscript '1', *i.e.*, $[\rho_1, p_1, \Phi_1] = [\{\rho_1(r), p_1(r), \Phi_1(r)\} Y_l^m(\theta,\phi) e^{i\omega t}]$.
By eliminating the variables ξ_h and ρ_1 using the above equations, one obtains:

$$\frac{1}{r^2}\frac{d}{dr}\left\{r^2 \xi_r\right\} - \frac{\xi_r g}{c^2} + \frac{1}{\rho}\left\{\frac{1}{c^2} - \frac{l(l+1)}{r^2\omega^2}\right\} p_1 - \frac{l(l+1)}{r^2\omega^2}\Phi_1 = 0, \tag{6.7}$$

$$\frac{1}{\rho_0}\left\{\frac{d}{dr} + \frac{g}{c^2}\right\} p_1 - (\omega^2 - N^2)\xi_r + \frac{d\Phi_1}{dr} = 0, \tag{6.8}$$

$$\frac{1}{r^2}\frac{d}{dr}\left\{r^2\frac{d\Phi_1}{dr}\right\} - \frac{l(l+1)}{r^2}\Phi_1 - \frac{4\pi G\rho_0}{g}N^2\xi_r - \frac{4\pi G}{c^2}p_1 = 0, \tag{6.9}$$

where, sound speed $c^2 = (\Gamma_1 p_0)/\rho_0$, the gravitational acceleration $g = (dp_0/dr)/\rho_0$, the Lamb Frequency $S_\ell^2 = [\ell(\ell+1)c^2]/r^2 = k_h^2 c^2$; k_h being the horizontal wavenumber, and the Brunt-Väisälä frequency $N^2 = g\left\{\frac{1}{\Gamma_1 p_0}\frac{dp_0}{dr} - \frac{1}{\rho_0}\frac{d\rho_0}{dr}\right\}$.

It can be seen that the Lamb frequency S_l is dependent on degree ℓ. On the other hand, N is the frequency with which a small fluid element will oscillate about its equilibrium position when displaced, and it is independent of ℓ. It may also be noted that the angular order m does not appear in these equations, hence for each ℓ, there is a $(2\ell+1)$-fold degeneracy of the eigenfrequencies. This is the result of neglecting rotation, magnetic fields and other sources of non-axisymmetry.

For a given solar standard model, these equations are solved numerically with appropriate boundary conditions, and only non-trivial solutions are considered for particular values of eigenfrequencies.

6.3.3 Local treatment and the Cowling approximation

Solutions of the differential equations (6.7-6.9) can be examined in a local treatment without going into detailed numerical solutions. For this, perturbations of the gravitational potential, Φ_1, being very small, can be neglected [90], while gravitational acceleration g still retains the effect of the unperturbed gravitation in the analysis. This is the *Cowling approximation*, which leaves us with the following two equations:

$$\frac{1}{r^2}\frac{d}{dr}\left\{r^2\xi_r\right\} - \frac{\xi_r g}{c^2} + \frac{1}{\rho_0}\left\{\frac{1}{c^2} - \frac{l(l+1)}{r^2\omega^2}\right\}p_1 = 0, \tag{6.10}$$

$$\frac{1}{\rho_0}\left\{\frac{d}{dr} + \frac{g}{c^2}\right\}p_1 - (\omega^2 - N^2)\xi_r = 0. \tag{6.11}$$

As the next simplifying assumption, the parameters, g, c, N, in the solar interior are considered to not vary much over one wavelength of perturbation, or remain locally constant (*the WKBJ approximation*). In this local approximation, the scale height $H = -\frac{\rho}{d\rho/dr} = \left(\frac{g}{c^2} + \frac{N^2}{g}\right)^{-1}$ is also constant.

On eliminating p_1 from the above set of linearised equations for non-radial oscillations ($\ell \neq 0$), we can describe the properties of the oscillation modes, and their frequencies by a single second-order differential equation for the radial component of displacement:

$$\frac{d^2\xi_r}{dr^2} = \frac{\omega^2}{c^2}\left(1 - \frac{N^2}{\omega^2}\right)\left(\frac{S_l^2}{\omega^2} - 1\right)\xi_r = -\mathcal{K}(r)\xi_r. \tag{6.12}$$

This equation represents non-radial oscillations in a crude approximation, and describes the overall properties of oscillation modes, but still provides a reasonably accurate determination of their frequencies.

It is evident from Equation (6.12) that the two characteristic frequencies, the acoustic or Lamb frequency, (S_ℓ), and the buoyancy, or Brunt-Väisälä frequency, (N), play very important roles in determining the behaviour of the modes in the solar interior.

6.3.4 Types of solar oscillations: the p- and g-modes

Equation (6.12) describes the behaviour of ξ_r depending on the sign of $\mathcal{K}$; oscillatory, when it is positive, and exponentially decreasing, or increasing, when it is negative. From the characteristic frequencies corresponding to (i) the acoustic S_ℓ, and (ii) buoyancy N, it is apparent that for these two types of oscillatory eigenfrequencies, or modes, the dominant restoring force depends upon location in the Sun, and on the perturbation frequency.

Thus, the solution is oscillatory for (i) modes with high frequencies (called p-modes): $|\omega| > |N|$, and $|\omega| > |S_\ell|$, or for (ii) $|\omega| < |N|$, and $|\omega| < |S_\ell|$. On the other hand, solutions grow, or decay exponentially for (i) modes with low frequencies (called g-modes): $|N| < |\omega| < |S_\ell|$, or for (ii) $|S_\ell| < |\omega| < |N|$.

For exploring these solutions further, let us assume that the vertical wavelength of perturbations is much smaller compared to r, and we look for oscillatory solutions of the form:

$$\xi_r \sim \rho_0^{-1/2} e^{ik_r r}, \text{ and } p_1 \sim \rho_0^{1/2} e^{ik_r r}, \tag{6.13}$$

where the $\rho_0^{1/2}$ factor takes care of the variable ρ_0, occurring in the differential equations under the Cowling approximation. This leads to the dispersion relation

$$k_r^2 = \frac{\omega^2 - \omega_a^2}{c^2} + S_\ell^2 \frac{N^2 - \omega^2}{c^2 \omega^2}, \tag{6.14}$$

where $\omega_a = c/2H$ is called the *acoustic cut-off frequency* below which acoustic oscillations do not exist, and $H = -\rho_0 (d\rho_0/dr)^{-1} = (g/c^2 + N^2/g)^{-1}$ is the density scale height, which is a constant.

The solutions are oscillatory if $k_r^2 > 0$, *i.e.*, k_r is real, and evanescent if $k_r^2 < 0$, or k_r is imaginary. Again, these solutions depend on S_ℓ, and N, which can be determined throughout the interior with r for a given solar standard model.

In order to see for what values of c, ω_a, *etc.*, the oscillations exist, one can set $k_r^2 = 0$, and solve the equation for ω^2, and plot its curves in the (k_h, ω)-plane with normalised coordinates, *i.e.*, ω/ω_a with $2Hk_h$ to examine the diagnostic diagram, where $k_h = \sqrt{\ell(\ell+1)/r^2}$ is the horizontal wavenumber. Let us consider the dispersion relation (Equation 6.14) in the two cases, as follows:

- *The high frequency regime*: At large ω, the dispersion relation is simplified to

$$\omega^2 = c^2(k_r^2 + k_h^2). \tag{6.15}$$

This is dispersion relation for acoustic waves, for which pressure gradient is the main restoring force. Therefore, the high frequency region is called the *acoustic*, or *p-mode* regime. Its lower boundary is at the cut-off frequency ω_a at $\ell = 0$. It approaches asymptotically to $\omega = S_\ell \approxeq ck_h$ for large ℓ. They travel with the sound speed c, and live longest where c is the lowest, *i.e.*, at the solar surface, since $c \sim T^{1/2}$. They propagate throughout the solar interior, but are evanescent outside the photosphere.

- *The low frequency regime*: The second oscillatory branch is found at small ω, for which one has

$$k_r^2 c^2 = S_\ell^2 \left(\frac{N^2}{\omega^2} - 1 \right), \tag{6.16}$$

which gives for large ℓ

$$\omega^2 = N^2 \frac{k_h^2}{k_r^2 + k_h^2} = N^2 \sin^2\theta. \tag{6.17}$$

Their frequencies depend only on angle θ between the propagation vector and the vertical direction. The propagation region is bounded by $\omega = N$ at large ℓ, and $\omega = S_\ell N/\omega_a$ at small ℓ.

For these waves, buoyancy due to gravity is the main restoring force, and N is the critical frequency. Therefore, they are known as *g*-modes. For $N^2 > 0$, the system is convectively stable to small perturbations, and $N^2 < 0$ in the convection zone where it is convectively unstable. Thus, these modes can propagate in the radiative interior and in the atmosphere but are evanescent in the convection zone.

The *g*-modes are expected to be most sensitive at the solar core, while the *p*-modes are most sensitive at the surface. The upper limit on solar interior *g*-modes lies below 1 cm/s while *p*-modes have amplitudes of a few metres/second.

6.3.5 Trapping regions of the modes

The properties of *p*- and *g*-modes can be further examined from the profiles corresponding to the characteristic frequencies S_ℓ and N for a given standard solar model. The sharp maximum in N near the very centre is associated with the increase in the He abundance in the region, where the nuclear burning

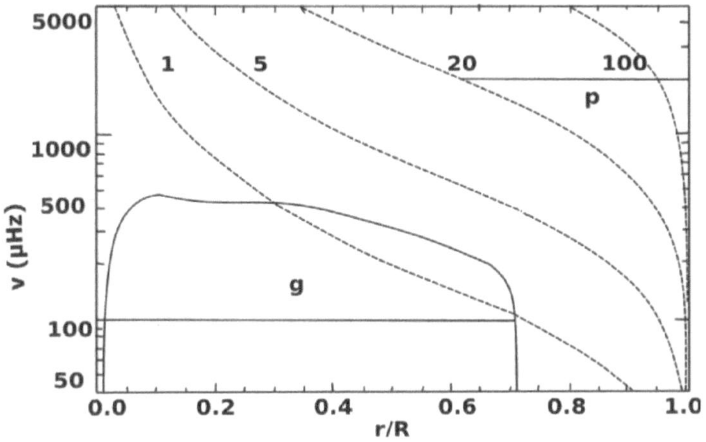

FIGURE 6.6: The propagation diagram for a standard solar model. The characteristic acoustic, or Lamb frequency S_l (dashed curves for degrees l=1, 5, 20, and 100), and the Brunt-Väisälä frequency N (solid curve) against fractional radius in a standard solar model. The horizontal lines show the trapping regions for a p-mode with ν=2000 μHz and ℓ=20, and for a g-mode with ν=100 μHz. Figure courtesy of Sarbani Basu [34].

is taking place. Here lighter material is on top of the heavier material, which adds to the convective stability, and hence, increases N.

On the other hand, the Lamb frequency S_ℓ is ℓ dependent. For $\ell = 0$ (radial modes), $S_\ell = 0$. These frequencies are illustrated in the propagation diagram for various values of ℓ in a standard solar model (Figure 6.6).

The horizontal lines in the figure indicate the trapping regions for a *g*-mode with frequency $\nu = 100\mu$Hz, and where a *p*-mode with degree ℓ=20 and $\nu = 2000\mu$Hz can propagate. S_ℓ tends to infinity as r tends to zero, and decreases monotonically towards the surface due to the decrease in c, and increase in r. The trapping regions for the *g*-modes and *p*-modes with different ℓ values are thus illustrated in this diagram.

6.3.6 The radial structures of the modes

Amplitudes with depth of some acoustic and gravity modes from a theoretical model are shown in Figure (6.7), where the eigenfunctions are scaled for clarity [84]. Radial extension of modes is delimited to resonant cavity by two turning points, outside which the mode decays. Outer turning point is the same, *i.e.*, the solar surface, for all modes, while the inner turning point is different for every (ℓ, n) pair.

The three acoustic modes, shown with frequency $\sim$ 3.3mHz (period T $\sim$ 5 min), are obtained for decreasing values of the radial number, n, while

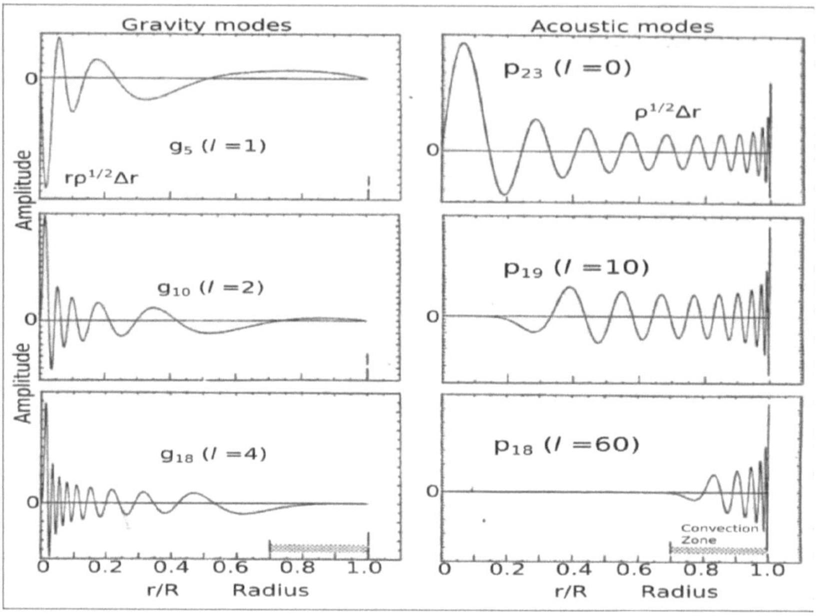

FIGURE 6.7: The radial structures of eigenfunctions associated with some g- and p-modes of low and high ℓ values with depth. With permission from Christensen-Dalsgaard [84].

increasing degree ℓ. As ℓ increases, these are increasingly confined towards the surface.

For the three gravity modes with frequency ~ 0.1 mHz ($T \sim 160$ min), and increasing (ℓ, n) gives structures increasingly confined towards the core, with upper reflection point near the base of convection zone.

6.3.7 Solar dispersion relation: the ℓ-ν diagram

Modes which resonate between the two turning points survive a destructive interference. This happens only when there is an integral number of vertical wavelengths fitting in the cavity defined by the two points:

$$\int_{r_0}^{r_1} k_r(r)dr = n\pi + \varepsilon,$$

where $k_r(r)$ the vertical wavenumber, and ε is a correction factor. The order n of the modes defines the dispersion relation between degree ℓ (through $k_r(r)$) and determines discrete frequency for every given ℓ.

Based on this relation, a diagram of frequencies ν as a function of degree ℓ can be plotted, where the power of solar modes appears as concentrated in

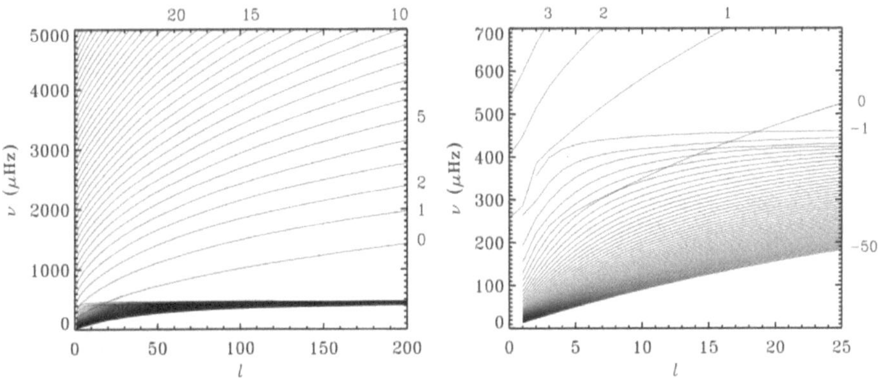

FIGURE 6.8: Frequencies $\nu_{n,l}$ obtained from a theoretical solar model for p- and g-modes at different radial orders n (indicated on R.H.S.). With permission from Christensen-Dalsgaard [84].

discrete ridges corresponding to every radial order n. It is called the $\ell - \nu$ diagram (Figure 6.8).

The observational $\ell - \nu$ diagram is obtained as follows: The temporal sequences of full disk velocity, or *Doppler images* are used to identify the global solar oscillation modes by decomposing into spatial spherical harmonics [161]. The coefficients of each spherical harmonic, as a function of time, are then *Fourier transformed* to generate the power spectrum as a function of three variables: ν, ℓ, and m. The resulting power spectrum is averaged over m to produce the 2-D $\ell - \nu$ power spectrum.

Only waves with specific combinations of period and horizontal wavelength, related to the Sun's interior structure, resonate within the Sun, and produce ridges of larger power in the period versus wavelength, or ℓ–ν diagram (Figure 6.9). These ridges correspond to p-modes of different radial order n. For the same ℓ, ridges represent mode with different n; for a given $\ell \neq 0$ penetration depth increases with frequency ν (as the number of ridges increases too).

6.3.8 Observational requirements: frequency resolution

The frequency resolution in the power spectrum is determined by the length of the time series, or the time duration over which the observations are made. For a time series of one day (24-hours) duration, one has the frequency resolution in the Fourier transform, $\Delta\nu = 1/T = 1/86400 \sim 11.6\mu$ Hz.

In order to obtain high spectral resolution, continuous solar time series observational data over long periods are required. This is particularly important to resolve mode frequencies at low ℓ's where the ridges of power in the

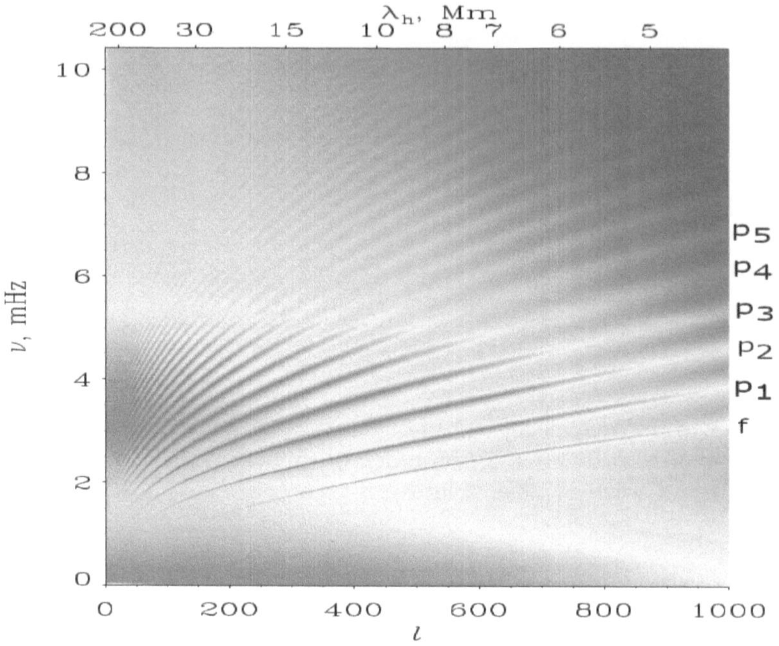

FIGURE 6.9: The $\ell - \nu$ diagram showing *p*- and *f*-ridges obtained from a long time series data of high-cadence MDI-SOHO Dopplergrams. (Courtesy of SOHO/MDI consortium. SOHO is a project of international cooperation between ESA and NASA.)

$\ell - \nu$ diagram are very narrow and crowded. As an example, let us consider observations using λ 600 nm solar line (width $\sim$ 0.01 nm). A velocity corresponding to $\sim$ 1 metre/sec would shift it by $\sim 10^{-6}$ nm [161]. Thus, individual oscillation mode having amplitudes $\sim$ 0.1 metre/sec would require observational measurement accuracy of parts per million of its width.

The long-time observations from a single ground-based site are affected by the data gaps due to the day-night cycle. These gaps introduce spurious, or ghost, frequencies in the power spectrum, rendering it nearly impossible to unambiguously identify the real solar modes. Therefore, it is essential to have near-continuous observations covering several days, months or even years. For example, a time-series of T=10 years would provide a frequency resolution $\Delta\nu \sim 3.18 \times 10^{-9}$ Hz, or 3.18nHz.

6.3.9 Observational requirements: temporal and spatial resolutions

The maximum value of harmonic degree ℓ that can be studied depends on the spatial resolution of the observations, whereas the maximum value of the frequencies ν that can be studied depends on the temporal cadence of the observations.

If the observations have a temporal cadence of $\Delta t = 1$ min, it gives a Nyqvist frequency of $1/(2\Delta t) = 8.3$ mHz, and only frequencies below this limit can be measured. As most of the p-mode power is contained below this maximum frequency, it is not required to have cadence faster than this.

Imaged data over the solar disk with spatial resolution $0 \leq \ell \leq$N, where N=number of pixels on the solar disk, determine the maximum value of ℓ that can be studied. Higher resolutions are required to detect larger ℓ modes. However, the atmospheric seeing of a ground-based observing station, leakages, foreshortening and other errors in the observations cause difficulties in estimating the maximum values of harmonic degree up to the Nyqvist ℓ.

6.4 Helioseismic observations from ground and space

There are two classes of observations from ground and space that are used for different goals of helioseismology. The observations can be either based on the line of sight velocity via Doppler shift, or intensity.

- Integrated sunlight measures the averaged velocity or intensity over the solar disk. This gives only low degree ($\ell \leq 4$) modes. In these observations, the signal corresponding to high ℓ modes having small scale sizes are essentially canceled out, while low ℓ modes produce radial motion that is in phase across much of the solar disk. This yields a detectable spectral shift in the unimaged sunlight.
- Imaged data over the solar disk with high spatial resolutions is required to obtain higher value of ℓ.

The long, continuous 24-hour coverage of the Sun is an essential requirement to determine the mode frequencies with greater precision in both imaged as well as un-imaged observations. This is particularly important at lower values of ℓ where the peaks in the spectrum are very narrow and crowded (Figure 6.9).

TABLE 6.3: Some ground-based networks and space-borne projects

	Ground-based network	Space-borne instrument
Un-imaged	BiSON IRIS	SOHO-GOLF
Imaged	GONG TON	SOHO-MDI SDO-HMI

There are three alternatives adopted in order to obtain near-continuous observations over long time periods for helioseismic studies (both imaged and un-imaged) (Table 6.3):

- Observations from the Antarctic during summer ([105], [199]). In principle, one can obtain observations spanning several months, but continuous observations extending over only a few weeks have been achieved because of severe weather conditions.

- Networks of ground-based observing sites: (i) the Birmingham Solar Oscillation Network (BiSON) [74]; (ii) the International Research on Interior of Sun (IRIS) [114]; (iii) the Global Oscillation Network Group (GONG) [231]; and (iv) Taiwan Oscillation Network (TON) [79].

- Space-borne instruments preferably placed in a sun-lit L1 Lagrangian point, such as, (i) SOlar and Heliospheric Observatory (SOHO) - Michelson Doppler Imager (MDI), Global Oscillations at Low Frequencies (GOLF) for both g and p-modes [130], and (ii) Solar Dynamics Observatory- Helioseismic and Magnetic Imager (SDO/HMI) [331].

One of the most successful ground-based helioseismology projects has been the Global Oscillation Network Group (GONG). It is a community-based international project designed in the 1980s and operational since 1995 for detailed study of the internal structure and dynamics of the Sun using helioseismic techniques [162]. GONG consists of six ground-based stations located around the Earth to obtain near-continuous observations of the Sun (Figure 6.10). Identical Doppler instruments are installed at each of the sites, which employ Michelson interferometers providing a narrow passband of 1 Å to measure the Doppler shift of a Ni I line 6768 Å. The network has been operating over an extended period of nearly 24 years. The data from individual sites are merged at the central facility in NSO, USA, to make full-day time series [367].

Long time series of helioseismic data have also been obtained from space-borne instruments, such as the Magnetic and Doppler Imager (MDI) instrument onboard the Solar and Heliospheric Observatory (SOHO) [327]. SOHO is a joint project of NASA and ESA, launched on 2 December 1995. It has been observing the Sun since then from the Sun-lit Lagrangian point $L1$, a point of gravitational balance located approximately 0.99 A.U. from the Sun and 0.01 A.U. from the Earth. December 2017 marked the 22nd launch anniversary of SOHO, making it the longest-lived heliophysics mission still operating. Like GONG, it has provided a nearly continuous record of solar and heliospheric phenomena over a full 22-year magnetic cycle (two 11-year sunspot cycles).

The Helioseismic and Magnetic Imager (HMI) on board the Solar Dynamics Observatory (SDO), launched in February 2010, is another important space instrument which is providing valuable data for helioseismology.

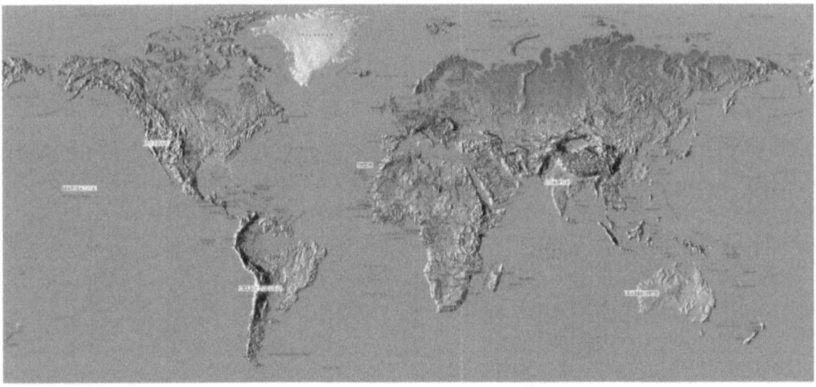

FIGURE 6.10: The six-site ground-based GONG network – Learmonth (Australia), Udaipur (India), Tenerife (Canarias, Spain), Cerro-Tololo (Chile), Big Bear (USA), and Mauna Loa (Hawaii, USA).

6.5 Direct and inversion techniques

From the discussions in the previous sections, it is obvious that solar oscillation mode frequencies depend on the structure of the equilibrium model. The accurately observed acoustic frequencies can, therefore, be used to infer the internal structure of the Sun. There are two complementary strategies for achieving this goal:

(i) *The Direct (Forward) approach*: It essentially adopts the method of comparing the observed mode frequencies with the theoretically computed eigenfrequencies of a given equilibrium standard solar model. The model is perturbed to obtain the eigenfrequencies of solar oscillations. Theoretically computed eigenfrequencies are then compared with the observed frequencies, which usually never turn out to be perfect and deviate from the observed frequencies.

Attempts are made to make small corrections to the parameters of the solar model having the largest uncertainties, *e.g.*, corrections to the ideal gas law, the fractional abundance of He, and heavier elements (Y,Z) in different layers of the Sun, the depth of the base of the convection zone, *etc.* Several results have been obtained from the direct modelling method, such as the chemical composition, depth of convection zone, internal mixing, *etc.* However, this method faces difficulty in determining adjustable parameters uniquely.

(ii) *The inverse approach*: This technique uses the observed mode frequencies to deduce the internal structure by inverting the observed oscillations. Most inversion techniques use a reference solar model to calculate the density and sound speed differences from the frequency differences (*reference-sun*). The computed solar models depend on assumptions about the physical properties of matter, in particular, the equation of state, the opacity, and the rates of nuclear reactions.

Hare and hound exercises were carried out in the initial stages, which provided the tests of inversions, where '*hare*' constructed a solar model, and calculated the frequencies of p-modes. These were perturbed by random errors, and then supplied to '*hounds*', who, using inversion techniques, calculated sound speed and density profiles. These profiles were sent back to '*hare*' for comparison with their model.

Two broad classes of inversion techniques are in use: (i) *Regularised Least Squares (RLS)*, and (ii) *Optimally Localised Averages (OLA)*. These are essentially complementary techniques, that have different aims [333]. RLS aims to find the $\delta c^2/c^2$ and $\delta\rho/\rho$ profiles that give the best fit to the data. On the other hand, the aim of OLA is to find linear combinations of the frequency differences that provide a localised average of the unknown function, while keeping the errors small. We do not attempt to provide their details here, and refer for details to ([148], [33]).

6.6 Results from inversion techniques

6.6.1 The sound speed and density

The BiSON, SOHO/SOI-MDI and GONG projects have obtained results for sound speed, density, fractional helium abundances, internal rotation rate, *etc.*, through inversions of measured frequency data (Figure 6.11). The quantity plotted is the deviation of the squared sound speed from solar models with radial distance from Sun's centre.

The differences between the SSM and the Sun are found to be very small, within around 0.4%, but are physically still very significant. Over most of the Sun, the match is found to be very good except at some problem areas indicative of more physics to be incorporated [35].

The largest deviations from the model are noticed just below the base of the convection zone and at the outer part of the core. These are locations of significant variation of the mean molecular mass μ. Because the lower He abundance increases the sound speed, and vice versa, it has been speculated that there would be a deficit (with respect to the model) of He just below the convection zone. It is expected that a small turbulent mixing can solve the problem. The strong rotational shear discovered by helioseismology at tachocline (Section 6.6.5), near the base of convection zone, is likely to induce the required mixing.

The sharp decrease of the sound speed as compared to SSM at the boundary of the core may indicate an over-abundance of He at the edge of the energy-generating core. The steep increase of sound speed towards the centre

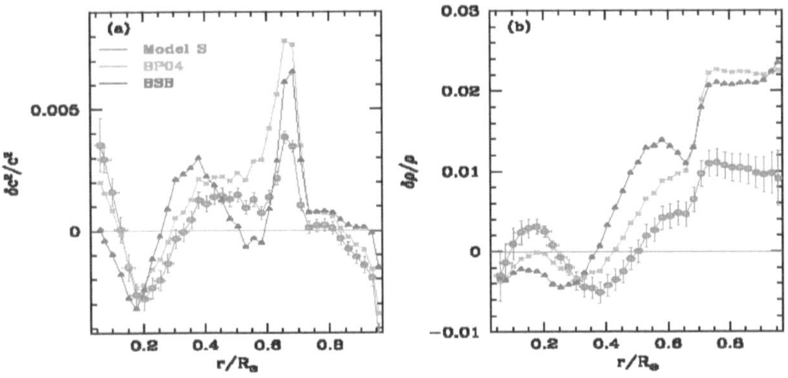

FIGURE 6.11: Relative difference in the squared sound speed and density between the Sun and three different solar models described in [34]. The differences are in the sense (Sun − Model). Results are obtained using the BiSON-13 data set. The vertical error-bars show 3σ uncertainties in the results caused by errors in the input frequencies. The horizontal error-bars are a measure of the resolution of the inversions. Figure courtesy of Sarbani Basu [34].

can be understood if He is less abundant in the core than in the SSM, possibly due to the use of incorrect nuclear reaction rates. Or it may be the result of macroscopic motions in the core, induced by the instability of ^3He burning.

The steep drop near the surface may be related to uncertainties of outer layers in the SSM. This requires better observations near the surface, but large degree ℓ-modes are difficult to detect due to the lack of very high spatial resolution observations.

6.6.2 Solar radius and the depth of convection zone

It is possible to measure the solar radius using the lowest ridge, or the fundamental ($f-$) mode, in the $\ell - \nu$ diagram. It has no radial nodes, and has a simple dispersion relation, *i.e.*, $\nu = \sqrt{gk_h}/2\pi$. The horizontal wavenumber is given by $k_h \simeq \sqrt{\ell(\ell+1)}/r$, and the f-mode frequencies are nearly independent of the internal structure of the Sun.

Using $g = Gm_\odot/r^2$, we find that $\nu \propto r^{-3/2}$. This gives r as it is possible to determine frequencies with very high precision. The problem with this method is that the optical depth, which defines the surface, must be defined precisely. Fixing the optical depth as τ=*2/3*, the radius of Sun r is estimated to be 695.7±0.026 Mm.

The depth of the convection zone is another critical distance. This is determined by the ratio between the mixing length, and the pressure scale height. The p-mode ridges of large ℓ are very sensitive to this ratio, which contributes to the determination of the base of the convection zone. With improving solar models, the estimate for its location is found as:

$$r_{base} = (0.713 \pm 0.001)r_{\odot}, \tag{6.18}$$

which corresponds to the depth of 199700 ±700 km, from the surface of the Sun.

6.6.3 Temperature of the central core

Due to the uncertainty of μ in the center of the Sun, helioseismology cannot give the central temperature $T_c \propto \mu c^2$ as accurately as it gives the speed of sound. The error estimate of temperature determined this way is about 2% around $T_c = 1.57 \times 10^7$K.

There are several other important results, which have been obtained from helioseismic studies, such as the abundance of $He \sim 0.249 \pm 0.003$, a significant increase in opacity of $\sim$ 15% at the base of the convection zone, *etc.*

6.6.4 Solar internal rotation

The rotation rate of the solar interior is a very important inversion problem [359]. Most solar standard models neglect solar rotation, while the observed frequencies are affected by the small, but non-zero solar rotation.

A standing wave arises from two waves traveling in opposite directions between two nodes. Their frequencies will be identical, if there is no rotation. But due to rotation, the waves are carried in opposite directions at the rotation velocity, v. Therefore, the wave moving in the sense of the rotation will have its frequency Doppler shifted by $\Delta\omega_+/\omega = v_+/v_p$, where v_p is the wave's phase velocity. The oppositely directed wave will undergo a reduction of $\Delta\omega$, *i.e.*, $\Delta\omega_-/\omega = v_-/v_p$. Hence the two waves will be separated, or split, in frequency by $2\Delta\omega$. Overall, at the surface, this would result in the entire standing wave pattern drifting slowly across the Sun in the opposite direction to that of the solar rotation.

Thus, if the Sun were to be spherically symmetric, there would be degeneracy in the azimuthal number m. Solar rotation causes a deviation from the spherical symmetry and breaks the degeneracy, *i.e.*, splits the frequencies due to dependence on m. This allows us to infer the rotation rate $\Omega(r,\theta)$ as a function of depth r [192], by measuring the splitting using the inversion technique:

$$D_{nlm} = \frac{\nu_{n,l,m} - \nu_{n,l,-m}}{2m} = \int_0^1 \int_0^1 dr\; d\cos\theta\; K_{n,l,m}(r,\theta)\Omega(r,\theta). \tag{6.19}$$

However, as the rotation rate is very small, less than 0.5 μHz, it is essential to have observations providing very good frequency resolutions.

Figure 6.12(a) shows profiles of solar internal rotation rate as a function of radius at different latitudes inferred by inversion of observed rotational

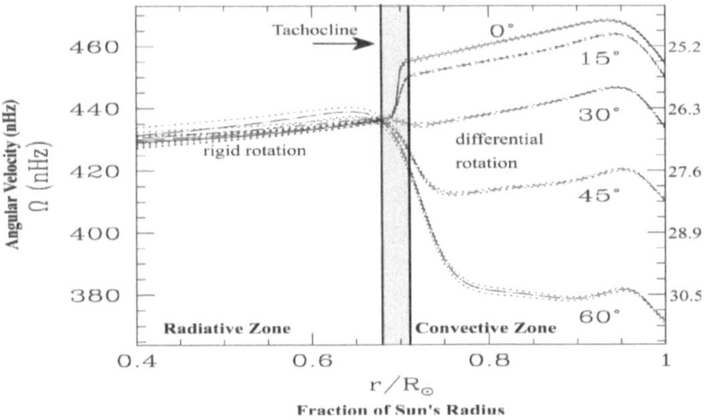

(a)

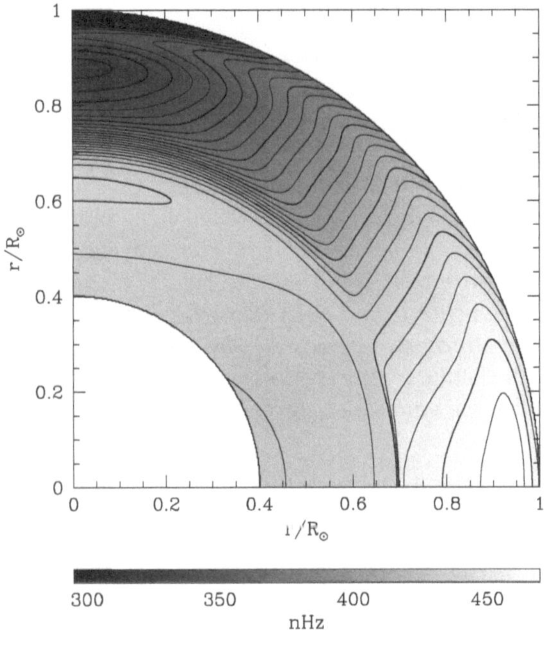

(b)

FIGURE 6.12: (a) Solar internal rotation rate as a function of radius at different latitudes inferred by inversion of observed rotational splittings. The dotted lines represent the 1σ error limits. The corresponding rotation periods (in days) are marked at the R.H.S. of the plot. (b) Contours of constant rotation rates at $5nHz$ intervals, while the thick contours are at 20 nHz, with the first thick contour near the equator marking 460 nHz. Figures are courtesy of H. M. Antia, drawn using 2D *Regularised Least Squares* (RLS) technique as described in [14].

splittings. The X-axis represents the solar equator while the Y-axis represents the rotation axis. These results are obtained from MDI data averaged over the solar cycle 23. Figure 6.12(b) is obtained by the 2-D inversion technique, shown here for only one quadrant, due to the symmetry of inversion results [14]. The results show that the observed surface differential rotation of the Sun persists through the convection zone (CZ). In the radiative interior, the rotation turns to nearly rigid body rotation, with a value that is intermediate between that of the solar equator and the pole at the surface.

Figure 6.12(a) marks the base of the convection zone at $r \sim 0.7R_{\odot}$. Shear layers are observed just below the equatorial region and around the base of the convection zone. Thus, the helioseismic calculations of the Sun's internal rotation profile show a rough separation into a rigidly-fast rotating core and the differentially-rotating outer envelope.

6.6.5 The interface, shear layer or "tachocline"

The transition from differential to solid rotation occurs near, or below, the base of the CZ over a very thin layer, referred to as the *tachocline* [345]. It is discovered between the radiative and the convective zone at $\sim 0.70R_{\odot}$, with a thickness of $0.04R_{\odot}$. This layer contains a substantial radial gradient of rotation velocity of opposite signs in low and high latitudes (Figure 6.12).

The tachocline has become interesting in recent years as more details about it have been discovered. It is now believed that the Sun's magnetic field is generated by a magnetic dynamo in the tachocline. Thus it could be the seat of the dynamo responsible for the solar magnetic cycle ([1], [359], [291]). The changes in fluid flow velocities across the layer, or shear flows, can stretch magnetic field lines of force and enhance them. This change in flow velocity gives this layer its alternative name, *the shear layer.* There also appear to be sudden changes in chemical composition across this layer. Another shear layer is found just under the equatorial surface (Figure 6.12b).

6.6.6 Solar internal magnetic fields

Apart from the solar rotation, there are other deviations from spherical symmetry, such as due to the solar internal magnetic fields. First estimate of toroidal magnetic field of 2 ± 1 MG, centred near the base of convection zone, was obtained using frequency splitting data [109]. This result was revised using GONG data, and an upper limit of 0.3 MG was reported on the toroidal magnetic field concentrated below the convection-zone base [32].

6.6.7 The changes related to solar cycles

An interesting feature of solar oscillations is the change in oscillation frequencies with solar activity related magnetic field variations [45], which affect the central frequencies, and frequency splittings of low degree modes, as well as intermediate degree modes. Frequency splittings also show solar-cycle related shifts [195].

An important result is the discovery of jet streams of plasma (more precisely, *torsional oscillations*) thousands of kilometres below the surface in the convective zone [379]. These jet streams form broad fronts at the equator, breaking into smaller cyclonic storms at higher latitudes. Torsional oscillations represent the time variation in solar differential rotation. They are alternating bands of faster and slower rotation.

So far there is no generally accepted theoretical explanation for the torsional oscillations. However, a close relation to the solar cycle is evident, as they also have a period of eleven years, known since their discovery in 1980 [183].

At low latitudes, just like the torsional oscillations, *zonal-flow bands* appear to migrate towards the equator with time, while at high latitudes these move polewards with time [12]. With the availability of more data, it has become clear that the zonal flows penetrate through most of the solar convection zone (for a review, see [185]).

6.6.8 Some results of g-mode detection

SOHO has made a fundamental discovery, and perhaps one of its biggest contributions in finding some evidence of the elusive low-frequency g-modes, after more than four decades of searching for it. Nearly two decades worth of full-disk solar data obtained from Global Oscillations at Low Frequencies (GOLF) instrument on-board SOHO was studied to search for possible collective frequency modulations produced by the periodic changes in the deep solar structure. Such modulations provided access to only very low frequency g-modes [121]. Around 100 g-modes of degree ℓ=1 and 2 are predicted in the oscillatory period between 9-48 hrs, but are not observed individually. But, when combined, they provide their asymptotic characteristics in good agreement. The inferred rotation frequency, reported as recently as in 2017, of the solar core is 1644$\pm$ 23 nHz (a period of one week). This is by a factor of 3.8 $\pm$ 0.1 faster than the rotation of the radiative envelope.

Although g-modes have been discovered in other stars, their discovery in our own star, the Sun, opens up new vistas of research, such as how differentially rotating layers of the Sun interact and what we can learn about the composition of the core based on its rotation.

Helioseismology has also helped in accurate measurements of the quadrupole (and higher-order) moments of the Sun's gravitational potential, which are found to be consistent with general relativity.

6.6.9 Limitations of global helioseismology

Although global helioseismology provides several important results about the solar interior, it has some limitation for the study of the near surface properties, *viz.*, sub-surface flows, emergence of active regions, effects due to transient activities, *etc.* The global modes extend over all longitudes, thus analysis of their frequencies provides essentially no information about the longitudinal variations of solar properties. They depend only on the component of rotation, which is symmetric around the equator.

The properties of global modes have little sensitivity to meridional or more complex flows, such as large convective eddies, which may be present in the solar convection zone. It became possible to explore the properties of relatively shallower layers underneath small spatial scales of ARs only after the advent of *local helioseismology*, which complements global helioseismology, and probes local perturbations in the sub-surface depths. For details of the advances in global and local helioseismology, one may refer to [221].

6.7 Local helioseismology

Local helioseismology, a term first used by Lindsey, Braun and Jefferies in 1993 [246], is a technique to interpret the full wave field, observed at the solar photosphere, not just the mode frequencies. Local helioseismology uses short wavelength, or high ℓ, acoustic waves measured on small patches of the solar surface. This usually corresponds to the scales of active regions (ARs), *i.e.*, much smaller areas than the whole solar disk. These studies exploit only the high ℓ modes, that are made possible by the availability of spatially resolved, imaged observations, using large format CCD detectors. Usually such studies require $\ell > 300$ p-modes, which are damped over time scales much shorter than the time required to circumnavigate the Sun. Therefore, their frequencies are local measures of the Sun's properties. Table 6.4 provides a comparison of the global and local helioseismology approaches.

TABLE 6.4: Global v/s local helioseismology

Global (whole Sun) helioseismology	Local (active region scale) helioseismology
1. Horizontal interference selects discrete values of ℓ.	1. No horizontal interference, can have any wavelength.
2. Entire sun is sampled.	2. Sampling of local areas.
3. Spherical harmonics describe global waves.	3. Plane wave approximation.
4. Does not get structure as function of longitude.	4. Longitudinal structure can be determined.
5. Does not get non-symmetric latitudinal structure.	5. Non-symmetric latitudinal structure can be determined.
6. Deals with $\ell \leq 180$.	6. Valid for $\ell \geq 300$.

Science objectives of local helioseismology are: dynamics of sunspot and their magnetic connectivity with the interior, imaging and prediction of sunspots at the far-side [57], 3-dimensional view of the solar envelope, which is important to understand large-scale flows, sub-surface weather (velocity flows), and their interactions in the solar interior [139]. To facilitate space weather forecasting, seismic images of the solar far-side are being produced nearly continuously by analysing data from the GONG and SOHO, and since 2001 the entire far side has been imaged with this data.

The three main techniques of local helioseismology related analysis are (i) the ring-diagram analysis [176], (ii) time-distance method [106], and (iii) acoustic holography [244]. There are also some studies using the Fourier-Hankel spectral analysis [58] and direct modelling [398].

6.7.1 The ring-diagram techniques

The ring-diagram analysis is one of the most powerful tools of local helioseismology. Its basic aim was to infer the speed and direction of horizontal flows under solar active regions. The analysis essentially deals with the high degree acoustic waves, assumed to be plane waves, travelling across shallow regions of the solar surface.

The primary data requirements of ring diagram analysis are high-resolution uninterrupted, long time series of photospheric Doppler images as provided by the ground-based $GONG$, and space-borne instruments, $SoHO - MDI$, and more recently, by the SDO/HMI. Small patches are selected, usually $16° \times 16°$ in size, around areas of interest. The three-dimensional Fourier transformation of the data is then carried out that requires equal sampling in temporal and spatial dimensions.

The main steps of the data processing for the ring-diagram technique are: (i) data cube extraction from Dopplergrams corresponding to the patch, (ii) remapping, and tracking of the data, (iii) interpolation and apodization, and (iv) filtering and three-dimensional Fourier transformation, providing power spectra [176].

Ring diagram analysis allows to derive the spatially (over the selected patch) and temporally (over the period of data) average properties of photospheric p-mode parameters, *e.g.*, mode amplitude, width, flow, *etc.*, at the scale of the patch, usually an active region. Further, the inversion of p-mode parameters obtained from the ring analysis can be used to understand various physical parameters with depth in the sub-photospheric layers of the local scales. Sound speeds, and adiabatic index can be compared within magnetically active and inactive regions [35].

The technique of ring-diagram was first developed to infer the horizontal flows and thermodynamic structure underneath the solar photosphere [176]. Using Doppler observations, flows of $100m/s$ were found from the equator to the pole. The presence of large-scale flows with longitudinal variations of about $20m/s$ was discovered in ARs [177]. Thereafter, several other important inferences have been derived using this technique [13], [263].

6.7.2 Time-distance analysis

The method of time-distance helioseismology was introduced in 1993 with the aim of measuring and interpreting the travel times of the solar waves between any two locations on the solar surface [106]. Direct measurement of the travel times and distances of individual acoustic waves, the predominant approach in terrestrial seismology, is expected to be more difficult in view of the number and stochastic nature of solar seismic sources. However, it is possible to extract time-distance information from temporal cross-correlations of the intensity fluctuations on the solar surface.

As is done in the ring-diagram analysis, the first operation in time-distance helioseismology is to track Doppler images at a constant angular velocity to remove solar rotation. The resulting data cube is Fourier transformed, and a filtering procedure is then applied to the data [107], first to filter out the frequencies below 1.5 mHz to remove the noise due to granulation, and super-granulation. The data are further filtered to select parts of the wave propagation diagram. The basic computation in time-distance helioseismology is the temporal cross-covariance between filtered signals at two points x_1 and x_2 on the solar surface,

$$C(x_1, x_2, t) = \frac{h_t}{T - |t|} \sum_{t'} \Psi(x_1, t')\Psi(x_2, t' + t). \tag{6.20}$$

This approach opens the way for seismic studies of local solar phenomena, such as the sub-surface inhomogeneities near sunspots. The seismic signature of inhomogeneities within the ray path connecting two surface locations gives a travel time anomaly. In order to infer the local structure and dynamics of the solar interior, one must then solve an inverse problem [200]. The time-distance technique can be used in the study of photospheric and sub-photospheric structures of the AR, and provides spatially and temporally resolved information [222].

6.7.3 Helioseismic holography: the far-side imaging

Helioseismic holography was introduced by Lindsey and Braun in 1990 [244] for the main purpose of far-side imaging, *i.e.*, a special case of phase-sensitive holography. The theoretical introduction to helioseismic holography is given in [245]. However, the basic concept was first suggested by Roddier in 1975 [322].

In simple terms, sunspots absorb helioseismic waves, which causes a seismic deficit that can be imaged at the antipode of the sunspot. Helioseismic holography uses the wavefield on the visible disk to learn about the active regions present on the far-side of the Sun. The basic idea is that the wavefield, *e.g.*, the line-of-sight Doppler velocity, observed at the solar surface, can be used to make an estimate of the wavefield at any location in the solar interior at a given instant in time.

6.7.4 The solar transient related changes

Activity related effects on p-mode parameters have been reported by many researchers ([75], [310], [178]). Using ring diagram analysis of the sites of intense magnetic fields, the sunspots, a significant amount of p-mode power absorption was reported [310]. Similarly, it is expected that the energetic phenomena, such as solar flares and CMEs that release a large amount of energy may be able to excite, or amplify the acoustic (p-) modes of oscillations.

The ring diagram technique provides properties of modes that are spatially averaged over the area of active region patches, and further, also temporally averaged over 24-hour periods of data cubes. Therefore, any effects related to the rather short-lived transient energetic events may not be expected to reflect clear signatures on the mode properties obtained from the ring diagram technique.

It is also notable that any mode amplification induced by transient energetic events would need to compete with the absorption effects associated with the sunspots. However, there have been some reports of power amplification in high-degree p-modes during the periods of high flare activity as compared to non-flaring regions of similar magnetic fields [9]. Evidence of substantial increase in mode amplitude and systematic variations in sub-surface flows, *i.e.*, the meridional and zonal components of velocity, was also inferred from comparison of the pre- to the post-flare phases for a very large, energetic flare [264].

Local helio-seismology has obtained the seismic images of solar flares. Time-distance analysis has been used to infer the sub-photospheric flows of ARs [106]. Solar quakes, or flare-related waves spreading away on the solar surface, have been reported by using time-distance analysis in several cases ([219], [220], [217]).

Acoustic holography is ideal for the detection of sources, and sinks of acoustic waves on the Sun. A region of lower acoustic emission in the 3 - 4mHz frequency band was discovered to extend far beyond the sunspots as acoustic moats into the quiet Sun [55]. It has obtained "power maps" which discovered acoustic sources of power associated with flares, or high-frequency emission, also termed '*acoustic glories*', surrounding active regions at the flare sites ([56], [102], [103]).

6.8 Extension to stars: asteroseismology

Asteroseismology refers to the study of the internal structure of stars through the interpretation of their oscillation frequency spectra in analogy with helioseismology (see [63] for a review). The success of helioseismology has led to efforts for other stars over the whole mass range. As the oscillations of the Sun are caused by turbulent convective motions near its surface, such oscillations are expected to be excited in other stars with outer convection zones. Solar-like oscillations are indeed predicted for the lowest-mass main-sequence stars up to objects with masses near some $1.6M_{\odot}$ (*e.g.*, [81], [85], [180]), as well as in red giants [108]. For basic theoretical formalism of stellar oscillation, one may also refer to [84].

Asteroseismology is the only available method to derive, in a near direct way, the internal structure of the distant stars with high precision. Reliable detection of oscillations on many low mass Sun-like stars have been made. Low-ℓ modes are of particular importance, as they reach deeper into the Sun, and other stars, and these are the only modes detectable on other Sun-like stars. Unambiguous detections of stochastically excited p-modes have been found after decades of efforts in many solar-like stars. The techniques of helioseismology are applicable to these stars once their frequency spectrum is determined. The search for oscillations in stars in the solar neighbourhood has been ongoing since the early eighties. The first indication of stellar power with a frequency dependence similar to that of the Sun was obtained in αCMi (Procyon, F5IV) [64], and first detection of individual frequencies was achieved from high-precision time-resolved spectroscopic measurements in 1995 for the G5IV star ηBootis ([212], [65]). Subsequently, these results were further confirmed ([70], [213]). One may refer for further reading to the book entitled *Asteroseismology* [3].

Other detailed asteroseismological results for probing the internal structure have been obtained for the *g*-modes in white dwarfs since the early 1990s. These are the compact end-products of stars with initial masses $< 9M_{\odot}$. Their *g*-modes have multi-periodicities of around 10 minutes.

The current status of asteroseismology of non solar-like main-sequence stars, *i.e.*, of stars with larger mass than the Sun, has not made much progress because asteroseismology for such objects is not straightforward. A prerequisite is the detection and identification of many oscillation modes, *i.e.*, the knowledge of their wavenumbers (n, l, m), which has been a major problem.

The best candidates according to the number of modes detected are the δ-Scuti stars, which pulsate in *p*-modes. For some of these stars, more than 30 oscillation frequencies have been detected from multi-site campaigns, such as WET and the DSN: δ-Scuti Network. The early attempts for such stars were very promising, but it became clear that unexplained amplitude and period changes occur in these stars. Moreover, the lack of accurate mode identifications, due to the incomplete observed frequency spectra, limits seismic applications.

The quest for solar-like oscillations in metal-poor stars, which are considerably less massive than the Sun, is another important problem for asteroseismic space missions. This is particularly so because asteroseismology not only probes interior stellar structure, but also derives age estimates [83]. Such low-mass stars are among the oldest objects not only in our Galaxy, but also in the Universe. Accurate age estimates of such objects in principle can provide a good age determination of the Universe, completely independent of any other existing methods.

A network of telescopes called the Whole Earth Telescope (WET) was set up around the Earth's equator in the late 1980s. A WET campaign on the star DOV PG 1159-035 implied a breakthrough in white-dwarf seismology and allowed to derive its mass, rotation rate, and internal stratification with precision, which led to other successful campaigns on white dwarfs.

There are several asteroseismology space missions, such as COROT, and Kepler, apart from the ground-based programs, such as, ESO 3.6m (HARPS), ESO VLT (UVES), and a network of several smaller telescopes carrying out such studies.

6.9 Summary

Helioseismology has become a unique diagnostic observational tool in the field of solar physics to study the solar interior, which provides proof of several theoretical interpretations. It has made a fundamental contribution to basic particle physics and eliminated uncertainties about the internal structure of the Sun. The most notable result of global helioseismology is the explanation of the solar neutrino problem by providing the proofs to the models of the solar interior [26]. It showed that the so-called neutrino flux problem of the Sun could not be caused by errors in solar (stellar) models, and was a fundamental problem related to particle physics that was ultimately resolved by invoking neutrino oscillations. Helioseismology inversions have led to the discovery of shear layer, or the tachocline, located at the base of the convection zone, where the transition from differential to solid-body rotation occurs. Extension of helioseismology to other stars has also been successful in providing information about the otherwise difficult problem involving our understanding about their interior. There are still several problems related to various types of stars and detection of the elusive g-modes that need to be followed up in the coming years.

Chapter 7

The Invisible Sun in Long Wavelengths

7.1 Introduction

According to solar radiation energy distribution, around 6.6% corresponds to UV and shorter wavelength photons, 44.7 % to visible, and 48.7% to IR and longer wavelengths (Figure 7.1). As discussed in Section 2.1 (page 25), the most important observing window for ground-based telescopes is in the wavelength range about 300-800 nm, *i.e.*, the optical window, in which most of the early astronomical knowledge was collected using photographic and spectroscopic observations.

The visible spectrum covers a wavelength range of only a factor of three, whereas the longer wavelength infrared spectrum extends over a factor of 1000. Considering the large span of wavelengths in IR, it is usually divided into three domains: Near IR (800-5000 nm), mid-IR (5000-25000 nm), and far IR (25000-10^6 nm). In the IR and NIR, there are several narrow windows between molecular absorption bands, in which solar radiation can reach to the ground-level (Figure 1.3).

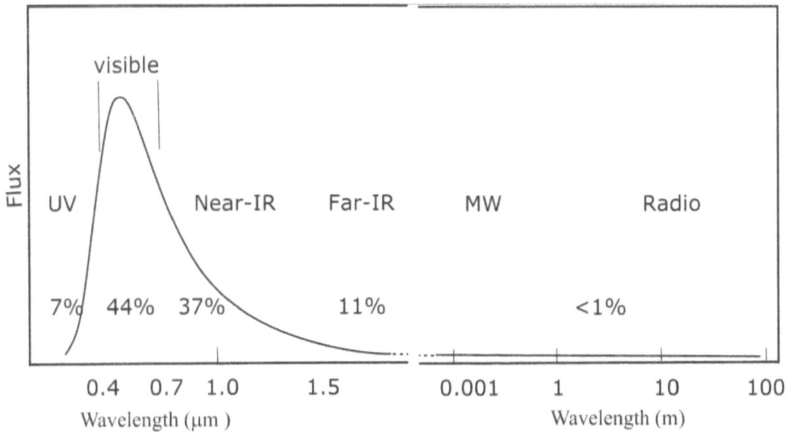

FIGURE 7.1: Solar radiation fluxes in various wavelength-bands.

Beyond 300 μm in the FIR band, there is a decreasing atmospheric absorption. It merges into the microwave radio region at about 1 mm, also known as *sub-mm band.* The span of the radio window, beyond IR, is also very large, covering three orders of wavelengths. The Earth's atmosphere is transparent to a large part of radio wavelengths as in the optical wavelengths. However, very long wavelength radio waves are reflected away by Earth's ionosphere, and hence, not accessible for observations from ground-based radio telescopes.

Observing in the longer wavelengths than the optical provides a good deal of new information about the Sun and the physical processes occurring in various layers of the solar atmosphere. Of these, the Sun has been studied extensively from the ground in radio, since the end of World War II, when military radars became available for use in astronomy. Although, the solar radio flux is minuscule as compared to the visible and IR, it provides very valuable information about the usually invisible corona from the ground. A large variety of solar transient activities, such as the radio bursts associated with the dynamical processes in the corona are also easily observed. On the other hand, it is rather more difficult to observe in the IR and FIR wavelengths due to several observational and technical issues.

7.2 Detection of IR radiation

A major difficulty in observing in IR wavelengths occurs due the very intense thermal background produced by the telescope itself and its surroundings. For black-bodies at temperature T, the emission has a peak at $\lambda_{max}T \approx 3000$ μm-K (from Wien's displacement law). Therefore, there is a peak emission in IR at about 10 μm due to the surroundings at $T \sim 300$K. This background undergoes changes, both in space and time, because of the changing ambient conditions. It can be much stronger, sometimes a million times or more, than the source itself (refer to [308], and [301]). Thus, it requires accurate subtraction of the constantly changing background by observing the source as well as the background. This is done usually by 'chopping', *i.e.*, rapidly switching between the source and a nearby patch of blank sky at 10-100 Hz by using an oscillating mirror, wobbling on and off the source.

Atmospheric absorption is the other main problem faced in the infrared, the major absorber being water vapour. IR observatories are, therefore, usually built on high mountain tops to minimise this absorption. However, even at mountain sites, IR in the range of about 25-300 μm is blocked by atmospheric water vapour and CO_2. Thus, it is required to use aeroplanes ascending to about 20 km, or balloons going to around 40 km, and even satellites for observing in these wavelengths.

7.2.1 IR detectors

In the near-IR, the early measurements were carried out by uncooled bolometers, which absorb radiation on a blackened surface, and convert it to heat. The resulting temperature is measured by a temperature dependent property of the material, such as the electrical conductivity. In order to overcome the limitation of their poor temperature resolution, and the effect of ambient temperature fluctuations, cooled germanium and silicon bolometers were used later. In order to reduce the thermal noise, these were operated at liquid helium temperatures (T$\sim$ 1-2 K).

With the development of new IR array detectors, it is possible to exploit the near-IR and mid-IR part of the solar spectrum in the range of around 1000-10,000 nm, from ground-based solar telescopes. In this IR regime, a large fraction of light coming from cosmic objects is transmitted through the Earth's atmosphere to ground level at several bands. The atmospheric transmission windows for the IR wavelength bands in the 1000–5000 nm range are traditionally known in stellar astronomy as J, H, K, L, and M with corresponding central wavelengths of around 1300, 1600, 2200, 3600, and 5000 nm.

In order to efficiently detect a photon, the detector pixel size should be about the size of the photon wavelength or larger. Thus, in the near-IR region, it is required to have array detectors with pixel sizes of order 10–100 microns. The quantum efficiency of silicon array detectors, both CMOS and CCDs, drops down to zero at around 1100 nm, and the silicon substrate becomes nearly transparent to longer wavelength photons. To overcome this difficulty, efforts are being made to develop detectors using other types of substrates suitable to the wavelengths of interest; for example, HgCdTe (1000 - 2200 nm), InGaAs (1000 - 1800 nm), InSb (1000 - 5000 nm), Si:X doped silicon arrays (2000 - 30 000 nm), Ge:X doped germanium arrays (28000 - 200000 nm), *etc.* Heterodyne receivers, or radiation waveguides, which are similar to radio detectors, are used for studying longer (0.1 - 1 mm) wavelengths [95].

7.2.2 Advantages and disadvantages of observing in IR

There are several advantages of observing in the IR wavelengths: for example, better atmospheric 'seeing', and scattering effects, and comparatively smaller instrumental polarization ([301], [205]). It has been known that seeing improves at the longer wavelengths. However, with the recent developments in adaptive optics (AO) techniques [319], this is no longer a serious issue, and solar telescopes equipped with AO have been successful in providing very good image quality even in the optical wavelengths. In fact, the telescope tracking accuracy and stability aspects are more important issues when spectroscopic and polarimetric studies are carried out, which usually need longer image integration time. The availability of large number of molecular lines and the ability to probe different layers of the solar atmosphere using the solar IR continuum are the other major advantages.

There are some disadvantages too in IR wavelength observations. The main scientific disadvantage is owing to the significantly lower signal (fewer photons) in this range as compared to the visible window. Another difficulty is that there are not many atomic absorption lines available in the IR band. One encounters larger diffraction limit, increased background noise levels, and atmospheric and transmission optics absorption.

Notwithstanding the difficulties in IR observing, there has been immense interest in IR observations due to the many important scientific advantages. In solar physics, the IR forbidden emission lines are very promising for the measurement of coronal magnetic fields, primarily due to the better Zeeman resolution in longer wavelengths IR as compared to visible wavelengths. For a detailed review of advantages and disadvantages of observing in IR, one may refer to [163].

7.3 IR emissions from the Sun

Solar coronal observations during total solar eclipses, and from space have clearly revealed the importance of magnetic fields in playing a central role in governing the observed properties of the corona. At the long IR wavelengths too, it has become possible to observe and study the dynamics of solar corona using ground-based instruments.

The solar IR emission, in the wavelength range of 1000 nm to 1 mm are important not only in studies of the solar disk, but also for studying the faint solar corona [306]. There are many wavelengths in this window that are suitable for studies involving opacity minimum, sunspot umbrae, chromosphere, chromospheric oscillations, prominences, and magnetic fields [250]. Therefore, it presents a new window of opportunity for exploring the Sun with ground-based solar telescopes.

IR solar radiation is formed in thermal equilibrium and provides true kinetic temperatures. Center-to-limb variations over the Sun's disk have been obtained in several wavelengths from the ground, using balloons or aircraft. This has contributed to improving solar atmospheric models up to the chromosphere.

The HeI 1083 nm line provides a useful ground-based method for studying coronal features on the disk of the Sun. By comparing Skylab X-ray telescope observations with HeI 1083 nm spectroheliograms, it was found that the coronal holes seen in the X-ray data also displayed a lack of HeI absorption against the disk of the Sun [165].

At 1.6 μm, opacity of the solar photosphere reaches a minimum. Therefore, it is possible to see $\sim$40 km deeper in this wavelength. This helps in the verification of convection models. Similarly, molecular bands of CO around

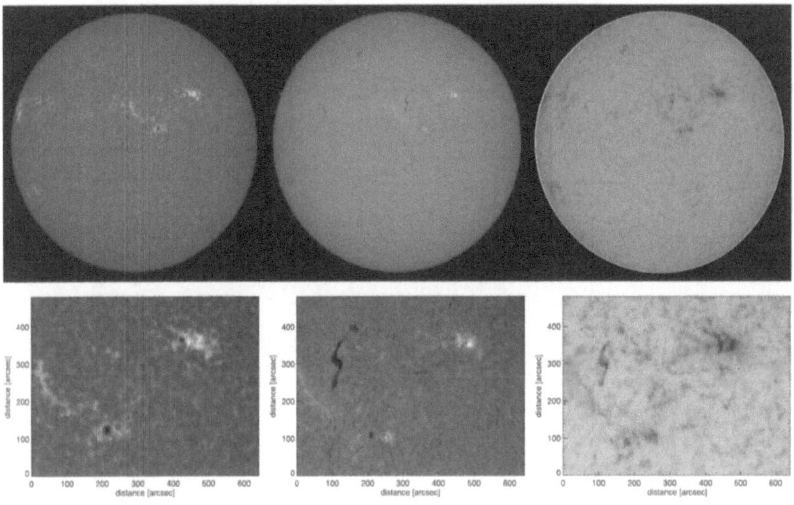

FIGURE 7.2: From left to right: Solar filtergrams in CaII K 393.37 nm, H_α 656.28 nm, and HeI 1083 nm showing the photospheric and chromospheric layers, and corresponding features like sunspot, plages, and filaments in He I. Adapted with permission from [44].

2.3 and 4.7 μm are good thermal probes of the temperature minimum region of solar atmosphere.

The progress in IR imaging techniques has allowed for mapping of the fine structures and active regions (Figure 7.2). The spectrum of the quiet Sun is mostly characterised by molecular lines. These are extremely temperature sensitive and help in probing the temperature and abundance fluctuations in granulation, or active regions. On the other hand, the structure of the corona can be probed using the fine structure lines [236].

The telescope polarization poses a major problem in the optical wavelengths in Zeeman polarimetry of solar spectral line. In IR at 12.5 μm, the telescope polarization is negligible, therefore, the magnetic field vector can be deduced directly [160]. Furthermore, there is the advantage that the Zeeman splitting increases as λ^2 of spectral lines, while the Doppler broadening increases only as λ. The IR wavelengths, such as FeI 1.5648 μm (g=3) line, are very promising for obtaining magnetic field vector of the solar surface. The simultaneous observation of select spectral lines at optical and infrared wavelengths allows for the determination of the magnetic field at different photospheric and chromospheric heights. It is thus possible to obtain magnetic field gradient with height in the solar atmosphere.

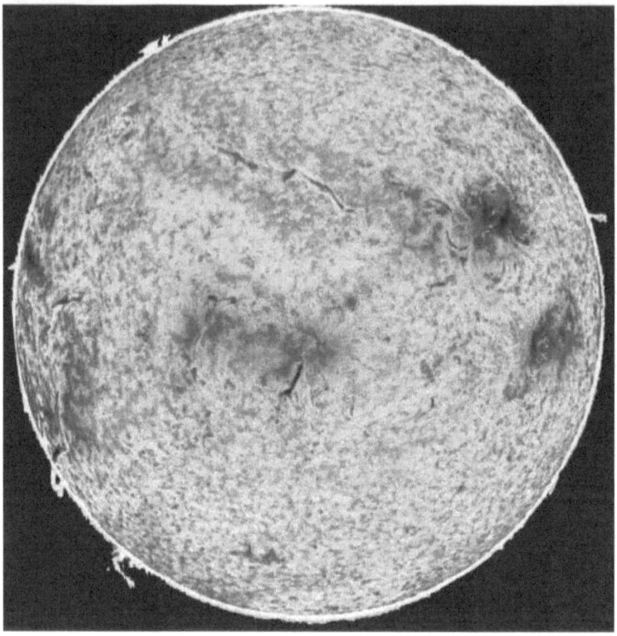

FIGURE 7.3: A full disk spectroheliogram in He I 1083 nm from the NSO/VSM instrument; solar north is up. Credit: NSO/AURA/NSF.

7.3.1 Imaging and polarimetry of solar atmosphere in IR

Figure 7.3 shows a full disk spectroheliogram taken in He I 1083 nm from the NSO/VSM instrument. The image shows polar coronal holes as regions of less absorption, and a low latitude coronal hole in the northern hemisphere. Dark absorption accompanies active regions, and the quiet Sun shows less absorption, but reveals the internetwork pattern. Limb prominences and limb emission are seen as bright regions off the solar limb, and a dark filament is visible on the disk in the northern hemisphere.

A Coronal Multi-channel Polarimeter (CoMP) instrument was integrated into the coronagraph at Sacramento Peak Observatory, and later, at the Mauna Loa Solar Observatory. This instrument helps to measure the polarization of light emitted by the solar corona and prominences, and allows to infer the strength and orientation of coronal magnetic fields [365]. The CoMP employs a combination of birefringent filter, and polarimeter to form spectroscopic images of the Sun simultaneously in two wavelengths, over an extended field-of-view of $2.8R_{\odot}$. It measures the polarization state at the Fe XIII 1074.7 and 1079.8 nm coronal emission lines, and the HeI 1083.0 nm chromospheric line. One may refer to [301] for a detailed review.

Oscillations in the He I 1083 nm line have been used to examine the dynamics of the chromosphere. CoMP instrument has also been used to obtain the Doppler images to perform time-distance seismology of the solar corona, which revealed the presence of propagating Alfvén waves in the solar corona. This instrument has already indicated an exciting opportunity for probing the structure and magnetic topology of the coronal plasma through coronal seismology [366]. Helioseismology in solar IR is also done using CO lines at 4666 nm, FeI 1564.83 nm, HeI 1083 nm and MgI 12318 nm.

Upgraded CoMP (UCoMP) is planned to be deployed at MLSO in 2019. It will provide total intensity, linear polarization, magnetic azimuth, and Doppler observations over a wide range of wavelengths from 530 nm to 1083 nm.

Another instrument, the Facility Infrared Spectro-polarimeter (FIRS) was installed at the Dunn Solar Telescope (DST) at Sacramento Peak (NSO/SP) [193]. Imaging and magnetic field measurements are routinely carried out simultaneously in the visible (Fe I 630.2 nm) and IR wavelengths (1564.8 nm line) in the photosphere, and He I 1083.0 nm line in high chromosphere.

Observations of the Sun at near-IR wavelengths brought major advances in our knowledge of magnetic fields in the solar atmosphere, including in the corona. The spectral lines which have been used are FeI 1564.83 nm [164], HeI 1083.0 nm, and FeXIII 1074.7 nm. But the amplitude of the Zeeman Stokes V signal for the 1074.7 nm coronal line is extremely small, approximately 10^{-4} of the intensity for a 1 G magnetic field [240]. Such small signals make these measurements extremely difficult.

7.3.2 Solar flares in IR

The IR spectrum holds promise for a better understanding of solar flares. IR observations have revealed that solar flares do emit in the IR and FIR continuum. Flare IR continuum observations are of particular interest as they provide a diagnostic of atmospheric electron content, with a temporal and spatial resolution adequate to discriminate between different flare energy transport models.

The first observations of a large X10 white-light flare obtained with the Dunn Solar Telescope (DST) focussed on flare dynamics observed in the NIR continuum at 1.56 μm at the opacity minimum [404]. Bright footprint emissions were observed at 5 and 8 microns. The IR sources were compact and rapidly variable, matching the hard X-ray footprints [205]. The IR continuum greatly exceeded the hot continuum extrapolated from the soft X-ray band. The first spectral mid-IR flare observation, in two broad bands at 5.2 and 8.2 μm also revealed the mid-IR sources to be compact and consistent with double-footpoint configurations, well associated with hard X-ray (HXR) and white light (WL) sources [300].

The main continuum thermal emission processes relevant to flare intensity enhancement in the mid- to far-IR (2-200 μm) spectral range has been interpreted as free-free emission on neutrals and ions [336]. It is found that the IR intensity evolution tracks the energy input to within a second and provides a very direct indication of the evolution of the atmospheric ionisation during flares. These observations encourage further flare observations in the future at the mm/sub-mm/IR wavelengths.

7.4 Detection and imaging in radio wavelengths

There are several advantages of observing in radio wavelengths, although the flux in radio wavelengths is extremely small as compared to the visible and IR. These observations have the advantage that they are not affected by cloud coverage, and dust poses no hindrance for radio waves passing through the Earth's atmosphere. The relatively long wavelength of radio waves, however, limits the angular resolution of the radio telescope.

The much lower spatial resolutions in radio wavelengths, as compared to the optical wavelengths, arises because λ_{radio} ranges in cm to m, whereas the $\lambda_{optical}$ is in the range of $\sim$ 400 nm. The ratio $\lambda_{radio}/\lambda_{optical} \sim 10^5 - 10^6$, affects the spatial resolution of a telescope, $\alpha \propto \lambda/D$. Therefore, for the same resolution, for example, as for an aperture $D = 1$ m of optical telescope, one needs a very large aperture, D $\sim$ 20 km, for radio observations. Obviously, the construction of such a large aperture single radio dish is not feasible. Thus, the resolution of a radio telescope is limited by the aperture, rather than by 'seeing' in optical wavelengths. To overcome this limitation, a variety of techniques are used to achieve higher resolutions in radio telescopes.

The detection technique of a radio signal is not the same as in an optical telescope. Unlike in optical detection, where properties of radiation as photons are used, in radio detection, one uses their wavelike properties. The radio waves excite an alternating field in the detector as an AC voltage:

$$V = V_0 \sin(\omega t - \phi). \tag{7.1}$$

The amplitude V_0 and phase ϕ of this voltage is analysed for constructing radio images. The phase is important, because it determines how waves are added together: if two waves have the same phase, their total amplitude will be the sum of the amplitudes of two waves, and if they are out of phase by π, then the total signal will be the difference of the two. This phase information becomes important for interferometry in radio astronomy.

The energy density in radio waves is usually extremely weak for astronomical objects, as compared to man-made signals. Therefore, flux density is measured in terms of a small unit, Jansky (Jy), where

$$1\ Jy = 10^{-26}\ W\ Hz^{-1}\ m^{-2}. \tag{7.2}$$

But, large solar radio bursts could give out as much as $10^8 - 10^9$ Jy. Therefore, in solar astronomy, a much larger unit, the *solar flux unit (sfu)* is used to describe the radio flux of the intensely bright Radio Sun, where 1 sfu $= 10^{-22}\ W\ Hz^{-1}\ m^{-2} = 10^4$Jy .

The antenna temperature T_A is defined by $p_\nu = k_B T_A$, where k_B is the Boltzmann constant, and T_A is the temperature of the resistor attached to the antenna, which would give the same power p_ν as the radio source, or a measure of power reaching the antenna. For an extended source like the Sun, one is interested in knowing the brightness distribution, expressed as the brightness temperature T_B, which is defined as the temperature of the black-body source that would give the same brightness distribution. The distribution of intensity of black-body radiation is given by the Planck spectrum:

$$B_\nu(\nu, T) = \frac{2h\nu^3}{c^2} \frac{1}{exp(h\nu/k_B T) - 1} \tag{7.3}$$

in $Wm^{-2}Hz^{-1}ster^{-1}$. At radio wavelengths, $h\nu << k_B T$, so that we can use the *Rayleigh-Jeans approximation* for brightness of a radio source, which is given in terms of wavelengths as:

$$B_\lambda(source) = \frac{2k_B T_B}{\lambda^2}. \tag{7.4}$$

For non-thermal sources, T_B is just a convenient measure of the signal strength which varies with wavelength.

7.4.1 Radio telescopes

Radio telescopes can be characterised by three attributes: (1) sensitivity, or the ability to detect weak signals, (2) angular resolution, or the ability to detect fine details, and (3) frequency response, the determination of the intensity of the radio emission as a function of frequency or wavelength. These telescopes consist of:

- *Antenna*: It is equivalent to an optical telescope for selecting the direction of source, collecting the radiation, and converting it to an AC signal. The antenna of a radio telescope is very sensitive to polarization of incident radiation. A polarized radio signal arises from the nature of the emission process, such as those involving magnetic fields, or by interactions with the medium between the source and the receiver.

- *Receiver*: It is the equivalent of a detector and analyser in an optical telescope, which selects the frequency of the radio signal and the bandwidth, defining the spectral resolution, processing the signal and recording it.

One of the simplest elements for detecting radio emission is the *Yagi Antenna*, also known as a Yagi-Uda array invented in 1926. It is named after one of the inventors, Hidetsugu Yagi (the other being Shintaro Uda). This type of antenna is common for VHF television and FM radio reception, whereas modern satellite television uses a high frequency, microwave dish antenna, much

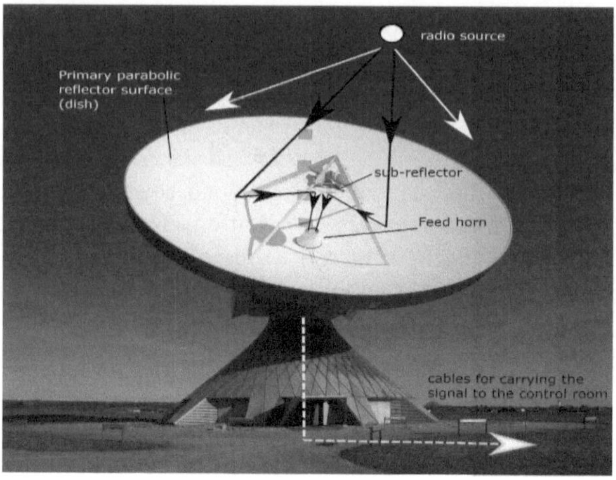

FIGURE 7.4: Radio parabolic dish telescope.

like modern radio telescopes. Yagi, is a directional antenna commonly used in communications in frequency ranges, HF (3–30 MHz), VHF (30–300 MHz), and UHF (300–3,000 MHz). This antenna consists of a number of linear dipole elements.

There are different instruments in use for detecting radio signals from cosmic sources, such as the Sun. Radio telescopes can be divided into two types: (i) filled, and (ii) unfilled aperture.

7.4.2 Filled aperture radio telescopes

A *parabolic dish telescope*, consisting of a parabolic reflector, is an example of a *filled aperture telescope*. Its closest analogue is an optical reflecting telescope, usually circular shaped, which collects the radio waves. The surface figure of the reflector should be accurate within a wavelength of the radio waves of interest, usually an open wire mesh << the wavelength.

The collected radio signal is focussed on a simple dipole, or a horn-shaped waveguide, called *the feed*, placed at the prime focus. The signal from the feed is transmitted by cable to the control room, where it is recorded. For shorter wavelength observations (less than a few metres), the instrument of choice is the parabolic reflector (Figure 7.4), usually mounted equatorially. The larger aperture steerable dishes usually have alt-azimuth mounting as is the case with the optical counterparts.

For large steerable dishes, the costs involved get very high. Therefore, many large radio telescopes use limited mechanical movement, being either a fixed transit telescope, or steerable in elevation only. An example is the Ooty Radio Telescope, India, which is an extended steerable parabolic cylinder along

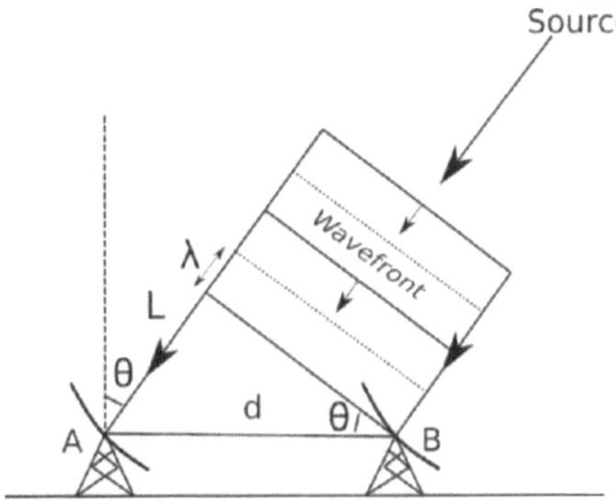

FIGURE 7.5: Schematic of two-element radio interferometry: two parallel beams arrive with a path-difference $d \sin\theta$, and interfere constructively, when the path difference is an integral number of wavelengths, i.e., $d \sin\theta = n\lambda$, n=0,1,2 ....

its axis. It is a 530-metre long and 30-metre cylindrical parabolic antenna, operating at a frequency of 326.5 MHz with a maximum bandwidth of 15 MHz at the front-end. It has a better resolution in the direction parallel to the long axis, than in the direction normal to the axis, giving rise to a flattened beam, or fan beam.

Another method of improving the resolution is by arranging many small parabolic dish units (or dipoles, or other simple aerial) in a two-dimensional array, all feeding their signals to a single receiver. This *filled aperture array* is analogous to multi-mirror telescopes in optical wavelengths.

7.4.3 Unfilled aperture radio telescopes

An array consisting of a small number of well-separated elements, where their spacing $d >> a$ size of an individual element, is an *unfilled aperture telescope*. It is analogous to an optical interferometer, which makes use of the fact that the resolution of a large filled aperture telescope depends only on its aperture dimension a, and not on the effective surface area. It helps collection of more signal, resulting in increased sensitivity. So, even if a large part of the effective surface is removed, the resolution is not affected. The simple example, analogous to the Michelson optical interferometer, is a two-element radio interferometer schematically shown in Figure (7.5).

The use of the Michelson interferometer was pioneered by Sir Martin Ryle and his group at the Cavendish Laboratory at the University of Cambridge (UK). Ryle was awarded the Nobel Prize in physics (1974) for his developments in theory and technique of aperture synthesis (simulating a large radio telescope by combining the signals from a number of small antennas spaced on the ground), based on the use of the Michelson interferometer.

The major limitation in observing in the radio wavelengths, as compared to the optical, due to the coarser angular resolutions, has been overcome due to this great innovation. With the huge interferometers spanning over $D \sim 10^4$ km, it is now possible to build large precision radio telescopes (with small *rms* surface errors $\sigma < \lambda/16$). Advances in radio interferometry techniques, and aperture synthesis, where several radio dishes are locked together electronically, have helped greatly in improving the achievable resolutions in radio wavelengths.

In optical astronomy, spectroscopy has played a significant role in the diagnostics of various physical parameters of cosmic objects using spectral lines. However, in radio astronomy the major proportion is continuum radio emission. The only important radio spectral line is the 21 cm line of neutral hydrogen predicted in 1944 by van de Hulst, and observed in 1951. Neutral hydrogen (HI) line results due to spin-flip transition, and has been used extensively in mapping galaxies. There are some radio recombination lines (between high-lying atomic states), and also some molecular lines, such as, CO, OH, *etc.*, that are promising.

Instead of using gratings to disperse incoming light to separate out wavelengths, in spectrography in radio astronomy, the signal is measured at different discrete wavelengths. The receiver of a radio telescope is tuned to select a narrow band centered at any desired frequency within the broad frequency band in which the radio antenna is sensitive.

7.5 Radio emission from the Sun

The discovery of radio emission from the Sun was made in the early 1940s, using the radars which became available after World War II. In February 1942, British radar stations observed a strong radio noise originating from the direction of the Sun in the metric wavelength range. It was associated with a big solar flare. During the same period, solar thermal microwave radio emission was also discovered, and a relationship between the intensity of the 10.7 cm emission, and the presence of sunspots on the disk was found.

The next important step was the development of radiospectrography, which allowed to continuously record the intensity of the solar emission as a function of frequency and time, first made in the frequency range of 40–70MHz. These initial discoveries provided the ground-based solar astronomers a new window into the electromagnetic spectrum for observing the Sun. This window has been very useful, and complementary to our knowledge of the Sun, which traditionally came from the optical window.

While the optical observations yielded knowledge on the solar structure, composition, and physical conditions of the Sun's photosphere and chromosphere, the radio observations provide information about the structure of the steady and rapidly varying outer atmosphere of the Sun, which is not directly accessible from the ground by optical observations. In fact, it was through the radio observations that the discovery of hot, tenuous coronal plasma was first made. For extensive reviews on the early developments, one may refer to [225], and the excellent book by Mukul Kundu [224].

There are essentially three components, with different characteristic times, associated with solar radio emissions. These are:

- the 'quiet' component contributed from the Sun, radiating as a thermal black-body and changing only slightly in time scales of years,
- the slowly varying component (SVC), related to solar active regions, or sunspots, in time scales of their life-times (days to weeks), or over the solar rotation,
- the transient, and highly variable 'radio bursts' with time scales of minutes to hours, related to energetic and explosive solar phenomena, such as flares, eruptive prominences, and coronal mass ejections (CMEs).

Total solar flux in radio emissions from the Sun, consists of all these aspects, and in fact, varies over $10^4 - 10^5$ times as compared with less than 1.01:1 in the optical window.

For solar radio observations, high time-resolution (*i.e.*, short time intervals) is usually essential. For detailed review on radio emissions, instruments, diagnostics, and flares in HXR and radio, one may refer to [305]

7.5.1 The black-body, or thermal, solar radio emission

Solar radiation flux in optical wavelengths is quite stable, with very small changes occurring over the solar cycle. In the 1 cm to 30 cm wavelength, however, the emission from a disturbed Sun, and for a quiet Sun, is significantly different from that of the black-body in the optical range. The slowly varying radio emissions are associated with brightness temperatures within the black-body limit of thermal radiation of solar atmosphere. The thermal radio continuum in cm and decimeter was found to vary slowly with the evolution of plages, and solar rotation.

The shorter, cm wavelengths originate nearly entirely in the dense chromosphere, while the meter and decameter waves are generated in the corona. The longer the wavelength, the higher is the region of radio emission in the solar atmosphere. The spectral density becomes much greater than at longer wavelengths for both a quiet and disturbed Sun. These elevated fluxes come mainly from the contribution of the solar chromosphere and the corona.

The radio observations in metre wavelength track plasmas ejected in flares and prominence eruptions propagating away through the corona. Their radio emission spectrum is generally not related to the optical counterpart, and arise due to non-thermal physical processes occurring in the ionized plasma of the upper solar atmosphere. Careful observations and identification of characteristic radio emissions help in understanding the solar eruptive phenomena that occur in the solar atmosphere. Figure (7.6) shows the schematics of emission mechanisms that dominate at different frequencies in solar atmosphere.

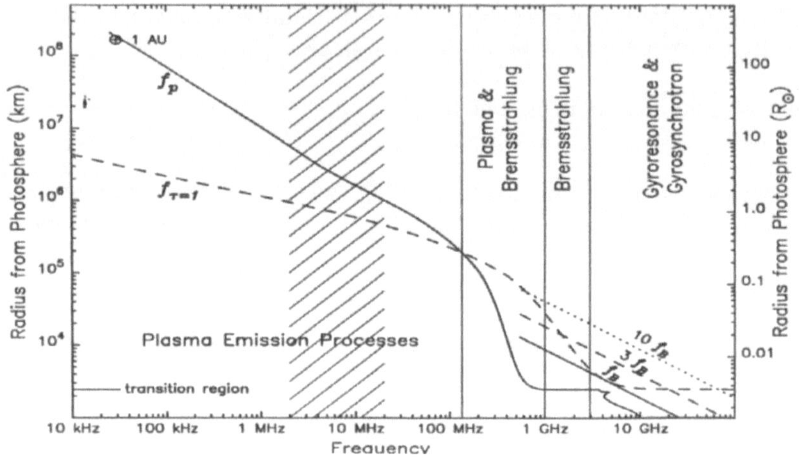

FIGURE 7.6: The solid curve specifies the type of emission mechanism that dominates at different frequencies in the solar corona and heliosphere based on the dependence of different emission mechanisms on the plasma parameters of temperature, density, and magnetic field strength. The regimes of dominant plasma emission, bremsstrahlung emission, and gyroresonance or gyrosynchrotron emission are marked as a function of frequency. The plot covers 7 orders of magnitude in frequency and in height. (Figure 15.1 page 639 from [18] by permission of M. Aschwanden.)

7.5.2 The non-thermal component of solar radio emission

The radio observations illustrate the fact that the chromosphere and corona are much hotter than the photosphere. The slope at longest wavelengths for the quiet Sun corresponds to very high temperatures (10^6 K) as compared to that for the main thermal black-body radiation. Sun's radio emission varies strongly with wavelength, because the radio emissions originate from various non-thermal plasma processes, and NOT just from the usual thermal (black-body) emission.

Radio emission at metre-, and decimetre-wavelengths from flares is quite different from emissions at centimetre and millimetre wavelengths. This has to do with the fact that at the shorter wavelengths the emission is caused by incoherent gyrosynchrotron radiation mechanisms, while at the longer wavelengths the emission comes from coherent plasma radiation. The 1-3 GHz frequency range appears to lie at the transition between the two types of emission mechanisms. The non-thermal processes are generally associated with the following processes:

(i) **Electron gyro-radiation, or cyclotron emission**: This is due to a non-relativistic electron moving under the influence of magnetic field. These electrons follow a helical orbit around the magnetic field lines, gyrating with frequency $f_{ge} = eB/(2\pi m_e)$ Hz, or $2.8 \times 10^{10} B$ Hz (B in Tesla).

(ii) **Gyrosynchrotron emission:** This is the emission when electrons are at mildly relativistic speeds (electron energies 100-300 keV). This form of emission is responsible for most radio emission from solar and stellar flares. There are several useful diagnostics that can be deduced from gyrosynchrotron emission, which is broadband emission, with optically thick emission at low frequencies (typically 2-5 GHz), optically thin emission at high frequencies ($>$ 10 GHz), and a peak around 5-10 GHz. The peak frequency is related to the magnetic field strength and number density. The optically thin spectral index is related to the electron power-law index, and the polarization provides the direction of the magnetic field. The sensitivity of emission to these parameters makes it a useful complement to other measures of the accelerated electrons, *i.e.*, hard X-rays.

(iii) **Synchrotron emission**: At highly relativistic speeds of gyrating electrons, the emission comes in narrow pulses at the cyclotron frequency, beamed along the direction of motion. This radiation is called the synchrotron radiation (Figure 8.6b). One requires electrons with energy of over 1 MeV for a brightness temperature of $\sim 10^{10}$K.

(iv) **Bremsstrahlung radiation (or braking radiation)**: It is produced due to accelerations caused by collisions between electrons and ions (Coulomb collisions) (Figure 8.6a, page 208). In head-on collisions, the electron can actually bounce off the proton, and lose most of its energy, or even reverse its direction. In this case, the energy of the photons emitted is typically in the X-ray range.

Radio emission, on the other hand, is due to relatively gentle, distant collisions of electrons with ions. The plasma consists of electrons and ions. An electron passing at high speed around the relatively immobile ions will experience a wiggling motion due to slight deviations as it passes each ion producing radiation in a broad spectrum (related to the energy lost or gained in each collision).

(v) **Plasma radiation**: Radio emissions occur in perturbed plasma layers of the solar atmosphere in plasma frequency, which is the frequency at which natural plasma oscillations take place. The associated electromagnetic emission these oscillations can produce is called the *plasma emission*. Figure 7.7 shows that increasing height in solar atmosphere corresponds to frequencies starting at the higher, and moving to the lower frequencies. The plasma frequency is related only to the number density of electrons in the plasma given as

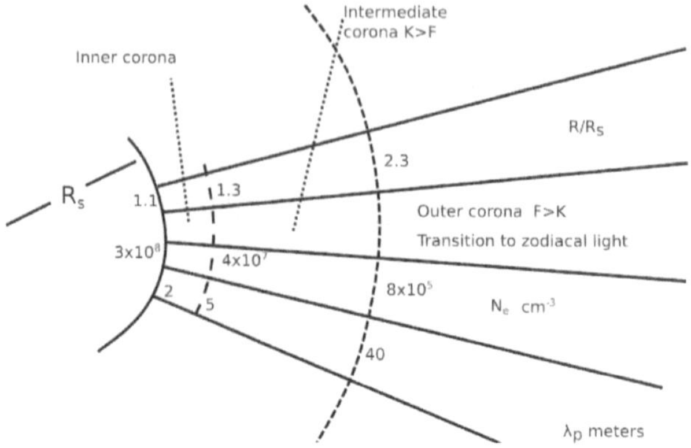

FIGURE 7.7: Plasma frequency corresponding to various solar atmospheric layers.

$$f_{pe} = \frac{1}{2\pi}\sqrt{\frac{n_e\, e^2}{\epsilon_0\, m_e}} \tag{7.5}$$

Using the values of the constants, this can be expressed as

$$f_{pe}(Hz) \approx 9 \times \sqrt{n_e(m^{-3})}. \tag{7.6}$$

Wavelength is often used to characterise the radio emissions, where λ (in metres) $= 300/f$ (in MHz). Thus, λ (1 mm) corresponds to a frequency of $f = 300GHz$.

Due to the large increase in electron density, radio emission becomes optically thick towards the lower chromosphere. Therefore, radio observations pertain only to the upper chromosphere and the coronal plasma. Let us start at the chromosphere and move outward in the solar atmosphere. In the solar chromosphere, the electron density drops from $2.5 \times 10^{17} m^{-3}$ to $10^{16} m^{-3}$ which corresponds to emission in plasma frequencies ranging from 4.5 GHz - 0.9 GHz (or wavelengths in the range 6.7 - 33.3 cm).

Going higher in the solar coronal atmosphere, *i.e.*, increasing radial distance, r, the temperature rises steeply to millions of degrees, while the electron density decreases according to the relation

$$n_e(r) = 1.55 \times 10^{14} r^{-6} + 2.99 \times 10^{14} r^{-16}, \tag{7.7}$$

where, density is in m^{-3}. At $2R_\odot$, the plasma frequency is ~14MHz (λ21 m). Plasma frequencies, with distance from the solar surface, with corresponding electron number densities are shown in Figure 7.7. Apart from this usual steady radio emission, a large contribution can come from transient non-thermal emission due to sudden, rapid changes in the coronal plasma.

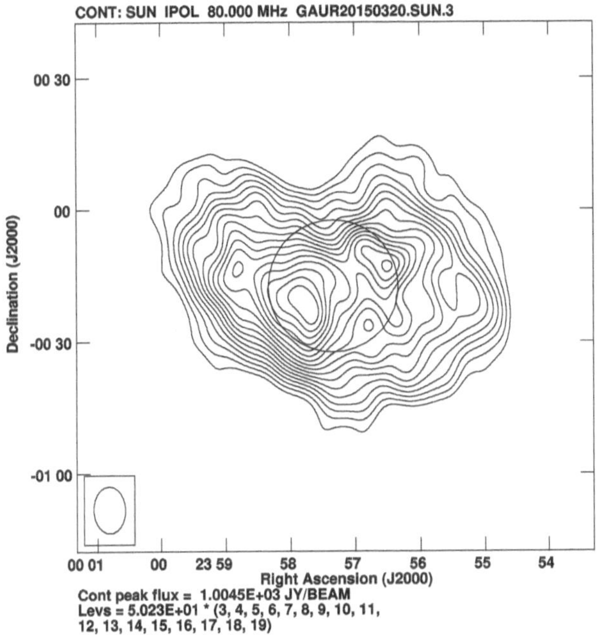

FIGURE 7.8: Contour map of the solar corona observed with the Gauribidanur radioheliograph on 20 March 2015 at 80 MHz. The circle in the center of the image represents the size of the Sun's photosphere. The ellipse at the lower left corner of the image indicates the size of the response pattern of the Gauribidanur radioheliograph. Credit: Indian Institute of Astrophysics, Bengaluru (India).

Figure 7.8 shows a radio heliograph map taken in 80 MHz showing the extended solar corona along with some sources of enhanced radio emissions associated with activity in the solar atmosphere. The intensity or flux density of the outermost contour in the image is about 1500 Jy.

Due to the decrease in the electron density with increasing height in the solar atmosphere, radio emission from the Sun in the above frequency range originates typically in the height range $\sim$0.1 to 0.8 $R_\odot$ above the Sun's photosphere. It is important to observe in this height range on a day-to-day basis, because the chain of solar activities leading to disturbances in the interplanetary medium and the terrestrial environment (commonly referred to as *space weather*) originate there.

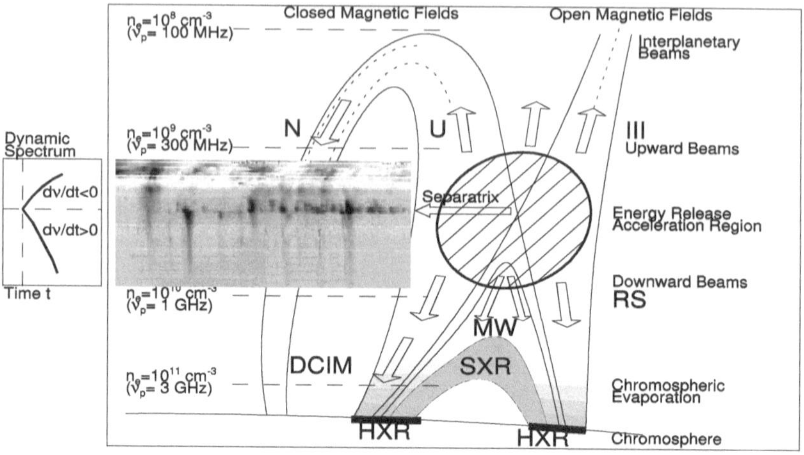

FIGURE 7.9: Schematics of active region emissions and radio burst types in the standard flare scenario. The acceleration region is located in the reconnection region above the soft X-ray-bright flare loop, accelerating electron beams in the upward direction. Electrons trapped on closed loops (the loops labeled MW) cause the gyrosynchrotron emission. (Figure 15.15 page 657 from [18] by permission of M. Aschwanden; refer for details to [19].)

7.5.3 Microwave emissions

The acceleration of electrons in flares from the reconnection region of magnetic fields into the emitting volume is called the '*injection*' process wherein the electrons can go directly to the chromosphere, generating hard X-rays foot-point sources, or they can go into a magnetic trap for a while, before ultimately escaping gradually. The coronal plasma is generally hot everywhere, and is even hotter in active regions, *i.e.*, the regions associated with sunspots. Here, the electrons trapped in closed magnetic loops of active regions emit in higher frequencies. Microwave emissions from flares are usually also associated with hard X-ray signature from the slowly escaping electrons. There are the trapped, huge population of electrons that have not yet escaped.

The shaded component in Figure (7.9) is the hard X-ray signature from the slowly escaping electrons. These electrons require only small accelerations in order to produce microwave emission much more readily than they do hard X-rays (which require large accelerations). Microwaves are thus far more sensitive to high-energy electrons than are the hard X-rays. The frequency of the microwaves is proportional to the magnetic field. Therefore, microwave images at many frequencies show a dispersion in position, essentially mapping out the coronal magnetic field.

7.5.4 Solar radio bursts

The development of radio-spectrographs allowed to continuously record the intensity of solar emissions as a function of frequency and time. At lower frequencies, metric burst types were discovered in the 1950s, through the use of spectrographs. Each type of radio burst has a unique signature in frequency and time. During strong solar disturbances, non-thermal radio emissions observed in metre wavelength can exceed the quiet levels by several orders of magnitude in bursts lasting seconds or minutes, and storms lasting over hours or days (Figure 7.10). A classification scheme has been developed by solar radio astronomers to categorise radio bursts into five main types.

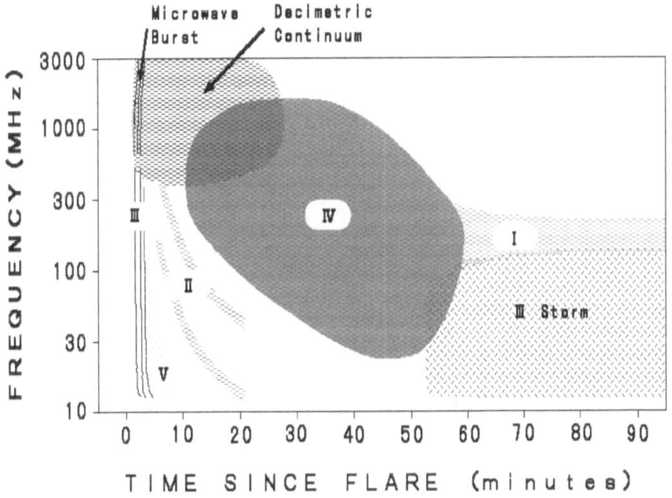

FIGURE 7.10: Classification scheme for solar transient bursts. Adapted from `http://sunbase.nict.go.jp/solar/denpa/hiras/types.html`.

Type I bursts are short duration, narrow band, non-drifting highly polarized bursts which have a short duration (a single burst of the order of one second). The cause of this emission appears to be plasma emission. A continuum emission embedded with many type I bursts lasting for days is known as Type I storm, which may be related to continuous reconnection above active regions.

Types II, III, IV and V are all generally associated with solar flare activity. Type II is the so-called slow drift, from high to low frequencies, generated during the expulsion of proton streams during a flare. They are caused by shock waves propagating outward into the corona to interplanetary space at 500-2000 km/s, and observed down to kilometric wavelengths.

Type III is fast drift bursts, from high to low frequencies, generated during the expulsion of electron streams moving at high speeds of $\sim c/3$ from low corona outwards during the flare. The description 'drift burst' arises from the fact that these particle streams excite progressively lower frequency radio waves as they travel out through the corona, propagating all the way to the Earth and beyond. Since the frequency of the radio waves is uniquely determined by the coronal electron density and magnetic field intensity at the point of generation, frequency versus time, recordings of these bursts can be used to deduce the velocity and apparent height of the exciters in the solar atmosphere.

The flare related events may be followed by Type IV broad-band continuum emission, due to either plasma emission or gyrosynchrotron emission, lasting for several hours. These are caused by a plasmoid, or loop filled with non-thermal particles. There may also be a moving type IV, caused by similar process, but entrained in a CME. Occasionally, the exciters which generate type III radio bursts also generate a continuum afterglow known as Type V emission in the outer corona, lasting over 1 to 2 minutes, with duration increasing as frequency decreases. This is generally confined to frequencies below 50 MHz. A schematic of active region emissions and radio burst types in the framework of the standard flare scenario is shown in Figure 7.9.

7.5.5 Radio diagnostics of coronal magnetic fields

One of the key measurements to analyse and understand solar activity (flares, CMEs, energetic particles, *etc.*) is of the magnetic field, in particular the coronal magnetic field. The radio domain is very promising to help in the otherwise difficult measurement of coronal magnetic fields. Magnetic field maps derived from radio observations consist of combining the radio imaging observations at a few discrete frequencies with X-rays, and EUV images and spectra. Both of these radiations, coming from two extreme ends of the electromagnetic spectrum, depend on temperature and density. But, it is only the radio emission that depends on the coronal magnetic field.

EUV and X-ray images at different wavelengths provide information on the temperature and emission measure maps in the active region which allows to compute the free–free brightness temperature as a function of frequency and to compare maps of free–free radiation to the observed radio images. The comparison of the radio maps computed with the extrapolated photospheric fields shows a good consistency with the observed brightness temperatures above the sunspots if gyroresonance emission is assumed [7].

The combination of radio, and X-ray or EUV observations, also allows to derive the three-dimensional coronal magnetic field above the sunspot and in its close surroundings ([62], [61]). Radio observations also provide ways to measure the longitudinal component of weak magnetic fields in both active and quiet regions of the Sun. The method consists in the measurement of the level of circular polarization of the free–free component, which will depend on the magnitude of the longitudinal magnetic field component. For further details one may refer to [132].

7.5.6 Some solar radio telescopes and radio heliographs

In order to obtain high-resolution radio images of the Sun, at scales comparable to those in optical wavelengths, it is obvious that a radio interferometer, and not just a single radio telescopes, is the main requirement. At present, there are several dedicated solar radio interferometers operating from mm to m wavelengths, such as Nançay Radio Heliograph (France), Nobeyama Radioheliograph (Nobeyama, Japan), *etc.* In addition, there are facilities, such as Giant Meter-wave Radio Telescope (GMRT), Near Pune, India, which are selectively used for solar observations.

The Sun is a highly variable and unpredictable radio source. Therefore, radio telescopes that are not dedicated to solar observations, such as VLA, NRH and GMRT, have only limited usefulness in campaigns. Dedicated long-term observations at high time resolution are required to obtain complete physics of solar dynamic phenomena, the best instrument being dedicated interferometers. In the solar radio spectrum, there are no lines, but structured features exist covering a wide frequency range. Therefore, imaging over a wide frequency range, particularly imaging spectroscopy, is a preferred approach.

In the following, we describe some radio telescopes, radio heliographs, and large radio arrays, which are used for solar radio imaging and spectroscopy.

Gauribidanur Radioheliograph: An example of radio heliograph, radio spectrograph, and a radio polarimeter is the Gauribidanur Observatory, located about 100 km north of Bangalore, India (Figure 7.11). It is operated by the Indian Institute of Astrophysics, Bengaluru, for daily observations of the Sun's corona in the frequency range 120 - 40 MHz. The heliograph, spread over an area of about 3 km, is being used to obtain two-dimensional images of the radio wave emitting structures in the corona (Figure 7.8). The relatively smaller spectrograph and the polarimeter are used to estimate the speed of the transient propagating disturbances from the Sun, and their associated magnetic field strength, respectively. These instruments, particularly the heliograph and the polarimeter, are unique in that currently there are no other similar dedicated solar radio observational facilities in the aforementioned frequency range, anywhere in the world (`https://www.iiap.res.in/?q=centers/radio`).

Culgoora Radioheliograph consisted of a series of aerials arranged in a circular array with a 3-km diameter at a site about 600 km north-west of Sydney. The array consisted of ninety-six, 13-m-diameter paraboloids and worked as a 3-km-diameter radio synthesis telescope for solar radio observations at a frequency of 80 MHz. The instrument was able to construct a two-dimensional image over a field of view corresponding to the entire solar corona (about 2° at 80 MHz). When it was decommissioned in 1986, the Australia Telescope Compact Array (ATCA) was built.

Radio Astronomical Telescope of the Academy of Sciences (RATAN-600) is a radio telescope located in the Caucasus Mountains, in Russia. Unlike most radio telescopes that utilise a dish, or an array of dishes, to focus radiation onto a receiver or receivers, RATAN-600 employs a ring of adjustable reflecting

FIGURE 7.11: A section of the Gauribidanur radioheliograph array. It is a T-shaped array with arms along the East-West and North-South directions. The primary receiving element is a log-periodic dipole antenna, and the array has 384 of them spread over a distance of about 3 km. Credit: Indian Institute of Astrophysics, Bengaluru (India).

panels, which can be angled to direct the radiation from any point in the sky to a central conical receiver. The ring is about 576 metres in diameter, and there is no solid structure. The overall effect of the arrangement is that of a partially steerable antenna, with the resolving power of a 600-m-diameter dish, making it the world's largest diameter individual radio telescope. It can make radio frequency observations in the frequency band ranging from 610 MHz to 30 GHz. The focus of the RATAN-600's scientific program is solar coronal radio observation at many wavelengths, and determination of coronal magnetic fields [48].

The e-Callisto Network of radio spectrometers. An international network of solar radio spectrometers, e-Callisto (Compound Astronomical Low cost Low frequency Instrument for Spectroscopy and Transportable Observatory) operates more than 159 instruments at over 98 locations worldwide. The Callisto spectrometer is a programmable heterodyne receiver designed in 2006 under the International Heliophysical Year (IHY) 2007, and the International Space Weather Initiative (ISWI) of the United Nations Committee on the Peaceful Uses of Outer Space launched in 2009. The main applications of Callisto is the observation of solar radio bursts. The instrument operates between 45 and 870 MHz using a broadband cable-TV tuner having a frequency resolution of 62.5 KHz. The network is able to continuously observe the solar radio spectrum for 24h per day throughout the year (`http://www.e-callisto.org/`).

Karl G. Jansky Very Large Array (The VLA) - The Very Large Array of the National Radio Astronomy Observatory near Socorro, New Mexico, USA, is an aperture synthesis instrument, and one of the most successful radio telescopes. It consists of 27 paraboloid antennas, each with a diameter of 25 m, which are located on a Y-shaped track. These antennas can be moved to a number of configurations along the three railway tracks, each of length about 21 km, thereby resulting in an antenna with an effective diameter of 35 km. If the VLA is used in its largest configuration and at its highest frequency, 23 GHz (1.3 cm), the resolution achieved is 0.1 arc-second, superior to many optical telescopes. In 2012, the sensitivity was increased by a factor of about ten. The VLA covers a wavelength range from 7 mm (45 GHz) to 4 m (74 MHz)—a factor of about 600.

VLA has been used to study the structure and evolution of radio emission from coronal loops and solar bursts with high spatial and temporal resolution. High-resolution observations at 20 and 91 cm have made it possible to locate and resolve sites of impulsive energy release at two different heights in the solar corona, thereby providing insights to acceleration mechanisms and plasma parameters of the energetic particles.

VLA data have also been compared with X-ray observations from the Yohkoh Soft X-ray Telescope, Extreme Ultraviolet observations from the SOHO, TRACE, and RHESSI satellites. A comparison of EUV and soft X-ray images taken with the SOHO Extreme Ultraviolet Imaging Telescope (EIT) and the Yohkoh SXT has been carried out with a VLA snapshot map at 91 cm wavelength.

MERLIN and VLBI - The highest resolution telescopes, such as British MERLIN telescope, in which the individual components are so widely separated in distance that they cannot be linked by cables, and hence, are coupled together by radio links, where signals are recorded separately at individual sites, and then combined later. Even higher resolution can be obtained with an extension of the aperture synthesis technique, called VLBI (Very Long Baseline Interferometry). With the VLBI technique the spacing between the antennas is restricted only by the size of the Earth. VLBI uses existing antennas (often on different continents), which are all pointed towards the same source. With VLBI techniques, it is possible to achieve resolutions of 0.0001 arc-second!

7.5.7 Solar transient in the interplanetary medium

Radio observations have been a crucial tool to understand and forecast some of the space weather effects. The radio flux measured at 10.7 cm (f10.7) is still being used as a proxy for the UV ionizing flux in the Earth's atmosphere, and is a key parameter for aeronomy and orbitography. Type II and type IV bursts at metric/decimetric wavelengths together with measurements of large and long duration soft X-ray flares are used as predictors of large energetic particle events.

Interplanetary scintillation (IPS) measurements of cosmic radio sources act as a useful tool for obtaining properties of solar wind plasma and the transient streams associated with CMEs. Imaging radio observations of coronal waves propagating on the solar disk or of bright extended radio sources observed at metric wavelengths can be used to detect the signatures of halo CMEs radially propagating towards the Earth. Radio telescopes, such as the Ooty Radio Telescope (ORT), and at the Solar-Terrestrial Environment Laboratory (STEL), Nagoya University, Japan, have exploited this technique to map the enhanced radio fluctuations in interplanetary medium by measuring the radio scintillations of compact radio objects such as pulsars and quasars.

7.6 Solar coronal oscillations – coronal seismology

In the past, there has been a lack of high-quality observations of coronal oscillations. However, more recent years have seen a marked change with space-based observations, particularly by the SOHO and the Transition Region and Coronal Explorer(TRACE) satellites. Ground-based radio telescopes too have observed various types of oscillations in radio wavelengths with periods of a few seconds to several tens of seconds.

Coronal plasma waves and oscillations have now been discovered in a wide variety of solar structures, such as coronal loops, polar plumes and prominences. These modes manifest themselves as standing waves (MHD oscillations) and propagating waves. The discovered phenomena can be classified in several distinct classes, such as EIT waves [358], compressible waves in polar plumes [289], and in coronal loops, flare-generated global kink oscillations of loops [282], and longitudinal standing oscillations within loops [383]. For more details on these waves and oscillations, and related topics, one can refer to the review [281], and articles in the book [214].

The theory of coronal MHD oscillations and waves, and the idea of exploiting observed coronal oscillations as a diagnostic tool for determining the physical conditions of the coronal plasma was first suggested in the 1980s [321]. But its applications to solar corona started only after high-resolution EUV imaging observations from space became available. These new discoveries have established a new discipline in solar physics that is now known as *coronal seismology*, and provides a new tool for obtaining basic physical properties of the solar corona.

The hot, inhomogeneous, magnetized coronal plasma supports a variety of magnetohydrodynamics (MHD) wave modes, which are associated with perturbations of macroscopic parameters of coronal plasma. These physical parameters are coronal plasma's density, temperature, bulk velocity, and the local direction/absolute value of the magnetic field. These perturbations in the coronal plasmas can be detected as fluctuations of intensity of coronal emission lines, their Doppler shifts, and non-thermal broadening. Based on basic plasma physics, the properties of observed MHD oscillations and coronal waves are well studied and identified. These include fast kink modes, fast sausage modes, slow (acoustic) modes, and torsional modes. A brief introduction of MHD formalism of solar plasmas is given in Section 7.7.

A major benefit of coronal seismology is the probing of physical parameters, such as Alfvén velocities and magnetic field strengths, which are very difficult to measure in the solar corona by other means. Coronal oscillations are increasingly employed for the diagnostics of the coronal magnetic field, and their physical damping mechanisms. Initial research in coronal seismology concentrated on measurements of oscillation periods, spatial displacements, damping times, and temperature and density diagnostics of oscillating structures. The excitation mechanisms of coronal MHD oscillations and waves, and excitation in magnetic reconnection outflow regions are some other important problems.

Coronal MHD oscillations and waves play important role in coronal heating. Attempts have been made on the detectability of various MHD wave types, and estimating the energies carried in the observed MHD waves and oscillations. It is reported that Alfvén waves could provide sufficient energy to sustain coronal heating, while acoustic MHD waves fall far short of the required coronal heating rates. Future exploration is likely to reveal additional aspects on the fundamental physics of wave excitation, wave propagation, waveguides, wave damping, phase mixing, and resonant absorption.

7.7 Solar MHD formulation: a brief introduction

MHD equations are derived assuming space and time scales to be larger than inherent scale-lengths, such as the Debye length, or the gyro-radii of the charged particles. Debye-length (λ_D) is essentially the screening or shielding length scale of a charge, where its electric potential decreases in magnitude by $1/e$. For example, in a plasma, with density $n_e = 10^6 m^{-3}$, $T_e = 10^5$K, λ_D turns out to be $\sim$ 10m. The gyro-radius $r_g = mv_\perp/|q|B$ (also known as Larmor radius, or cyclotron radius) is the radius of the circular motion of a particle, with mass m, charge q, in the presence of a uniform magnetic field B, and $v_\perp$ is the component of its velocity perpendicular to the direction of magnetic field.

The MHD equations are used to study the magnetic properties and behaviour of electrically conducting fluids, such as the stability, instabilities, waves and oscillations. The main physical quantities characterising an electrically conducting fluid, having a charge density ρ^*, are the bulk plasma velocity field $\vec{v}$, the current density $\vec{j}$, the mass density density ρ, and the plasma pressure p. The flowing electric charge in the plasma is the source of the electric and magnetic fields, $\vec{E}$ and $\vec{B}$. All quantities generally vary with time, t.

The magnetized plasma is governed by two sets of equations: (i) Maxwell's equations, and (ii) the hydrodynamic (or, fluid) equations, consisting of basic conservation laws of mass, momentum and energy conservation. In what follows, we discuss these equations, and some of the essential concepts used in solar plasmas. For more details, one may refer to the excellent book by Francis Chen [78].

7.7.1 Maxwell's equations

The set of four electromagnetic equations, which were modified by Maxwell, who included a displacement current in the Ampere's law, are given as under (in MKS, or SI units):

1. Ampere's law: Steady currents or time varying electric fields may produce magnetic fields.

$$\nabla \times \vec{B} = \mu \vec{j} + \frac{1}{c^2}\frac{\partial \vec{E}}{\partial t}. \tag{7.8}$$

2. Gauss' law of magnetism (no monopoles).

$$\nabla \cdot \vec{B} = 0. \tag{7.9}$$

3. Faraday's law: Time varying magnetic fields produce electric fields.

$$\nabla \times \vec{E} = -\frac{\partial \vec{B}}{\partial t}. \tag{7.10}$$

4. Gauss' law: Electric fields can be produced by positive and negative charges.

$$\nabla \cdot \vec{E} = \frac{1}{\epsilon}\rho^*. \tag{7.11}$$

where ϵ is the permittivity of free space, μ is the magnetic permeability in vacuum ($4\pi \times 10^{-7} Hm^{-1}$), $\rho^* = e\,(n^+ - n^-)$ is the charge density, n^+, n^- are the number of ions and electrons, respectively.

7.7.2 MHD approximations

The first and main assumption in MHD approximations is non-relativistic fluid motions, where the plasma velocity is much smaller than the speed of light, *i.e.*, $v << c$. This results in making the *displacement current term* (the last term) in the Ampere's law (Equation 7.8) negligible, that can be dropped. Accordingly, Ampere's law simplifies to

$$\nabla \times \vec{B} = \mu \vec{j}. \tag{7.12}$$

In the second MHD approximation, the plasma in the solar atmosphere is considered to be mostly neutral, *i.e.*, $(n^+ - n^-) << n$, where n is the total number density. From Faraday's law (Equation 7.10), one has $E = LB/T = VB$, in order of magnitudes. Therefore, from Gauss' law (Equation 7.11), we get $E/L = (e/\epsilon)(n^+ - n^-)$, which implies that we can neglect the charge density in Equation (7.11), provided that $n >> (\epsilon\ V\ B)/(Le)$. This is the condition for *charge neutrality*.

7.7.3 The hydrodynamic equations

The set of hydrodynamic equations are:

1. *The equation of mass conservation* states that matter is neither created nor destroyed. Density at a given point increases only if mass flows into the surrounding region.

$$\frac{\partial \rho}{\partial t} + \nabla \cdot (\rho \vec{v}) = 0. \tag{7.13}$$

2. *The equation of momentum conservation*: A moving hot plasma under the influence of gravity of the Sun and coronal magnetic field will experience a Lorentz force $\vec{j} \times \vec{B}$, apart from that of the thermal pressure p, and gravity $\vec{g}$, therefore,

$$\rho \frac{d\vec{v}}{dt} \equiv \rho \left(\frac{\partial}{\partial t} + \vec{v} \cdot \nabla \right) \vec{v} = -\nabla p + \vec{j} \times \vec{B} + \rho \vec{g}. \tag{7.14}$$

   In a case where currents are flowing along $\vec{B}$, i.e., $\vec{j} || \vec{B}$, so the Lorentz force term vanishes, and the ratio of pressure and gravity terms provide

$$\frac{p_0/L_0}{\rho_0\ g} >> 1 \Rightarrow L_0 << \frac{p_0}{\rho_0\ g} = H, \tag{7.15}$$

   where H = scale height.

3. *The energy equation* is given by

$$\frac{\mathrm{d}}{\mathrm{d}t} \left(\frac{p}{\rho^\gamma} \right) = 0, \tag{7.16}$$

   where $\gamma = 5/3$ is the ratio of specific heats for an adiabatic equation of state. This energy equation is, of course, only applicable in the absence of shocks or heat conduction as it assumes that the entropy of a fluid element does not change.

In Equation (7.14), the Lorentz force can be interpreted by using Ampere's law for electric current (Equation 7.8), as follows,

$$\vec{j} \times \vec{B} = \frac{\nabla \times \vec{B}}{\mu} \times \vec{B}. \tag{7.17}$$

Using the vector identity, $\frac{1}{2}\nabla(\vec{B} \cdot \vec{B}) = (\vec{B} \cdot \nabla)\vec{B} + \vec{B} \times (\nabla \times \vec{B})$, one gets

$$\vec{j} \times \vec{B} = \frac{\left(\vec{B} \cdot \nabla\right) \vec{B}}{\mu} - \nabla \left(\frac{B^2}{2\mu}\right), \tag{7.18}$$

where, the 1st term on the R.H.S. is the magnetic tension force parallel to $\vec{B}$, which is important for curved fields, and is directed towards the center of curvature of B. Larger restoring force would act if the curvature is larger. The 2nd term is the magnetic pressure force (important when $|B|$ changes along the field). It is the gradient of a scaler. Magnetic pressure and tension represent two types of restoring forces associated with distinct waves modes: the *Alfvén waves* and the *magneto-acoustic waves*.

The ratio of pressure to the magnetic (Lorentz) terms in the equation of motion is defined as *plasma* β

$$\beta = \frac{p}{B^2/2\mu} \tag{7.19}$$

such that when $\beta << 1$, Lorentz force dominates, as in the solar corona. In such a case, the L.H.S. term in the equation of motion is approximately equal to the Lorentz force term, which gives $v \approx v_A = B/\sqrt{\mu\rho}$, *i.e.*, any perturbations or waves would propagate at a characteristic velocity, called the *Alfvén velocity*.

In MKS units, one may use the relation $\beta = 3.5 \times 10^{-21} N\ T/B^2$, where B is in Tesla ($10^4$G), N is number density in m^{-3}. Also, the Alfvén velocity $v_A = 2 \times 10^9\ B/N^{1/2}$. Typical values of various parameters for the Sun are listed in Table 7.1

TABLE 7.1: Typical parameters for solar atmosphere

Parameter	Photosphere	Chromosphere	Corona
N (m^{-3})	10^{23}	10^{20}	10^{15}
T (K)	6000	10^4	10^6
B (G)	$10 - 10^3$	100	10
β	$10^6 - 1$	10^{-1}	10^{-3}
$v_A (km/s)$	0.05-10	10	10^3

Equation of magneto-hydrostatic equilibrium: If $v << v_A$, then *L.H.S.* << *R.H.S.*, and one obtains

$$0 = -\nabla p + \vec{j} \times \vec{B} + \rho\vec{g}, \tag{7.20}$$

which describes magneto-hydrostatic equilibrium.

- Case (i) - when $L << H$, then $\rho\vec{g} << \nabla p$ term, which gives $0 = \nabla p + \vec{j} \times \vec{B}$ - *Equation of magnetostatic equilibrium.*
- Case (ii) - when $L << 2H/\beta$, then the Lorentz force term is much larger than the pressure and gravity terms, and one gets $0 = \vec{j} \times \vec{B}$ - *force-free field*. This is a good approximation for solar corona, and is satisfied by the following solutions:
    1. $\vec{B} = 0$ (a trivial solution),
    2. $\vec{j} \equiv \nabla \times \vec{B} = 0$, *i.e.*, $\vec{B} = \nabla\varphi$= potential field, where φ is the scalar magnetic potential, which gives $\nabla^2\varphi = 0$ using solenoid condition $\nabla \cdot \vec{B} = 0$. Potential fields thus satisfy Laplace's equation, so that the general solution can be given in terms of associated Legendre polynomials as $\varphi = \sum_{i=0}^{\infty} \sum_{m=-l}^{l} (a_{lm}r^l + b_{lm}r^{-(l+1)})P_l^m(\cos\theta)e^{im\phi}$.
    3. $\nabla \times \vec{B} = \alpha\vec{B}$, where α = constant, or a function of r. This corresponds to a stressed magnetic field configuration with non-zero current $\vec{j}$ parallel to $\vec{B}$ everywhere.
    Operating throughout by $\nabla \cdot$ on $\nabla \times \vec{B} = \alpha\vec{B}$, we get $\nabla \cdot (\nabla \times \vec{B}) = \nabla \cdot (\alpha\vec{B}) \Rightarrow 0 = \alpha\nabla \cdot \vec{B} + \vec{B} \cdot \nabla\alpha$.
    Using Ampere's law, $\nabla \cdot \vec{B} = 0$, we get $\vec{B} \cdot \nabla\alpha = 0$. This implies that α is a constant along each field line, although it may vary from one field line to another. If α=0, then the magnetic field reduces to the potential case.
    This approximation is generally used for extrapolation of solar magnetic fields at higher layers from the observed photospheric fields used as boundary condition.

7.7.4 The generalised Ohm's law

Plasma moving at a non-relativistic speed $\vec{v}$ in the presence of a magnetic field $\vec{B}$ is subject to an electric field $(\vec{v} \times \vec{B})$ in addition to the electric field E at rest. Then, the generalised Ohm's law states that current density is proportional to the total electric field,

$$\vec{j} = \sigma\,(E + \vec{v} \times \vec{B}) \tag{7.21}$$

where, σ is the electrical conductivity in $mho\ m^{-1}$, and $\vec{v}$ is the plasma velocity. *Thus, Ohm's law provides a link between the electromagnetic fields, and currents through Maxwell's equations and the plasma fluid equations through the fluid velocity,* $\vec{v}$.

In solar MHD, the primary variables are $\vec{B}$ and $\vec{v}$. Therefore, we can eliminate the secondary variables, electric current density $\vec{j}$ and electric field $\vec{E}$ from the set of Maxwell's, and the hydrodynamic equations. This is achieved by making use of Ohm's law (Equation 7.21):

$$\frac{1}{\mu}\nabla \times \vec{B} = \sigma(\vec{E} + \vec{v} \times \vec{B}) \Rightarrow \vec{E} = -\vec{v} \times \vec{B} + \frac{1}{\mu\sigma}\nabla \times \vec{B} \tag{7.22}$$

where we have substituted for $\vec{j}$ from Maxwell's equation for Ampere's law (Equation 7.8). Next, we can substitute for $\vec{E}$ in Maxwell's equation for Faraday's law (Equation 7.10), to obtain

$$\frac{\partial \vec{B}}{\partial t} = -\nabla \times \vec{E} = -\nabla \times (-\vec{v} \times \vec{B} + \eta\nabla \times \vec{B}) \tag{7.23}$$

Here, $\eta = 1/(\mu\sigma)$ is magnetic diffusivity, which depends on T and ρ. Assuming it to be a constant, we get

$$\frac{\partial \vec{B}}{\partial t} = \nabla \times (\vec{v} \times \vec{B}) - \eta\nabla \times \nabla \times \vec{B} \tag{7.24}$$

Now, using the vector identity for $\nabla \times \nabla \times \vec{B} = \nabla^2\vec{B} - \nabla(\nabla \cdot \vec{B})$, and from $\nabla \cdot \vec{B} = 0$, we get the *Magnetic Induction Equation*:

$$\frac{\partial \vec{B}}{\partial t} = \nabla \times (\vec{v} \times \vec{B}) + \eta\nabla^2\vec{B} \tag{7.25}$$

This equation implies that magnetic field in a plasma can change due to *advection* (R.H.S. - 1st term), and *diffusion* (R.H.S. - 2nd term). Physics of MHD fluids are derived from the induction equation combined with the equation of motion.

The ratio of advection and diffusion terms is defined as the *Reynold's Number*, a dimensionless quantity:

$$R_m = \frac{advection}{diffusion} = \frac{L^2}{\eta\tau} = \frac{LV}{\eta}, \tag{7.26}$$

where we have used the dimensions for the L.H.S. ($\approx B/\tau$), and R.H.S. ($\approx B/\tau$ and $\eta B/L^2$) of the induction equation. L, V are the typical length scale, and plasma velocity. We can infer the dimension of $\eta \approx L^2\tau^{-1}$.

It is evident that R_m is an important quantity, which indicates the importance of advection, or diffusion terms, for a given plasma system. We consider the following two cases:

- Case (i) - When $R_m << 1$, then $\eta >> 1$, and the diffusion term dominates over advection, therefore, magnetic field diffuses throughout the plasma. In this case, the induction equation is approximated by

$$\frac{\partial \vec{B}}{\partial t} = \eta \nabla^2 \vec{B}. \tag{7.27}$$

This implies that the magnetic field diffuses away on a time-scale of $\tau_D = L^2/\eta$.

- Case (ii) - When $R_m >> 1$, then $\eta << 1$ as usually is the case for the Sun, since L and V are typically large. In this case, the diffusion term is negligible as compared to advection, and the induction equation can be approximated by

$$\frac{\partial \vec{B}}{\partial t} = \nabla \times (\vec{v} \times \vec{B}) \tag{7.28}$$

This is Alfvén's *frozen-flux theorem: In a perfectly conducting fluid, where* $R_m \to \infty$, *magnetic field lines move with the fluid, i.e., the field-lines are frozen in the plasma.* In such a situation, the magnetic field configuration is determined by the flow field. This is particularly important in solar active regions, where strong flows are evident.

The advective term is usually much larger than the diffusion term in most of the Universe. Therefore, most plasmas have high Reynold's number, implying that magnetic field is frozen in plasma, except where large gradients in B are present.

Let us consider some typical examples in solar plasmas:
(i) *For sunspots* - taking $L = 10^4 km (10^7 m)$, characteristic velocity $V \sim 10^3 m/s$, and $\eta \sim 10^3 m^2 s^{-1}$, gives $R_m = 10^7 >> 1$, and $\tau_D \sim 10^{11} s$, or ~ 3000 years.
(ii) *For solar corona* - taking $L = 10^5 km$, $V = 10^3 m/s$, and $\eta = 1 m^2 s^{-1}$, implies $R_m \sim 10^8 >> 1$.
(iii) *For solar flares* - the observed rapid energy release rates require very short diffusion time-scales $\tau \sim 100s$. This can be achieved only if very small length-scales $L \sim 10\ m$, or large η are involved. On the other hand, in solar coronal atmosphere, the electrical conductivity is very large, *i.e.*, resistivity is very small.

This makes it possible to store energy by moving the fluid, or the source of the magnetic field, while not allowing the reconnection of magnetic field-lines. The stored magnetic energy can then become available for release, when triggered by some instabilities, only if the conditions for ideal MHD somehow break down (*i.e.*, resistivity is no longer very small - *anomalous resistivity!*) allowing magnetic reconnection to take place. It is expected that large η in neutral sheets may occur, where classical concepts of electrical resistivity may not hold.

7.7.5 MHD instabilities

Stability of plasmas is an important field, where one analyses the equilibrium of the plasma, i.e., whether a small perturbation acting on it will grow, oscillate, or be damped out. There are various types of instabilities, such as ideal MHD instabilities (linear and nonlinear), and resistive MHD instabilities (addition of dissipation in ideal MHD).

Some of the most common observed instabilities in experimental plasmas are listed in Table 7.2. Recent solar observations have been able to identify many of the instabilities observed in laboratory plasmas also in solar coronal plasmas, which has been the mainstay of coronal seismology (Section 7.6, page 188).

TABLE 7.2: Common MHD instabilities.

Ideal MHD	$m = 0$ sausage	$m = 1$ kink	$m \geq 2$ kink
Resistive MHD		$m = 1$ tearing mode	$m \geq 2$ tearing mode

7.7.6 Magnetic field lines and flux tubes

Magnetic field lines are curves with tangent in the direction of the field $\vec{B}$, defined in three-dimensional space by $dx/B_x = dy/B_y = dz/B_z$, or in two dimensions as $dx/B_x = dy/B_y \Rightarrow dy/dx = B_y/B_x$.

Magnetic flux tubes are surfaces generated by a set of field lines intersecting a simple closed curve. Magnetic flux crossing a section of area A is defined as $F = \int \vec{B} \cdot d\vec{A}$. Since $\nabla \cdot \vec{B} = 0$, the flux F is constant along a flux tube (Figure 7.12a).

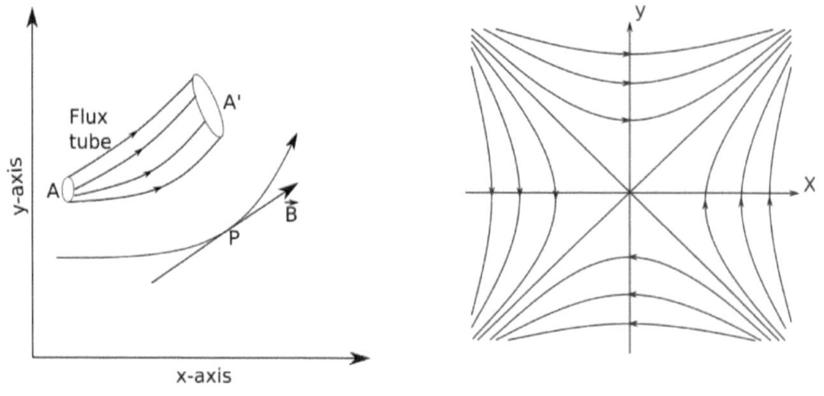

(a) (b)

FIGURE 7.12: (a) Magnetic field-line, flux tube, and (b) a null-field configuration.

Let us consider magnetic field geometry for a simple 2-D field having its x-, and y-components: $B_x = y,\ B_y = x$. Then, the magnitude of the field is $|B| = \sqrt{B_x^2 + B_y^2} = \sqrt{x^2 + y^2}$. The magnetic field lines must satisfy $dy/dx = B_y/B_x \Rightarrow dy/dx = x/y \Rightarrow ydy = xdx$.

On integrating, one gets the equation for the field lines as $x^2 - y^2 =$ *constant* (Figure 7.12b). The direction of the field can be obtained using the slopes. At the origin, x=0, y=0, one has $B_x = 0,\ B_y = 0$, and $|B| = 0$. Such a point is called a *neutral* or *null* point.

7.8 Summary

Recent observations in the IR spectrum have been carried by several groups, extending the existing capabilities. For example, spectro-polarimetry of IR spectral lines, having high magnetic sensitivity, has demonstrated that a variety of magnetic field strengths exist in addition to kilo-gauss flux tubes. Infrared spectroscopy has enhanced our understanding of the temperature structure and the dynamics of the solar atmosphere from the upper photosphere to the lower chromosphere. IR spectro-polarimetry of coronal spectral lines provide a new window into the critical magnetic processes which drive flares.

Following the early findings, radio observations have provided new insights in the understanding of solar activity and of its link with the interplanetary medium. Many results have emerged from coordinated ground-based radio observations with multi-wavelength observations from space. Radio emissions can remotely detect structures or sites of particle acceleration where magnetic energy release occurs that would not be visible otherwise.

The location of the burst radio sources, measured at several frequencies, provides information on the large-scale coronal magnetic field patterns. Many observations confirmed the early findings that the gyroresonance emission is usually the dominant mechanism above sunspots. Maps of the magnetic field were obtained for several active regions together with plasma temperature and emission measure maps when multifrequency radio observations or additional X/UV observations were available.

Measurements of the longitudinal component of weak magnetic fields in both active and quiet regions of the Sun were also obtained from radio observations. Major advances on the measurement of coronal magnetic fields are expected in the future from broadband imaging spectroscopy in the microwave domain.

Spatially resolved observations in the metric/decimetric domain, in combination with X-ray and gamma-ray observations provide useful information on the time scales of electron acceleration, the location of the electron acceleration sites and the link between the characteristics of the energetic particles (*e.g.*, spectra, electron to proton ratio) and the production sites of energetic electrons. It is confirmed from the observations of radio-spikes and of radio and X-ray bursts that the flare acceleration sites are located near the top of magnetic loops, possibly in the cusp region (X-type reconnection point). Radio observations, in flares and in CMEs, are consistent with the general concept of magnetic reconnection, and support reconnection models for magnetic energy release.

Chapter 8

The *Invisible* Sun in Short Wavelengths

8.1 Introduction

Although the Sun produces high energy γ- rays in the $p - p$ chain of nuclear-fusion processes operating in the central core of the Sun, these super high-energy photons are converted to photons of lower energy as they make the journey from the core to the Sun's photosphere. The black-body spectrum of the solar electromagnetic radiation contains most of the flux in the visible and IR in the long wavelengths (Figure 1.3, page 11). Towards the shorter wavelengths of UV, EUV, and X-rays, there is a very abrupt decrease in the flux. However, the high temperature outer layers of solar atmosphere, *viz.*, the transition region (TR) and corona emit UV, EUV and X-ray radiation (Section 3.5.2, page 69). Additionally, transient enhancement of these high-energy radiations, and sometimes, even γ-rays, occurs during the energetic explosive events in these outer layers of the solar atmosphere.

The high-energy processes in the solar atmosphere include the rapid release of energy stored in unstable magnetic configurations, the conversion of the stored energy into the kinetic energy of hot plasma, and accelerated particles (primarily electrons, protons and ions). The acceleration of electrons is revealed by hard X-ray and γ-ray bremsstrahlung, while the acceleration of protons and ions is revealed by γ-ray lines and continuum.

Ultraviolet, visible and infrared radiations are produced by changes in the energy states of the electrons in atoms, whereas, γ- and hard X-rays are produced in much higher energy processes that involve the transitions in atomic nuclei or in mutual interactions of elementary particles. Gamma ray astronomy involves studies of radiation having energies above 100 keV ($\lambda <$ 0.01 nm). These observations in the highest frequencies give information on very energetic processes.

Due to the proximity of the Sun, the high-energy emissions are observed as several orders of magnitude more intense than from any other cosmic source. The high-energy particles and radiation are transported through the solar atmosphere, and into interplanetary space. The energetic charged particles encounter Earth's magnetosphere, which they cannot penetrate through

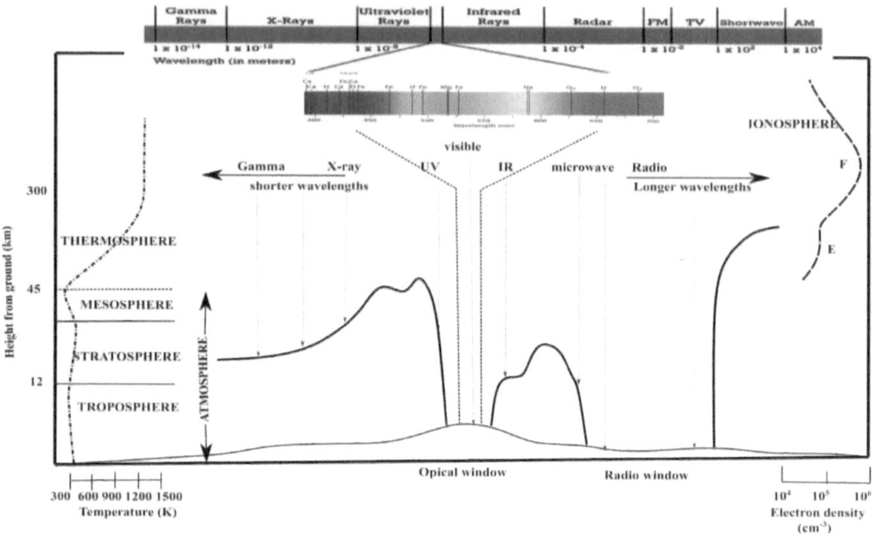

FIGURE 8.1: Passage of electromagnetic radiation through Earth's atmosphere. There are 'windows' in optical and radio wavelengths, while the shorter and longer wavelength radiations get absorbed at varying altitudes. Various layers of atmosphere are marked at the left, and the ionospheric E- and F-regions are marked at the right-hand side.

directly due to the Lorentz force. On the other hand, the high-energy UV, EUV, X-, and γ-ray photons are absorbed due to different processes at various altitudes during their passage through the Earth's atmosphere (Figure 8.1).

At wavelengths under 300 nm, absorption by atmospheric ozone prevents the radiation from reaching the ground. The atmospheric ozone is concentrated in a thin layer at a height of about 20–30 km. Thus, the ozone layer protects life on the Earth from harmful ultraviolet radiation. At still shorter wavelengths, the main absorbers are O_2, N_2, and the free atoms which get ionized in the process. Nearly all of the radiation under 300 nm is absorbed by the upper parts of the atmosphere leading to the creation of the ionosphere. As a result of the absorption of high-energy photons high in the atmosphere, their observation can only be made by using instruments carried up by rockets, balloons, or satellites. Owing to the Earth's atmosphere, astronomical observation of high-energy photons is essentially a space-age science [125].

8.2 The early era of observing from space

The photographic detection of Sun's UV radiation goes back to 1840 by A. Becquerel, and later during 1920-1930, using balloon observations from around 20-km height by K. Kiepenheuer. Some of the work in this period was based on UV absorption by atmospheric ozone and ionospheric studies. However, the first solar UV spectra were obtained as recently as 1946 by Naval Research Laboratory workers, who used the captured German V2 rocket flight equipped with instrumentation. These observations showed that the observed UV emission flux from the Sun was significantly lower below around 300 nm than what is expected from the black-body radiation. It was inferred that most of this variation at lower UV wavelengths corresponded with solar plage regions. In this region, several of the strong UV spectral lines occur in the narrow transition region of the Sun (interface between the chromosphere and corona), corresponding to various highly ionized elements, at high temperatures of $10^4 - 10^6$ K. Since the continuum emission from the photosphere and chromosphere is much weaker in these wavelengths, the spectral lines in UV provide a very useful diagnostic tool to infer the properties, such as plasma density and temperature, of the transition region.

X-ray astronomy began soon after World War II with observations of the Sun using the V-2 rockets. X-ray telescopes were first used in astronomy to observe the Sun, because of the high brightness of the Sun in X-rays. X-ray emission in the wavelengths below 10 nm from Sun's corona was first detected in 1949, using rocket-borne X-ray payloads [126]. Balloon-borne X-ray instruments were also used which led to the first detection of hard X-rays and γ-rays from solar flares in 1958 [303]. The early X-ray telescopes used small focussing elements. X-rays and photographic films were used as the detector/imaging medium at first, but later on CCD detectors replaced them.

Study of astronomical objects at the highest energies of X-rays and γ-rays began in the early 1960s. The first orbiting astronomical observatory, SOLar RADiation (SOLRAD 1) was launched in 1960, and studied radio fade-outs caused by solar X-ray emissions. Later, a series of 8 Orbiting Solar Observatories (OSO) were launched, starting with OSO 1 in 1963, with the primary mission to measure solar UV, X-ray, and γ-ray emissions. But it was only during the 1970s that it acquired momentum with the satellite-based observations.

Solar imaging in X-rays received a big boost with SkyLab. It was the first space station of NASA, launched on May 14, 1973. It spent six years orbiting the Earth until its decaying orbit caused it to re-enter the atmosphere. It made time-resolved UV and X-ray observations of the solar transition region, and the solar corona. UV and X-ray solar photography was also carried out. X-ray spectrography of solar flares, and active regions were other important objectives. The major discoveries from SkyLab were the first observations of coronal holes, and coronal transients, now known as the coronal mass ejections (CMEs).

NASA-Solar Maximum Mission (SMM), launched in 1980, was the next major satellite dedicated to observe γ-rays, X-rays and UV radiation from solar flares and solar corona during the period of high solar activity of Cycle 21. SMM faced technical problems only a few months after its launch, but became operational again in 1984, when the Space Shuttle Challenger mission retrieved, repaired, and re-released it into orbit. Eventually, the SMM mission ended in 1989 when it re-entered Earth's atmosphere.

The Japanese Hinotori satellite (ASTRO-A), the first X-ray astronomy satellite of Japan, was launched on February 21, 1981. It pioneered hard X-ray imaging of solar flares, and carried out observations of many solar flares, until it ended in 1991. In the last five decades of X-ray astronomy, the sensitivity of detectors has seen a marked increase, and the ability to focus X-rays has made rapid progress in obtaining high-quality images, following the early missions.

8.3 UV and EUV detection and imaging

The radiation domain of UV radiation is wavelengths between 10 and 400 nm. The UV wavelengths are distinct from optical domain in the sense that below 300 nm, these are completely blocked out by the ozone layer in Earth's atmosphere, and ground-based observations are not possible. The shorter wavelength region from 10 to 91.2 nm is called the extreme ultraviolet (EUV, XUV). Balloons and rockets, along with satellites, have been used to explore this domain of the electromagnetic spectrum with photometers or spectrometers. In the UV range, instruments have their roots in optical telescopes.

In the shorter, EUV range, the technology is strongly related to that utilised in X-ray astronomy. EUV detectors use photon counting devices, such as proportional counters (as used in X-ray astronomy), and photomultipliers (as used in UV range). The first orbiting EUV imaging telescopes were based on grazing-incidence (Section 8.4.2, page 203), which is necessary for wavelengths below $\lambda \sim 30$ nm. However, for EUV wavelengths, the incidence angles can be larger, around $10°$, as compared to X-ray telescopes ($1 - 2°$), because the reflectivity is higher in the EUV.

8.4 X-ray detection and imaging

The soft X-ray domain merges into the EUV at the limit of $\lambda \sim 10$ nm. The X-ray domain is further subdivided in 10–1 nm (soft X-rays) and 1–0.002 nm (hard X-rays). X-ray photons are often described by their energy instead of wavelength. An approximate relation between wavelength λ and photon energy E is given by $\lambda E \approx 1$ nm-keV. Therefore, the corresponding ranges in terms of photon energies are for soft X-rays $E < 1keV$, and for hard X-rays $1keV < E < 0.5MeV$. Several different technologies for detection of X-ray radiation are used, such as ionization chambers, Geiger counters or scintillators. CCDs or CMOS sensors are used for X-ray imaging.

8.4.1 X-ray detectors

Proportional counters were the main type of detectors for energies less than about 20 keV. They consist of a gas-filled tube, entering in which a photon ionizes a gas atom releasing a free electron, which creates more electrons, as it accelerates towards a central anode wire placed in the tube. This cascade of electrons gives rise to a pulse on reaching the anode from which one registers arrival of an X-ray photon and its energy.

For energies above 20 keV, X-rays pass through the tube without being absorbed. For such higher energies, solid-state detectors have been used, where the incident photon ionizes an atom in the crystal lattice, producing a high energy free electron, which can excite many other electrons. The impurity atoms present in the lattice capture some of these electrons, and produce flashes of visible light, or 'scintillations', that are detected by photomultiplier tubes attached to the 'scintillation counter'.

Semi-conductors have been used more recently as high-energy detectors. Bragg crystal spectrometers are also used for high-energy detection, in which the regular lattice of crystal acts as a three-dimensional diffraction grating. These crystals can also be used as polarimeters, because the scattering from lattice atoms also introduces polarization. One may refer to [338] for details about the recent developments in X-ray observing techniques.

8.4.2 X-ray imaging

The basic elements of any optical telescope are (i) the optics for focussing, or collimating, that collects the radiation entering the telescope, and (ii) the detector, on which the focussed, or collimated, radiation is collected and analysed. For imaging X-rays, however, there are difficulties related to its short wavelength, and high energies involved.

The first problem is that optics in this range need to be figured at much better than the wavelengths involved, *i.e.*, about 0.1 nm. The other difficulty is that the X-rays are so energetic that they cannot be reflected, as they tend to pass through most matter, including any mirrors. X-rays mirrors are, therefore, designed to be kept at very low angles from the plane of reflection (typically 10 arc-minutes to 2°). These are known as glancing (or grazing) incidence mirrors.

The use of grazing incidence relaxes the requirement of accuracy in figuring the mirrors by a factor of the order of the sine of the grazing angle. However, accuracy of the order of several nm is still required. On the other hand, the effective area of the telescope is also reduced by the same amount in grazing incidence telescopes. Therefore, they are not very sensitive for high resolutions.

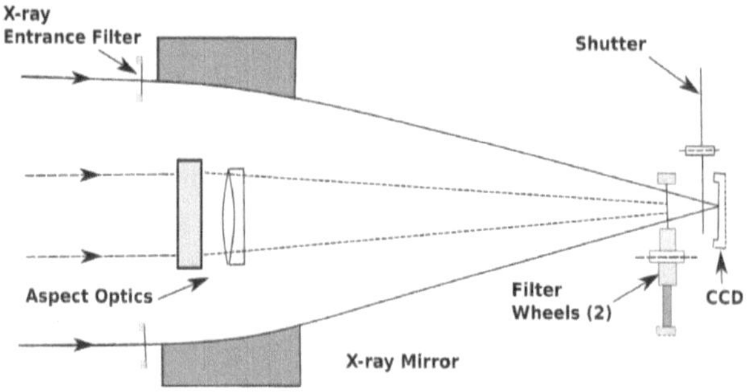

FIGURE 8.2: Schematic diagram of Yohkoh-SXT telescope. With permission from [368].

In 1952, Hans Wolter outlined three different configurations for such telescopes, referred to as Wolter type I, II, and III telescopes, each having its advantages and disadvantages [397]. The first solar X-ray imaging observations were made from a rocket-borne experiment in 1965, employing the most common of these, a Wolter Type I grazing-incidence optics mirror [135]. Figure (8.2) shows the Wolter Type I configuration that was used in the Japanese Yohkoh Solar X-ray telescope.

Wolter's key innovation was that by using two mirrors it is possible to create a telescope with a usably wide field of view. In contrast, a grazing incidence telescope with just one parabolic mirror could focus X-rays, but only very close to the centre of the field of view as it would suffer from extreme coma.

8.4.3 Focussing X-ray mirrors versus collimation

Nearly all the existing telescopes employ some variation of the Wolter Type-I design, with two grazing incidence reflections from the inner surfaces of near cylindrical mirrors. These have the limitations of narrower fields of view (typically $< 1°$) than visible or UV-telescopes. The mirrors are made of ceramic or metal foil, coated with a thin layer of a reflective material, typically gold or iridium.

In the other technique of collimating optics, a flat aperture patterned grille is used in front of the detector. It offers a larger field of view as compared to focussing technique, and can be employed at higher energies, where grazing incidence optics become ineffective. Imaging in collimation technique is reconstructed by post-processing of the signal. The drawbacks of this design are less sensitivity and low imaging quality, leading to poorer identification of source position.

As compared to the collimated optics, focussing optics allows (i) high resolution imaging, (ii) high telescope sensitivity, as the radiation is focussed on a small area, and (iii) higher signal-to-noise ratio. This technology is limited in energy range by the inverse relation between critical angle for total reflection and radiation energy. The limit in the early 2000s with Chandra and XMM-Newton X-ray observatories was about 15 keV X-rays, but it was pushed up to 79 keV using multi-layered coated mirrors.

8.5 The γ-ray detection and imaging

The distinction between X-rays and γ-rays is again rather arbitrary. It is known that annihilation of positron and electron leads to production of two γ-ray photons. Therefore, one can take the rest mass of electron, *i.e.*, 0.5 MeV and above as the photon energy to define γ-rays. The γ photons have energies much larger than the X-rays, and therefore, they are much harder to observe.

The low-energy γ-rays can be detected using technology similar to that used for hard X-rays, such as scintillation detectors and semi-conductors. In this case, however, there is a difference that the incident γ-ray photon may produce an electron-positron pair, rather than being absorbed by the atom, and produce electron-ion pair. In order to detect the resultant high-energy electron-positron pair, which may pass through the scintillation detector, it may be required to use a second detector.

Thus, γ-ray observations are made with scintillation detectors composed of several layers of detector plates, where eventually it is transformed into visible light, which is then detectable by photomultipliers. The energy of a γ-ray photon is determined from the depth to which it penetrates the detector. The analysis of the tracks of electron-positron pair provides the information about the energy and direction of the incident γ-ray. The directional accuracy is low, and in γ-ray astronomy the resolution is far below that in other wavelength regions.

Imaging in γ-rays is more difficult than it is for X-ray imaging due to the higher energies involved, and less effective even with grazing incidence telescopes resulting in even lower resolution images.

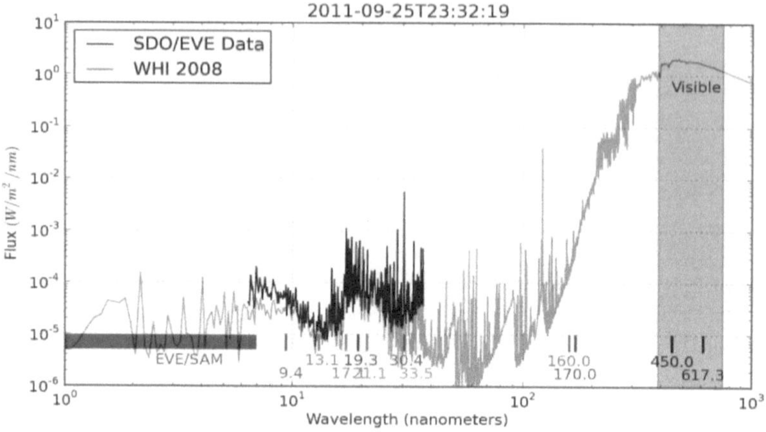

FIGURE 8.3: Solar UV spectrum from SDO/EVE. The EVE data (dark) overlays the Whole Heliosphere Interval (WHI) 2008 reference spectrum of the Sun [399]. Markers along the bottom indicate wavelength coverage for other SDO instruments such as EVE/SAM imager, AIA, and HMI. Note the transition from absorption lines (for $\lambda > 200$ nm) to emission line ($\lambda < 200$nm). *Courtesy of SDO/EVE science team.*

8.6 High-energy emissions from the Sun

Absorption lines are dominant in solar spectrum down to 210 nm. At shorter wavelengths the intensity is reduced to correspond to the effective temperature of 4700° K. Below 150 nm, emission lines start to dominate the spectrum, instead of absorption lines as seen in the visible [401]. The strongest is the hydrogen Lyman α line, centred at 121.57 nm (Figure 8.3). At the shorter wavelengths, solar spectrum becomes highly variable, implying non-uniform distribution of the emission sources in the solar atmosphere. This non-uniformity is both spatial and temporal.

8.6.1 The transition region

Moving above the chromosphere, temperature increases rapidly from around $20,000°$K to 10^6°K within a short height span of around a few hundred kms. Density falls too in this transition region that separates the hot corona from the cooler chromosphere (Figure 8.4). This increase in temperature is much more than the radiative heating of the chromosphere; one of the outstanding problems in solar physics.

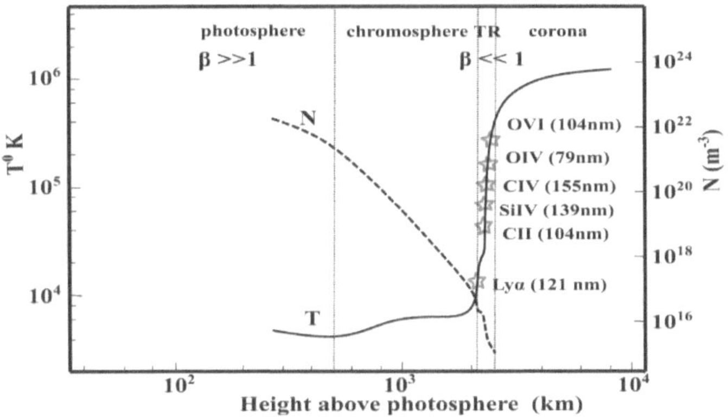

FIGURE 8.4: The solar transition layer: interface between the chromosphere and the corona.

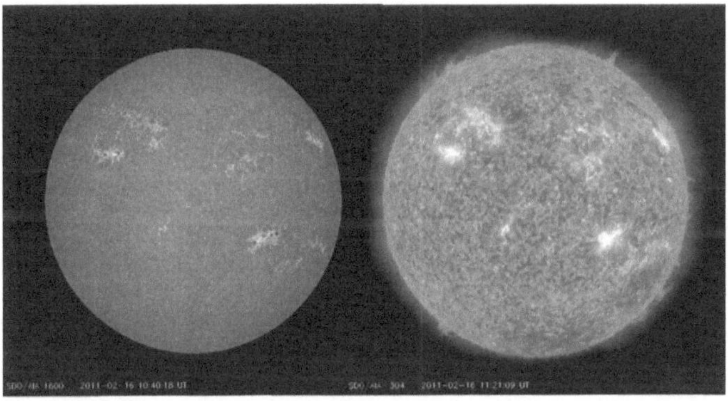

FIGURE 8.5: The solar transition layer: SDO/AIA images of (i) lower TR at 160 nm ($\sim$ 10,000°)K, and (ii) TR at 30.4 nm (50,000°K). *Courtesy of NASA/SDO and the AIA, EVE, and HMI science teams.*

The million degree temperatures of the outermost hot plasma is adequate to ionize hydrogen. Therefore, it is difficult to observe the transition region in H_α any more. It is dominated by ionized elements such as C IV, O IV, and Si IV (each with three electrons stripped off), and hence emits UV and EUV (Figure 8.4). The emissions in wavelength band below 120 nm, the extreme ultraviolet (EUV), come both from neutral atoms and from these ions in very high ionization levels. The temperature of the EUV emissions cover the range from $6 \times 10^4 - 2 \times 10^7$°K. As these radiations are absorbed in Earth's atmosphere, the TR can only be observed from space.

The Atmospheric Imaging Assembly (AIA) on-board Solar Dynamics Observatory (SDO) is routinely providing simultaneous full-disk images of the corona and transition region above the solar limb (Figure 8.5). Narrow-band imaging is done in seven EUV band-passes centred on specific lines: FeXVIII (9.4nm), FeVIII, XXI (13.1 nm), FeIX(17.1nm), FeXII, XXIV (19.3nm), FeXIV (21.1 nm), HeII (30.4nm), and FeXVI (33.5nm).

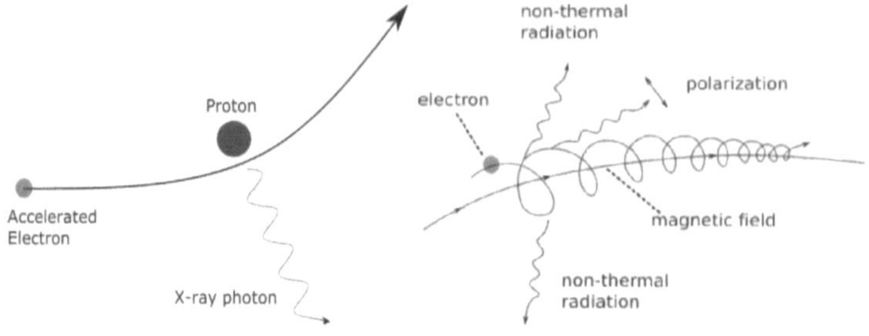

(a) (b)

FIGURE 8.6: Non-thermal emission mechanisms: (a) Bremsstrahlung, and (b) synchrotron radiation.

The high-energy emission from TR and corona occurs also due to non-thermal processes, such as bremsstrahlung, and synchrotron emission. When an electron hits another particle in a Coulomb collision, it emits bremsstrahlung (Figure 8.6a). In fully ionized plasma, this physical process is also known as *free-free radiation* of thermal electrons. The change in direction and momentum causes a particle with some keV energy, or greater, to emit X-ray photons.

8.6.2 Solar corona in X-rays

The thermal soft X-ray spectrum has an exponential shape, whereas the non-thermal hard X-ray spectrum shows a power-law at increasing energy. The hard and soft X-rays from astrophysical plasmas, including the Sun, are a mixture of *bound-bound* atomic transitions (soft X-rays), nuclear line emission (hard X- and γ rays), and electron bremsstrahlung. In the hard X-ray range (10-1000 MeV), the emission is purely electron bremsstrahlung. Therefore, HXR images provide direct information about the generation, transfer and confinement of high-energy electrons.

The observed solar X-ray spectrum consists of two basic components: (i) thermal radiation of hot electrons in soft X-rays in the range 1-10 keV (or $\sim$ 1-10 nm), (ii) non-thermal hard X-rays radiation beyond 100 keV by electrons accelerated to relativistic velocities. In both the cases, the dominant radiation mechanism is bremsstrahlung of electrons moving among ions (mostly protons) of the ambient gas.

Soft X-ray emission is particularly sensitive to high-temperature plasma, and is thus an important diagnostic of the coronal temperature distribution and elemental abundances and their evolution. A systematic survey of high-temperature emission in solar active regions has shown that spectrally resolved observations are crucial for constraining the various coronal heating models [388]. Recently, observations of the solar emission in the 0.5–5 keV range of soft X-rays has been carried out and its spectrum is obtained [71].

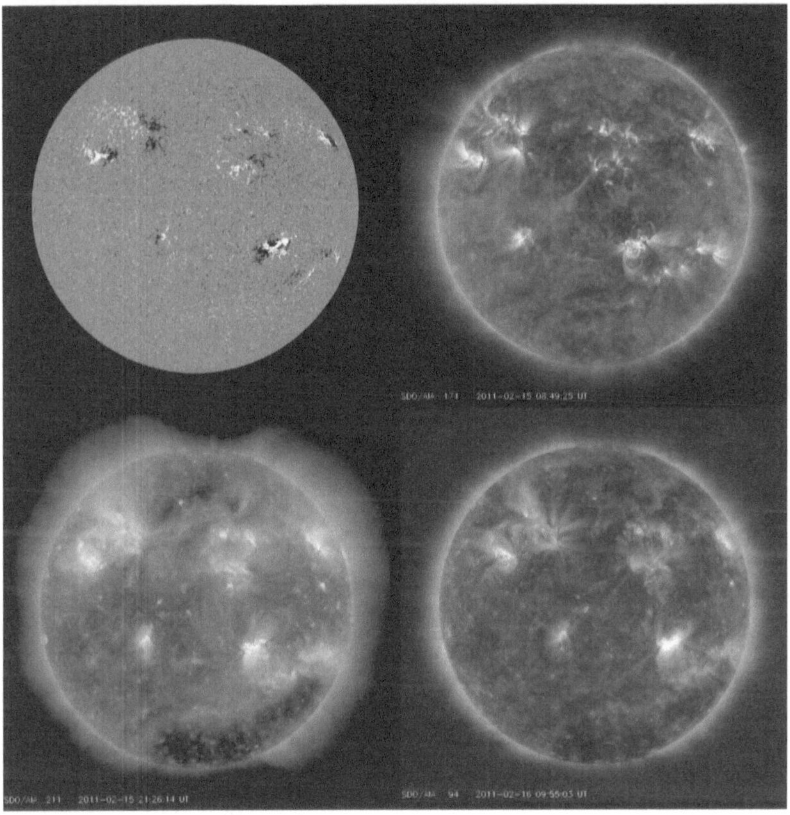

FIGURE 8.7: The Sun's X-ray images at 17.1, 21.1 and 9.4 nm wavelengths taken by SDO-AIA instrument. The first image is corresponding SDO/HMI magnetogram. The active-regions, having enhanced magnetic fields, are the regions of enhanced brightness (hotter temperature) in X-rays. *Courtesy of NASA/SDO and the AIA, and HMI science teams.*

Energetic events, such as solar flares, increase the EUV and soft X-ray (0.1 nm - 10 nm) spectrum quite considerably. There are sudden intense enhancements of high-energy emissions during these short periods of the transients (Section 8.7, page 210). In extremely energetic events, even hard X-rays and γ-rays are emitted by non-thermal processes. Charged particles ejected during these transients spiral around the magnetic field lines, and generate radio emission by synchrotron process (Figure 8.6b), while soft X-ray emission results due to heating of plasma to high temperatures within the magnetic loop below the reconnection site of acceleration. A composite spectrum from soft X-ray to γ-ray solar emission is shown in Figure (8.12), on page 216.

The Atmospheric Imaging Assembly (AIA) on-board Solar Dynamics Observatory (SDO) is routinely providing simultaneous full-disk images of the corona at several wavelengths, such as FeIX 17.1 nm for quiet corona, FeXIV 21.1 nm for active region corona, FeXVIII 9.4nm for flaring corona, *etc.* (Figure 8.7; refer also to Table 8.2).

The observed fraction of energy is determined by the total flare irradiance to exceed the soft X-ray emission (< 27 nm) by a factor of 5 in some large flares. In the lower layers, at the base of the magnetic field-lines, H_α flare ribbons are produced by recombination of electrons and protons that are accelerated toward the chromosphere from the reconnection site. The EUV spectrum is routinely measured by the EUV Variability Experiment (EVE; [400]) on board the Solar Dynamics Observatory (SDO; [302]).

8.7 Multi-wavelength observations of solar flares

Flares produce radiation across the electromagnetic spectrum, although with different intensities. There are typically three stages to a solar flare:

- *Precursor phase* during which the release of stored or excess magnetic energy in the active region is triggered. SXR emission is detected in this stage.
- *Impulsive phase* in which a large fraction of the energy released goes into 10–20 keV electrons: the electrons that precipitate into the denser regions of the solar atmosphere and produce hard X-ray bursts (HXBs) and H_α brightening, and the electrons that escape produce type III solar radio bursts.
- *Gradual or decay phase* of gradual decay of SXR and in H_α.

The duration of these stages can be as short as a few seconds or as long as several minutes and even hours in extended long duration flares. A schematic representation of the different phases of a typical solar flare as observed in electromagnetic and particle radiation were first presented by S. R. Kane [203] that greatly helped in understanding the physics of flares, and in the construction of composite flare models.

8.7.1 Coronal signatures of flares in X-rays

The ordinary corona has density and temperature that are not suggestive of hard X-ray emission. The coronal X-ray emission contains a thermal part dominating at low energies, and a weak non-thermal part above about 8 – 10 keV. Flare energy released in the solar atmosphere is mostly thermalised and some parts are heated to high temperatures emitting in soft X-rays. In the absence of further energy release the plasma then cools by thermal conduction to the chromosphere. The non-thermal emission in the corona is usually soft, consistent with the idea of a thin target. Thus, the accelerated electrons lose only a small fraction of their energy and continue to propagate towards the chromosphere. Solar X-ray flares are monitored regularly by the Geostationary Operational Environmental (GOES) satellite (`www.noaa.sec.gov`).

The soft X-ray (SXR) flux builds up gradually and peaks a few minutes after the impulsive emission. The flaring soft X-ray loops are about ten times hotter than the quiescent non-flaring coronal loops. Thus the plasma is heated up to several times $10^{7\circ}K$, adequate to strip almost all electrons from the iron atoms (neutral iron has 26 electrons) and the radiation comes from the inner shells of species such as $FeXXIII - XXVI$. This implies that there are also line emissions in the X-ray band, *e.g.*, at 0.1778 nm ($FeXXVI$), 0.185 nm ($FeXXV$), 0.3177 nm ($CaXIX$), 0.5039 nm (SXV), 0.917 nm ($MgXI$), and 1.346 nm ($NeIX$). From the density-sensitive lines, loop densities have been estimated to be $10^{17} - 10^{18}$ electrons m^{-3}.

The emission of hard X-rays (HXR) with a non-thermal energy distribution was first observed to be located at foot-points of coronal loops. It was reported that 95% of the $\geq$ 150 keV X-ray emission in impulsive flares originates at altitudes $\leq$ 2500 km, that is, at the level of the chromosphere [204]. This upper limit in altitude satisfies the collisional thick-target model in which precipitating electrons lose their energy in the dense, cold target. The emission presumably originates from flare-accelerated non-thermal electrons precipitating into the lower corona and chromosphere along the legs of the coronal loop. The hard X-ray sources are nearly co-located with the magnetic foot-points of the soft X-ray loops, and Hα emission in the chromosphere, tracing the base of a magnetic arcade (Figure 8.8), also see [262]. This is further supported by a similar time profile of radio waves at centimetre wavelengths, also produced by the non-thermal electrons.

First hints about where is the primary source of flare located came from a HXR source above the SXR loop observed in Masuda flare [256] (Figure 8.9). The altitude of coronal HXR sources, 5000 – 25,000 km, is compatible with observed time delays of HXR peaks emitted at the foot-points. The HXR loop-top and foot-point sources observed during flares, and SXR flare loops are consistent with the flare scenario of primary energy release in the hot corona, and the energy transport to the chromosphere and chromospheric evaporation.

8.7.2 High-energy γ-ray emission during transient events

The X-ray detectors, and γ-ray spectrometers on-board the SMM satellite, launched by NASA on 14 February 1980 [49], and Hinotori, the first Japanese satellite to study solar X-rays and γ-rays launched on 21 February 1981 as

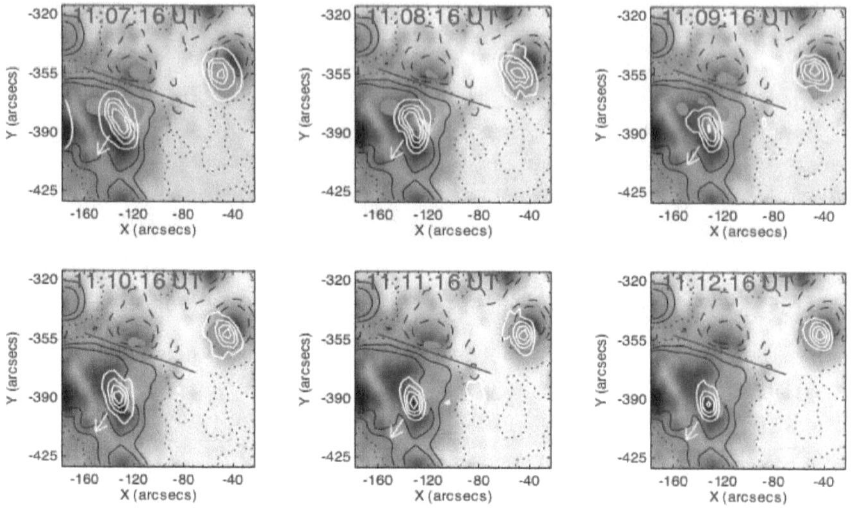

FIGURE 8.8: A mosaic of white-light flare (lighter patches) during the impulsive phase of the X17/4B flare on 28 October 2003, match with RHESSI HXR sources (white contours) in the energy range 100–200 keV, at 20, 40, 60, 80% of the maximum value. Superimposed dashed (solid) black contours represent negative (positive) magnetic flux at ±800, ±1500G levels, while the dotted contours mark the magnetic neutral lines.

Astro-A [116], during the 1980s were dedicated to studies pertaining to the solar X- and γ-ray radiations. Dynamic spectra of many flares were analysed which showed the delayed lines [318].

The solar flare γ-ray spectrum appears as a superposition of electron bremsstrahlung, line radiation, and pion decay emission. Above about 1 MeV, line radiation dominates the spectrum. This is illustrated in Figure (8.10), which shows the spectrum of a flare, in the range of 0.28-9MeV, observed by the Solar Maximum Mission (SMM) Gamma Ray Spectrometer [318]. Various γ-ray lines, as marked in Figure (8.10), are identified in this spectrum : (1) 0.511 MeV (annihilation line), (2) 1.24 MeV (^{56}Fe), (3) 1.38 MeV (^{24}Mg), (4) 1.63 MeV (^{20}Ne), (5) 2.223 MeV (neutron capture line, attenuated for this flare that was close to the limb), (6) 4.43 MeV (^{12}C), (7) 6.13 MeV (^{16}O), and (8) 7.12 MeV (^{16}O). The lines at 4.43 and 6.13 MeV, marked as (6'), and (7'), respectively, are instrumentally produced escape peaks. The spectrum from the threshold (0.28 MeV) to 1 MeV resembles a power law continuum, and lines are evident above 1 MeV up to about 7.5 MeV (the cutoff of the instrument), and above 10 MeV (not shown here), pion decay γ-rays and electron bremsstrahlung would contribute.

The neutron-capture line at 2.223 MeV, marked '5' in Figure (8.10), is the strongest of the many individual narrow and broad γ-ray lines generated

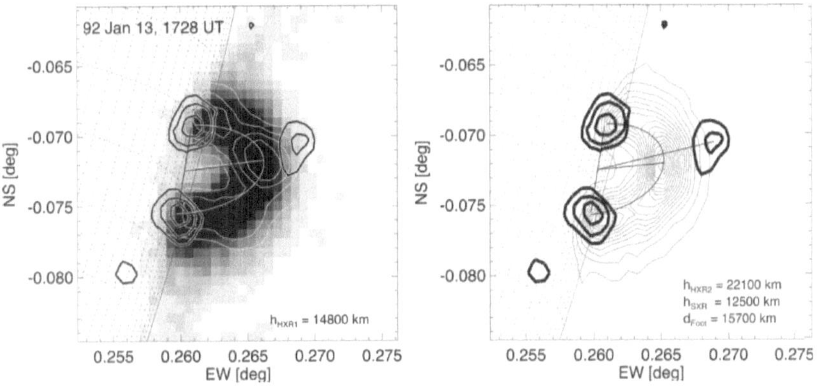

FIGURE 8.9: The Masuda flare: The overlays of Yohkoh/SXT (greyscale in left frame and thin contours in right frame) and Yohkoh/HXT (14-23 keV, white contours in left frame; and 23-33 keV, black contours in both frames. [20]). (Figure 13.34, page 600 from [18] by permission of M. Aschwanden.)

by accelerated ions [311]. As is the case with flare-accelerated electrons, ions accelerated by a flare would also be expected to interact at foot-point locations. On the other hand, ions accelerated by a CME shock would be expected to be widely dispersed and/or displaced from the flare site. Possibility for the latter was suggested by the detection of the 2.223 MeV line in an above-the-limb flare event [378]. Accelerated $\sim$ 10 to $>$ 100 MeV nucleon^{-1} protons and alpha particles colliding with the solar atmosphere produce fast neutrons that then thermalise in the photosphere before capture by ambient hydrogen to form deuterium, with the emission of an extremely narrow line ($<$ 10 eV FWHM). Since the neutrons travel only a short distance during the $\sim$100 s thermalisation time, images in the 2.223 MeV line indicate the site(s) of the original nuclear interactions. The analysis of time history of the 2.223 MeV line can provide the information, such as the abundance of ^3He and the depth in the atmosphere where these reactions take place.

Thus, in exceptionally energetic flare events, hard X-ray, and even γ- foot point sources may be produced by high-energy particles due to nuclear reactions in the chromospheric (thick-target) layer. Protons and other nuclei accelerated from about 10 MeV to $\geq$ 100 MeV per nucleon collide with the nuclei in the dense chromosphere, and produce neutrons. Neutrons thermalise and, after traveling less than 1000 km, are captured by an ambient proton to form deuterium. The γ-ray lines between 0.8 and 20 MeV are emitted due to atomic nuclei excited by impinging ions.

These reactions are *spallation reactions* which break heavier nuclei into lighter nuclei in excited state, followed by a de-excitation reaction which releases high-energy γ-rays or hard X-rays. These emissions may also be produced in the chromosphere during flares by reactions including electron-positron

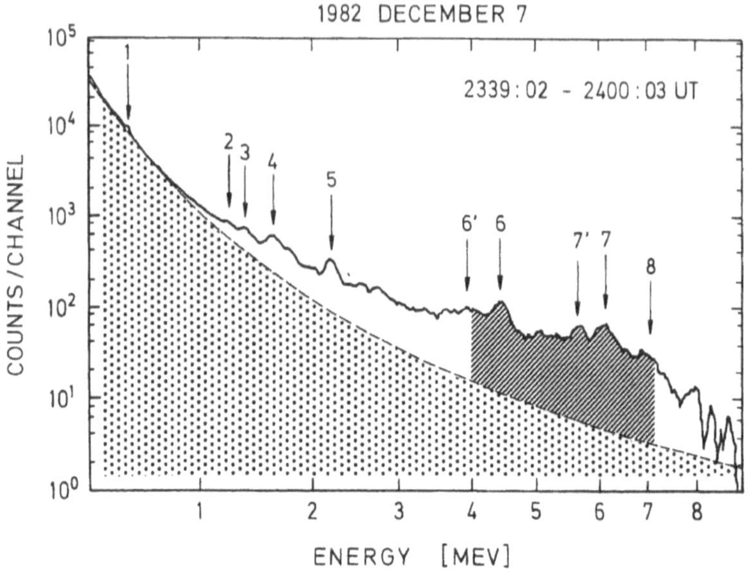

FIGURE 8.10: The energy spectrum of 7 December 1982 flare, marked with significant lines and their corresponding origin. With permission from [318].

annihilation, or by production of deuterium by proton-neutron reactions. Mechanisms of production of high-energy γ-radiation and neutrons during solar flares are discussed in [318], and schematically shown in Figure (8.11).

Free neutrons are produced when accelerated ions interact with the ambient solar atmosphere. These neutrons can subsequently escape from the Sun to interplanetary space, while those remaining in the solar atmosphere can be captured by nuclei before decaying, and produce lines such as the 2.233 MeV line in the reaction $^1H(n,\gamma)^2H$. Contrary to the de-excited lines, the 2.223 MeV line emission is delayed due to the finite capture time of neutrons on protons. As the 2.223 MeV line is produced in the photosphere, it is by far the strongest line for on-disk flares, while it is attenuated in limb flares. Gamma-rays of energies $> 10MeV$ are also produced as a result of pion production, which requires protons with energies $> 100MeV$ through the prompt decay of neutral pions, or from the decay of charged pions to positrons and electrons. Figure (8.12) shows a composite spectrum of a large flare produced from soft X-rays to high-energy γ-rays by various mechanisms.

Observations of γ-ray line emission show that ions are accelerated up to $\sim$ MeV–GeV energies in large solar flares. The solar energetic particles (SEPs) directly observed near 1 A.U. with comparable energies appear to be accelerated much higher in the solar atmosphere by widespread shock waves driven by fast ($> 1000km - s^{-1}$) coronal mass ejections (CMEs). Figure (8.8) shows location of the HXR sources images which spatially match with white light

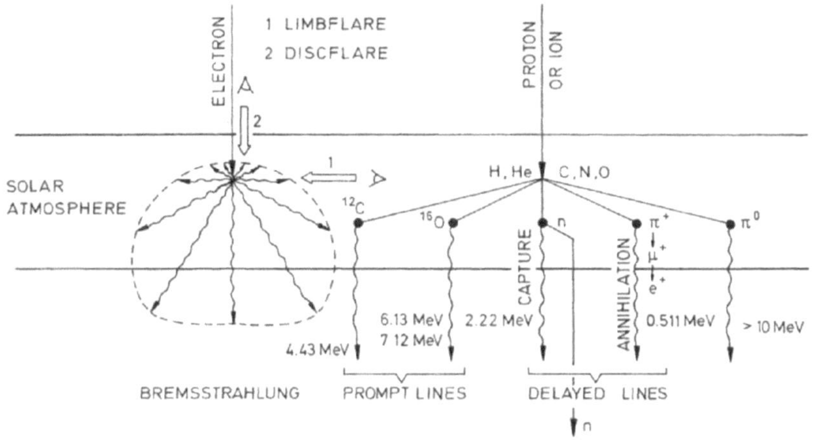

FIGURE 8.11: Schematic of production mechanism of γ-rays and neutrons during limb- and disk-flares. Here, the dashed oval shows the anisotropic emission of photons from electron bremsstrahlung. The arrows indicating the γ-rays and neutrons produced by ion interactions do not represent directivity of radiation. Figure with permission from [318].

patches of the energetic flare. A similar spatial match is also seen in an overlay of RHESSI γ-ray emission on a TRACE 195 Å image [189].

8.7.3 RHESSI and FermiGRO missions

Until recently, γ-ray line studies have been limited to spatially integrated spectroscopy. Since 2002, however, NASA's Reuven Ramaty High Energy Solar Spectroscopic Imager (RHESSI; [243]) has provided the first capability for solar γ-ray imaging, as well as high-resolution spectroscopy of ion-produced γ-ray lines and X-ray imaging spectroscopy of the bremsstrahlung radiation from energetic electrons. RHESSI was designed to image solar flares in energetic photons from soft X-rays ($\sim$ 3 keV) to γ-rays (up to $\sim$20 MeV), and to provide high-resolution spectroscopy. The RHESSI imager is highly sensitive to flare X-rays even from microflares [159].

The RHESSI mission contributed to most of the recent progress in understanding the high-energy aspects of solar flares. It made one of the most important discoveries of γ-ray foot-point structures, where in all the largest flares $\geq$ 10 MeV ions are localised at the vicinity of the flare. Also, in the largest flare, RHESSI is able to distinguish two ion-associated foot-points that are closely parallel to the electron-associated foot-points, establishing that flares accelerate ions and electrons in a similar manner.

Another γ-ray observatory began in 2008 with the launch of NASA's Fermi γ-Ray Observatory (FermiGRO) that allowed observations of prolonged

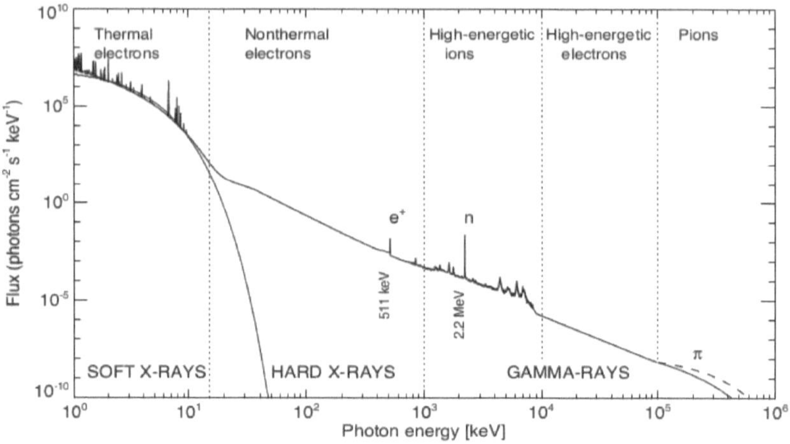

FIGURE 8.12: Composite spectrum of a large flare: soft X-rays (110 keV), hard X-rays (10 keV-1 MeV), γ-rays (1 MeV - 100 GeV). The energy spectrum is dominated by (i) thermal electrons (in soft X-rays), (ii) bremsstrahlung from non-thermal electrons (in hard X-rays), (iii) nuclear de-excitation lines ($\approx$ 0.5–8 MeV γ-rays), (iv) bremsstrahlung from high-energetic electrons ($\approx$ 10–100 MeV γ-rays), and (v) pion-decay ($\geq$ 100 MeV γ-rays). Also notable are the prominent electron-positron annihilation line (at 511 keV) and the neutron capture line (at 2.2 MeV). (Figure 14.1, page 608 from [18] by permission of M. Aschwanden.)

high-energy γ-emission during Solar Cycle 24. The Fermi observatory is comprised of two instruments: the Large Area Telescope (LAT) designed to detect γ-rays from 20 MeV up to more than 300 GeV [21] and the Gamma-ray Burst Monitor (GBM), sensitive from $\sim$ 8 keV up to 40 MeV [270].

During a powerful X5.4 solar flare observed on March 7, 2012, FermiGRO detected up to 4 GeV γ-ray emission. It was so far the highest energy radiation associated with an eruption on the Sun. This flare produced an intense burst of γ-rays making the Sun briefly become the brightest object in the γ-ray sky. The most likely scenario suggested for this energetic event was a continuous acceleration of protons in the solar corona that penetrated the lower solar atmosphere, and produced pions that decayed into γ-rays [6]. This discovery heralded the role of FermiGRO as a powerful new tool for understanding solar outbursts. Measurements of the γ-ray emission from solar flares provide an opportunity to investigate the mechanisms of impulsive energy release and acceleration [2].

8.7.4 Flare classifications

X-ray classification of flares is based according to their soft X-ray flux data from monitoring satellites (GOES10 &12 satellite data) (Section 8.9.3, page 222). They are quantitatively identified using 5 min averages of solar X-ray flux in 0.1-0.8 nm, and 0.05-0.4 nm wavelengths (Figure 8.13).

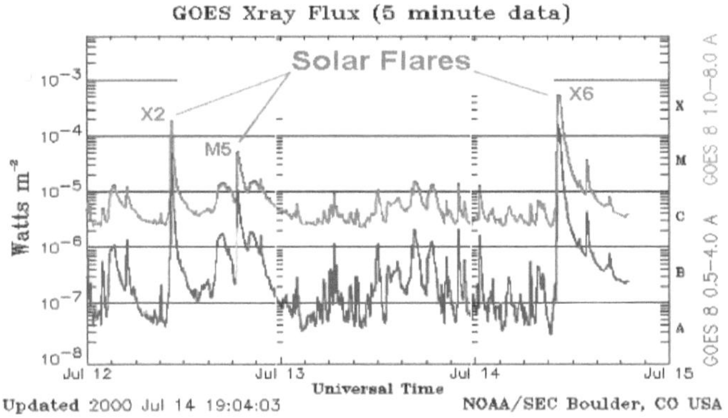

FIGURE 8.13: The X-ray classification of solar flares: the three flares marked here are X2, M5, and X6.

There are 3 main categories: (i) The X-class flares are major events that can trigger planet-wide radio blackouts, and long-lasting radiation storms, (ii) the M-class flares are medium-sized, and cause brief radio blackouts affecting Earth's polar regions, and minor radiation storms, and (iii) the C-class flares are small events with few noticeable consequences on the Earth. Even smaller events are classified as B- and A-class. Each category has nine subdivisions ranging from 1 to 9. One of the largest observed X-ray flares on November 4, 2003, was classified as X27, close to the upper limit of GOES observations, but was perhaps not quite accurate.

TABLE 8.1: Flare classification in various categories

H_α Class	Area ($10^{-6} A_\odot$)	Flux at 5000 MHz (SFU)	SXR (0.1-0.8nm) Class ($W-m^{-2}$)
S	< 100	5	C2 (2×10^{-6})
1	$100-250$	30	M3 (3×10^{-5})
2	$250-600$	300	X1 (1×10^{-4})
3	$600-1200$	3000	X5 (5×10^{-4})
4	> 1200	30000	X9 (9×10^{-4})

Flares were traditionally observed in chromospheric H_α telescopes from ground-based observatories much before the X-ray observations began from space. The optical classification of flares was based on both the area of flare during the maximum phase, and the brightness in H_α. The flare areas were calculated in terms of the solar hemisphere area $A_{\odot} = 3.0 \times 10^{12} km^2$, while the flare brightness was more qualitative, as assigned by the observer, using the suffix F (faint), N (normal) and B(bright) along with the flare importance, for example, a large bright H_α flare may be identified as class 2B. A typical equivalence of various classifications is listed in Table 8.1.

8.7.5 Modelling the flares

From the observations of flares accumulated over several years, many models have been proposed for describing solar flares, such as *the standard, or CSHKP model*, and the *Eruptive Flare model* by Shibata (1996) [334]. The CSHKP model was developed by Carmichael (1964), Sturrock (1966), Hirayama (1974) [179], Kopp and Pneuman [216]. These models attempt to describe various observational signatures of flares obtained in different wavelengths, from chromospheric H_α to soft and hard X-ray, microwave, and radio bursts produced at successively higher layers of solar atmosphere [203]).

Masuda et al. reported [256] observations of a limb flare of January 13, 1992, from Yohkoh-SXT and HXT and suggested a flare model scenario that consists of four basic elements: a particle accelerator above the top of a magnetic loop (consistently imagined at the peak of a cusp), a coronal source visible in SXR and HXR, collision-less propagation of particles along the magnetic loop and HXR foot-points in the chromosphere. This flare is popularly known as the Masuda flare (Figure 8.9, page 213).

According to these observations, a flare scenario is given where the primary source of flare energy is related to reconnection of magnetic field lines in corona at the top of a loop. High-energy particles are accelerated relativistically away from the reconnection point, that emit loop top HXR, microwaves, radio bursts, and the finite resistivity of plasma may cause Joule heating of gas, leading to $10^7 K$ temperatures. Some particles may escape the Sun in the form of solar cosmic rays (Figure 8.14).

Solar flares affect all layers of the solar atmosphere. Some of the non-thermal electrons channel down the loop and hit the chromosphere and emit foot-point HXR by electron-ion bremsstrahlung. Beams of accelerated protons entering dense atmosphere cause nuclear reactions and emit γ-ray spectral lines. Chromosphere is heated and material rises into the loop giving a slow, gradual increase in SXR. This upwelling of heated material is chromospheric evaporation, and occurs in the decay phase of the flare.

The coronal source often appears before the main flare hard X-ray increase, but is well correlated in time and spectrum with the foot-points. These observations suggest that there is a strong coupling between corona and chromosphere during flares. Non-thermal hard X-ray emission during impulsive solar flares is highly variable, often showing activity peaks and dips with durations ranging from seconds up to several minutes. This behaviour can be observed in the largest X class flares as well as in much smaller B and C class flares. The non-thermal photons usually follow a power-law distribution in energy.

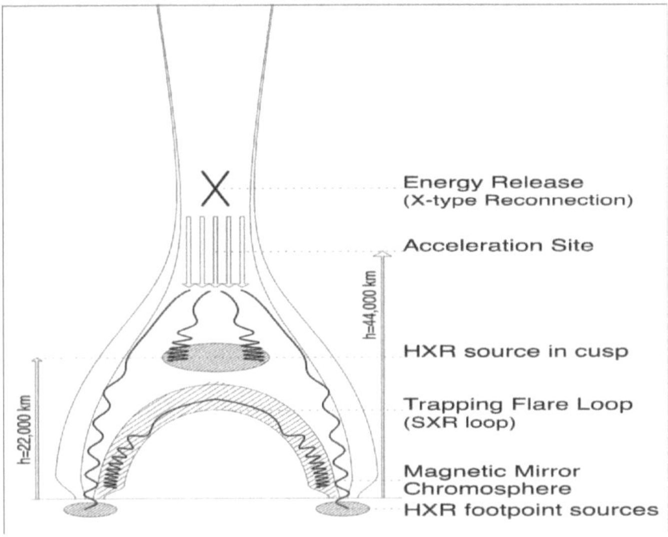

FIGURE 8.14: Diagram of a possible flare scenario for the Masuda flare, constrained by the height (h = 22 Mm) of the above-the-loop top hard X-ray source and the height (h = 44 Mm) of the acceleration source inferred from electron time-of-flight measurements [20]. (Figure 13.35, page 601 from [18] by permission of M. Aschwanden.)

8.7.6 Impact of solar flares and CMEs on space weather

Solar flares strongly influence local space weather in the vicinity of the Earth. Flares are associated with the ejection of plasmas and particles through the Sun's corona into outer space; and they emit radio waves. X-rays and UV radiation emitted by solar flares can affect Earth's ionosphere and disrupt long-range radio communications. The radiation risks posed by solar flares are a major concern for manned missions to Mars, the Moon, or other planets.

The soft X-ray flux of flares increases the ionization of the upper atmosphere, which can interfere with short-wave radio communication. Increased SXR flux can heat the outer atmosphere, and thus increase the drag on low orbiting satellites, leading to orbital decay. Hard X-rays and energetic particles can damage the spacecraft and instruments they carry. Direct radio emission at decimetric wavelengths may disturb the operation of GPS, radars and other devices that use those frequencies.

Flares can also produce transient streams of highly energetic particles in solar wind, known as a solar proton event. In the case of Earthward direction of ejection, associated particles penetrate into the magnetosphere causing auroras. Most proton storms take at least two hours from the time of visual detection to reach Earth's orbit, giving astronauts as little as 15 minutes to reach shelter.

Massive solar flares may be accompanied by CMEs. The CMEs are important in triggering geomagnetic storms that can disable satellites and knock out terrestrial electric power grids for extended periods of time.

8.7.7 Forecasting of solar flares and CMEs

Current methods of flare prediction and CMEs are difficult and problematic, and there is no certain indication when an active region will produce a flare. Many properties of sunspots and active regions correlate with flare productivity. For example, magnetically complex regions called δ-spots (inverted polarity structure) generally produce the largest flares.

A simple scheme of sunspot and its magnetic classification, or related to fractal complexity, is commonly used as a starting point for flare prediction. Predictions are usually stated in terms of probabilities for occurrence of flares above M or X GOES class within 24 or 48 hours. Alerts are issued on a daily basis by NOAA, USA, and other agencies. One can check websites of ground- and space-based missions to check the state of the Sun (Also, `www.spaceweather.com`)

8.8 The coronal heating problem: observational evidence

There are, in general, two possible mechanisms suggested for the outstanding problem of coronal heating, *viz.*, the MHD wave heating and the magnetic reconnections (Section 3.5.4, page 71). Observational evidence for these suggestions is emerging from the recent UV, EUV and X-ray observations. The EUV observations at the temperatures of 10^5°K (i.e., in the thin transition region) have shown that there are localised hot spots that explode and shoot material upward at speeds of hundreds of km/s. These hot upward plasma jets occur above the lanes of the magnetic network. It has been claimed that the jets would carry enough energy to heat the corona but the observations are still inconclusive.

UV and EUV observations of the SOHO and TRACE satellites have shown that there are large numbers of small explosive events occurring all over the Sun. They may be occurring at the rate of 300-400 events per second during even the quiet phases of the solar cycle. These are known as microflares or nanoflares having energies in the range 10^{16} J - 10^{19} J. The Skylab mission in 1973 first revealed the presence of X-ray bright points (XBPs), and that their distribution is uniform over the whole Sun.

The brightest micro/nanoflares lie above enhanced elements of the magnetic network. It is suggested that the lower corona is not only heated but continuously replenished by chromospheric material that has been heated to coronal temperatures. A substantial part of the energy may come with the heated plasma from below. A possible scenario is that the new magnetic fluxes emerge from the Sun in the supergranular cells, which is then carried to their edges by convective motion. They finally reconnect with the magnetic fluxes from the adjacent cells and release energy to power the microflares observed in the overlying low corona.

There has been debate whether the small-scale flares are adequate to account for the coronal heating. There are observations both in support of this interpretation, and otherwise. However, SOHO and TRACE have conclusively demonstrated a correlation between the magnetic field and coronal heating. The variability of the small-scale magnetic elements observed in the photosphere (so-called *magnetic carpet*) does correlate with temperature fluctuations in the corona. Furthermore, observations of the temperature distribution forming polar plumes within the coronal holes seem to correlate with photospheric fine-structure associated also with the supergranular structure and magnetic network.

The coronal heating problem is not yet completely solved, but there are already several observational indications that the final picture may be emerging.

8.9 Some past solar space missions

In the previous sections, we have discussed the observations from several space-borne instruments, which have contributed to furthering our understanding of the invisible upper layers of solar atmosphere, comprising the transition region and the corona. In the following, we describe some of these major space missions, which followed after the early missions, starting with SkyLab in 1973.

8.9.1 Yohkoh (Solar-A) mission

Japan's Yohkoh (Sunbeam), or Solar-A, satellite was launched in 1991 to observe solar flares at X-ray wavelengths ([223], [368]). Yohkoh carried four instruments, *viz.*, the Soft X-ray Telescope (SXT), Hard X-ray Telescope (HXT), Bragg Crystal Spectrometer (BCS), and Wide Band Spectrometer (WBS). Yohkoh SXT was an X-ray telescope with a glancing incidence X-ray mirror of 1.54-m focal length forming X-ray images in the range 0.25–4.0 keV on a CCD sensor of 1024×1024 pixels with angular size of 2.45 × 2.45 arc-sec per pixel. Its field of view of 42 × 42 arc-min was larger than the whole solar disk. The schematic diagram of Yohkoh-SXT with optical set up and key elements is shown in Figure (8.2). Yohkoh HXT was a Fourier-synthesis X-ray imager sensitive to photons with energies from 14 keV to 93 keV. The angular resolution was about 5 arc-sec, image synthesis field of view was 2' × 2', and cadence was 0.5 s.

Yohkoh was successful in observing the X-ray Sun during the entire solar cycle, however, it lost the track to the Sun due to an annular eclipse in 2001, and eventually ended in 2005 due to its re-entry to Earth's atmosphere. In its more than a decade observing the Sun, it was the first spacecraft to have continuously observed the Sun in X-rays over an entire sunspot cycle. Yohkoh made numerous significant discoveries about the solar corona, solar flares and space weather. It established the canonical magnetic reconnection model on a much firmer observational foundation, and trans-equatorial interconnecting loops related to a class of CMEs. It discovered X-ray jets, and dimmings and established the fact that sigmoids were more likely to be associated with solar eruptive events, such as CMEs.

8.9.2 Hinode (Solar-B) mission

Hinode (also known as Solar-B) also focussed on observing soft- and hard X-rays from the solar corona. It was the follow up mission to Yohkoh, with the joint collaboration of Japan, USA and UK. It was launched on 22 September 2006, to explore the interaction between the solar magnetic fields and its corona. The Solar Optical Telescope (SOT) on board Hinode was a 50-cm Gregorian optical telescope with an angular resolution of about 0.2 arc-sec over the field of view of about 400×400 arc-sec^2 which produced images of the solar photosphere and chromosphere in different interference filters. The SOT spectro-polarimeter (SP) produced very sensitive vector magnetograms of the photosphere.

8.9.3 GOES satellites

The Geostationary Operational Environmental Satellite (GOES) system, operated by the United States National Environmental Satellite, Data, and Information Service (NESDIS), supports weather forecasting, severe storm tracking, and meteorology research. The GOES-14 satellite provides continuous monitoring of the solar integrated flux in soft X-rays (SXR). The GOES X-ray flux plot contains 5-minute averages of solar X-ray output in the 0.1–0.8 nm and 0.05–0.4 nm passbands. The GOES 12–15 spacecrafts each carry a sophisticated Solar X-ray Imager (SXI) to monitor solar X-rays for the early detection of solar flares, coronal mass ejections (CMEs), and other phenomena that impact the geospace environment. GOES data has been used for identifying the solar transient activity and its association with various physical parameters characterising the non-potential nature of active regions.

8.9.4 SOHO satellite

A joint mission of NASA, and the European Space Agency (ESA), Solar and Heliospheric Observatory (SOHO) was launched on 2 December 1995 to observe from the Sun-lit Lagrangian point L1 [327]. It has been one of the most successful, and longest lasting, solar observing satellites. As of 2019, only the LASCO coronagraph is still routinely producing data, with the other instruments all having been switched off. SOHO has provided continuous coverage of the Sun at many wavelengths.

The three main scientific objectives of SOHO were (i) investigation of the outer layer of the Sun, which consists of the chromosphere, transition region, and the corona. CDS, EIT, LASCO, SUMER, SWAN, and UVCS are used for solar atmosphere remote sensing, (ii) observations of solar wind and associated phenomena in its vicinity at L1 (CELIAS and CEPAC are used for "in situ" solar wind observations), and (iii) probing the internal structure of the Sun using three different instruments for helioseismology (Michelson Doppler Imager (MDI), Global Oscillations at Low Frequencies (GOLF), and Variability of solar Irradiance and Gravity Oscillations (VIRGO)).

The MDI instrument on board SOHO used the photospheric absorption line Ni I 676.78 nm to obtain full disc magnetograms, Dopplergrams and white-light images (MDI archive: `http://soi.stanford.edu/sssc/datasets.html`). The MDI data have spatial resolution of ~2"/pixel and cadence of 1 minute. The accuracy for the Doppler and magnetic fields observations are 20 m/s and 30 Gauss, respectively.

With the help of the instrument called LASCO (Large Angle and Spectrometric Coronagraph) on-board SOHO, which studies the structure and evolution of the corona by creating an artificial solar eclipse (see also Section 2.7, page 48, and Figure 2.10c), SOHO has discovered more than 3000 comets going around the Sun, including some Sun-grazing comets. SOHO had been the main source of near real-time solar data for space weather prediction. It has observed a large number of CMEs. Along with the GGS Wind, Advanced Composition Explorer (ACE) and DSCOVR, SOHO is one of four spacecraft in the vicinity of the Earth-Sun L1 point.

8.9.5 Transition Region and Coronal Explorer (TRACE)

Its major goal was to better understand how magnetic fields and plasma (superheated gas) act in the Sun's environment. The spacecraft, during its life (1998-2010), helped understand the high-resolution geometry and dynamics of the upper solar atmosphere and the nature of the hot corona. TRACE was also the first mission to study the Sun during an entire solar cycle.

8.9.6 STEREO mission

In order to obtain stereoscopic images of the Sun and solar phenomena, such as coronal mass ejections (CMEs), two identical spacecrafts, COR1 and COR2, were launched in October 2006 under a Solar Terrestrial Relations Observatory (STEREO) mission into orbits around the Sun, one ahead of Earth in its orbit, the other trailing behind, whose separation angle has varied with time. COR1 is an internally occulted coronagraph and is one of the STEREO SECCHI suite of remote sensing telescopes. It took observations of CMEs from 1.3 to 4 solar radii in three different polarization angles every five minutes. The last contact with STEREO-B was in September 23, 2016. These STEREO coronagraph data were very useful to study the CMEs beyond one solar radius by providing their stereoscopic observations [361].

8.9.7 The RHESSI mission

The Reuven Ramaty High Energy Solar Spectroscopic Imager (RHESSI; [243]) is a NASA small explorer mission, launched on 5 February 2002, with the primary aim to explore the basic physics of particle acceleration and explosive energy release in solar flares. RHESSI was designed to image solar flares in soft X-rays ($\sim$3 keV) to γ- rays (up to $\sim$20 MeV), and to provide corresponding spectroscopy. Following communication difficulties, RHESSI ceased science operations on 11 April 2018, and was decommissioned on 16 August 2018.

RHESSI provided spatial resolution of 2" at X-rays from $\sim$4 keV to $\sim$100 keV, 7" to $\sim$400 keV, and 36" for γ-ray lines, and continuum emission above 1 MeV. The first ever image of a solar flare in γ-rays was obtained by RHESSI [191]. The raw format data (i.e., modulated Fourier components) is used to construct HXR maps for the events using "clean" method as described in [190], and provided under the SolarSoft (SSW) package. RHESSI observations have immensely contributed to change our perspective on solar flares, particularly on high-energy processes in flares.

8.9.8 The Ulysses mission

Nearly all solar satellites observed the Sun from the plane of the ecliptic. Therefore, their observations provided details of solar equatorial regions, and not of the polar regions. To overcome this aspect, the Ulysses probe (1990-2009) was launched with the primary goal of studying the Sun's polar regions. It travelled to Jupiter, and then was 'slingshot' into an orbit far above the plane of the ecliptic. It made three-dimensional observations of solar wind and magnetic field strength at high solar latitudes. Its major discovery was that the solar wind from high latitudes was moving at about 750 km/s, which was slower than expected, and large magnetic waves emerging from high latitudes that scattered galactic cosmic rays (Figure 9.5, page 240).

8.9.9 Coronas-Photon mission

Coronas-Photon (Complex Orbital Observations Near-Earth of Activity of the Sun) was a Russian (Roskosmos) solar-terrestrial science mission, launched on 30 January 2009. It aimed for the study of the solar radiations from Extreme UV (EUV) up to high-energy γ-ray radiation. The Russian partners in this mission were from Poland, India, and Ukraine. It carried many instruments including a solar photometer in X-rays (SphinX; [133]), and a RT-2 γ-telescope contributed by TIFR/ICSP/VSSC, India. All scientific instruments on the satellite were turned off on 1 December 2009 due to some power problems.

8.10 Recent solar space-missions

8.10.1 Solar Dynamics Observatory (SDO)

To continue after the SOHO, another space mission fully dedicated to studies of the Sun, the Solar Dynamics Observatory (SDO), was launched on February 11, 2010, under NASA's "The Living With a Star (LWS)" program. Unlike the SOHO, SDO is observing the Sun from a circular geosynchronous orbit (located at 102° West), about 36,000 km above Earth. As a result, it faces periods of several weeks, when it is briefly blocked by the Earth on each day of observing the Sun.

It contains three main instruments: (1) the Atmospheric Imaging Assembly (AIA) for obtaining multi-wavelength images of the Sun, (2) the Helioseismic and Magnetic Imager (HMI) for solar Doppler and vector magnetic fields, and (3) the Extreme ultraviolet Variability Explorer (EVE) for solar irradiance studies (`https://sdo.gsfc.nasa.gov/`).

The AIA is an array of four Cassegrain telescopes that together provide full-disk images at 1 arc-sec resolution (4096×4096-pixel images) in 10 wavelengths of ultraviolet (UV) and extreme ultraviolet (EUV) windows every 10 seconds [235]. Each f/20 telescope has a 20 cm primary mirror and an active secondary mirror, having effective focal length of 4.125 m. These are designed to observe solar emissions from the transition region and corona. Table 8.2 gives the details of various wavelength channels of AIA along with the corresponding regions of solar atmosphere at a characteristic temperature.

8.10.2 Interface Region Imaging Spectrograph (IRIS)

Interface Region Imaging Spectrograph (IRIS) (2013-present) focusses on the interface (transition) region of the Sun, to obtain information on how solar material moves through the Sun, as well as contributing to the heating

TABLE 8.2: The primary ions observed by SDO/AIA

Channel nm	Primary ions (s)	Atmospheric region	Char. Log(T)
450	Continuum	Photosphere	3.7
170	Continuum	Temp. min., photosphere	3.7
30.4	He II	Chromosphere, TR	4.7
160	Ca IV-cont	TR, upper photosphere	5.0
17.1	Fe IX	Quiet corona, upper chromosphere	5.8
19.3	Fe XII, XXIV	corona and hot flare plasma	6.2, 7.3
21.1	Fe XIV	active region corona	6.3
33.5	Fe XVI	active region corona	6.4
9.4	Fe XVIII	flaring corona	6.8
13.1	Fe VIII, XXI	TR, flaring corona	5.6, 7.0

of the corona. This region emits most of the UV radiation. IRIS obtains high-resolution UV spectra and images of the Sun's upper chromosphere and transition region, specifically on the non-thermal energy. It detected in 2016 tiny solar 'bombs' (micro- and nano-flares) that may contribute to solar coronal heating.

8.10.3 Deep Space Climate Observatory (DSCOVR)

The Deep Space Climate Observatory (DSCOVR) was launched on February 11, 2015. It is at the Sun–Earth L1 Lagrangian point, 1.5 million km from Earth. It has the main aim of monitoring variable solar wind conditions, providing early warning of approaching CMEs, and maintaining the accuracy and lead time for space weather alerts and forecasts.

8.10.4 Parker Solar Probe

NASA's Parker Solar Probe was launched on August 12, 2018, and is the first-ever mission to observe the solar corona from within the Sun's atmosphere, about 6.5 million km from the Sun's surface. Its four instrument suites are designed to characterise the dynamic region of the corona by measuring particles and electric and magnetic fields. Each of these are specially designed to withstand the harsh radiation and temperatures they encounter at such close distances from the Sun. These instruments are Solar Wind Electrons Alphas and Protons investigation (SWEAP), Wide-Field Imager for Parker

Solar Probe (WISPR), and FIELDS instrument for the scale and shape of electric and magnetic fields in solar atmosphere, and for observing waves and turbulence in the inner heliosphere. The Integrated Science Investigation of the Sun (IS$\odot$IS) is for measuring particles across a wide range of energies.

8.10.5 Aditya-L1 mission

The Indian Space Research Organisation (ISRO) has planned to launch a satellite named *Aditya-L1*, carrying a coronagraph for studying the dynamics of the solar corona. It is expected to be launched in 2020. Aditya-L1 will provide observations of the soft and hard X-ray, UV, and the corona in visible emission lines and NIR. In addition, particle payloads will study the solar wind particle flux emanating from the Sun.

8.11 Summary

The most obvious consequence of the high temperatures in TR and corona is the ionization of the atoms. This has a major impact on the MHD of the upper atmosphere because the electric conductivity becomes high and the diffusivity drops dramatically. For an observer, the suite of emission lines provides an unique method to probe the transition region and the corona. Space-borne instruments have made use of these lines to make specific diagnostics of TR and corona, which have helped in pushing the frontiers of the physics of these layers from speculation to better clarity.

Various thermal and non-thermal processes operate in the corona during the quiet, and active phases of flares and CMEs, giving rise to enhanced emissions across the entire range of high energy of the electromagnetic spectrum. Observations of these emissions have helped in our understanding of the physics of energetic phenomena occurring in the solar atmosphere.

Inside the Sun the energy transfer is totally dominated by radiation and convection. In the atmosphere the radiation takes over. In the coronal plasma it is also important to include the thermal conductivity. Far from the Sun, in the solar wind, the thermal conductivity is a non-trivial issue because in the collisionless magnetized plasma the thermal conductivities of different particle species are different and difficult to determine from theory. Finally the heating is necessary to drive the supersonic expansion of the solar wind. Some of these aspects shall be discussed in the next chapter.

Chapter 9

Solar Wind, Space Weather, and the Heliosphere

9.1 Introduction

The solar system objects, including the planets that revolve around the Sun, are immersed not in vacuum, but in a space that is filled with charged particles and radiation emanating from the ever-expanding solar corona. This is the *solar wind* blowing outwards from the Sun that carries away around a million tons of solar plasma per second. It constitutes the *interplanetary medium*(IPM) reaching beyond Pluto, and finally meeting the interstellar plasma, where a pressure balance stops its further flow. The space of IMF, called the heliosphere, is a bubble of solar influence in space produced by the solar wind as the Sun moves through the interstellar space around the centre of the Milky Way.

Within this IPM, various temporal and spatial changes take place due to solar activity and different types of transient, energetic phenomena occurring in the outer solar atmosphere. In addition, there can also be temporal changes that may occur in-situ due to the localised physical processes occurring in the medium itself. The term *space weather* covers the temporal condition of the interplanetary medium, and the environment surrounding the planets including their magnetospheres, ionospheres and exospheres. Space weather has increasingly become an important branch of space physics due to the increasing dependence on satellites, space stations, communication and observing instruments, and astronauts operating in outer space.

There were some indications already in the 19th century, when Carrington's flare was discovered in 1859 (Section 1.6.6, page 14), that there was some connection between the Sun's activity and magnetic disturbances on the Earth. Unfortunately, this aspect of solar-terrestrial connection could not be explored further due to the strong objection by Lord Kelvin that any such connection was only a coincidence, and was fictitious, thus delaying the birth of space weather by many decades.

9.2 Solar wind: a historical perspective

In 1910, Arthur Eddington suggested the existence of the solar wind, without naming it, in a footnote to an article on Comet Morehouse. But, it was Kristian Birkeland who first inferred in 1916, from his studies of geomagnetic and auroral activities, that the ejected material consisted of both ions and electrons. Birkeland proposed that "From a physical point of view, it is most probable that solar rays are neither exclusively negative nor positive rays, but of both kinds". Chapman proposed in 1929 that solar flares would emit plasma clouds, and could cause magnetic disturbances if the cloud hit the Earth's magnetic field. But the question arose: how could such a cloud escape the intense gravity of the Sun?

Around the 1930s, it was determined that the temperature of the solar corona must be a million degrees from the way it stood out into space, as seen during total eclipses. Later, spectroscopic work confirmed these extraordinarily high temperatures (Section 1.6.7). In the mid-1950s, Sydney Chapman calculated the properties of a gas at such temperatures. He determined that it should be a superb conductor of heat, and it must extend out beyond the orbit of Earth [76].

FIGURE 9.1: The two distinct plasma and dust tails are clearly visible in this picture of Comet Hale-Bopp due to solar wind and the radiation pressure of the Sun. Photo Credit: A. Dimai and D. Ghirardo (Col Druscie Obs., Astronomical Association of Cortina, Italy.)

Ludwig Biermann pointed out in 1951 that a comet, headed towards or away from the Sun, displayed its tail always pointed away from the Sun. Biermann postulated that the Sun emits a steady stream of particles that pushes the comet's tail away [46] (Figure 9.1). Alfvén suggested that this flow must be magnetized plasma. This was confirmed later in 1959 by the direct observations made from outside the Earth's atmosphere by Soviet Lunik III and Venus I spacecrafts. The continuous flow of magnetised plasma, as solar wind, was further confirmed by the US Mariner II during 1962-67. However,

Wilfried Schröder has claimed that it was Paul Ahnert who first related solar wind to comet tail direction, based on observations of the comet Whipple-Fedke [5].

In the following, we briefly discuss some theoretical models, which indicated that solar corona could not be in static equilibrium, but must be a dynamic component of the Sun.

9.2.1 Chapman's model of static corona

Chapman (1957) developed a model for static corona, based on thermal conduction. He obtained an estimate for its temperature at the Earth's orbit, *i.e.*, 1 A.U. from the Sun, and then proceeded to calculate its density under the assumption of hydrostatic equilibrium. Accordingly, the heat flux is given by $Q = \kappa \nabla T$, where thermal conductivity $\kappa = \kappa_0 T^{5/2}$.

In the absence of heat source or sinks, $\nabla . Q = 0$, which provides

$$\nabla .(\kappa \nabla T) = 0 \Rightarrow \nabla .[(\kappa_0 T^{5/2}) \nabla T] = 0. \tag{9.1}$$

In spherical coordinates, this can be written as

$$\frac{1}{r^2}\frac{d}{dr}\left\{ r^2 \kappa_0 T^{5/2} \frac{dT}{dr} \right\} = 0. \tag{9.2}$$

On integration, Equation (9.2) leads to

$$r^2 \kappa_0 T^{5/2} \frac{dT}{dr} = c \Rightarrow \int T^{5/2} dT = \frac{c}{\kappa_0} \int \frac{dr}{r^2} \Rightarrow \frac{2}{7} T^{7/2} = -\frac{c}{\kappa_0 r}. \tag{9.3}$$

The constant c can be set by ensuring temperature $T = T_0$ at the base of solar surface, *i.e.*, $r = r_0$, and $T \to 0$ as $r \to \infty$. This determines $c = -(2\kappa_0 r_0 T_0^{7/2}/7)$.

Thus, temperature T at a distance r from the Sun is

$$T = T_0 \left(\frac{r_0}{r} \right)^{2/7}. \tag{9.4}$$

Assuming $T_0 = 2 \times 10^6 K$ at $r_0 = 1.05 R_\odot$ as the base of corona, Equation (9.4) gives $T \sim 4 \times 10^5 K$ at the Earth's location ($1\ A.U. = 214\ R_\odot$). This is close to the measured value, although it was not known in 1957, and was actually observed from space at a later stage.

Assuming a hydrostatic equilibrium for corona, one has from the equation of momentum conservation,

$$\frac{dP}{dr} = -g\rho, \tag{9.5}$$

where $\rho = n(m_e + m_p) \approx n m_p$ is the plasma density, m_p, and m_e are proton and electron masses, respectively.

Since electrons and protons both contribute to the pressure, we have $P = 2nk_BT = 2\rho k_BT/m_p$. Substituting for P in Equation (9.5), and integrating for density ρ, one obtains

$$\rho = \rho_0 exp\left(-\frac{mgr}{k_BT}\right), \tag{9.6}$$

where ρ_0 is the density at $r \sim R_\odot$. In Equation (9.5), now substituting for pressure, $P = \rho k_BT/m_p$, and T from Equation (9.4) provides

$$\frac{d}{dr}\left\{2\frac{\rho k_BT_0}{m_p}\left(\frac{r_0}{r}\right)^{2/7}\right\} = -\frac{GM\rho}{r^2}. \tag{9.7}$$

Simplifying, and integrating for density ρ from ρ_0 (at $r = r_0$) to $\rho(r)$ at r,

$$\int_{\rho_0}^{\rho}\frac{d\rho}{\rho} = \frac{2}{7}\int_{r_0}^{r}\frac{dr}{r} - \frac{GMm_p}{2k_BT_0r_0^{2/7}}\int_{r_0}^{r}\frac{dr}{r^{12/7}}, \tag{9.8}$$

or,

$$\rho(r) = \rho_0\left(\frac{r}{r_0}\right)^{2/7} exp\left[\frac{7}{5}\frac{GMm_p}{2k_bT_0r_0}\left\{\left(\frac{r_0}{r}\right)^{5/7} - 1\right\}\right]. \tag{9.9}$$

Substituting for $\rho(r)$ from Equation (9.9), and for T from Equation (9.4), in $P = 2nk_BT = 2\rho k_BT/m_p$, the expression for pressure at any point r is obtained as

$$P(r) = P_0 exp\left[\frac{7}{5}\frac{GMm_p}{2k_bT_0r_0}\left\{\left(\frac{r_0}{r}\right)^{5/7} - 1\right\}\right]. \tag{9.10}$$

From Equations (9.9) and (9.10), one gets $\rho \to \infty$ as $r \to \infty$, which is unphysical. Similarly, as $r \to \infty$, one gets $P \to constant \sim 10^{-4}P_0$. This is far greater than $P_{ISM} \sim 10^{-15}P_0$.

From these considerations, Eugene Parker concluded that it is not correct to consider the corona in a static equilibrium. In other words, the corona must be dynamic - either expanding, or collapsing.

9.2.2 Parker's dynamic model of expanding corona

Eugene Parker realised the heat flowing from the Sun in Chapman's model, and the comet tail blowing away from the Sun in Biermann's hypothesis, had to be the result of the same phenomenon, which he termed the "solar wind". In 1958, in an elegant paper, Parker showed that even though the Sun's corona is strongly attracted by solar gravity, it is still very hot at large distances due to its large conductivity. Since gravity weakens as distance from the Sun increases, the outer coronal atmosphere can escape supersonically into the interplanetary space [296]. Parker developed a dynamic model due to the problems inherent in the assumption of hydrostatic equilibrium in Chapman's model.

The high temperature and high thermal conductivity of coronal plasma imply that the corona can be assumed to be approximately isothermal. As $P_\odot >> P_{ISM}$, there must be a steady symmetric outflow from the Sun. This can be given by the momentum conservation equation describing the acceleration of gas due to a pressure gradient and gravity:

$$\rho\frac{dv}{dt} \equiv \rho v\frac{dv}{dr} = -\frac{dP}{dr} - \frac{GM_\odot \rho}{r^2}. \quad (9.11)$$

Assuming a perfect gas, pressure $P = R\rho T/\mu$, where R is the gas constant and μ is mean atomic mass. Thus, the pressure gradient term in Equation (9.11) can be rewritten as

$$\frac{dP}{dr} = \frac{R\rho}{\mu}\frac{dT}{dr} + \frac{RT}{\mu}\frac{d\rho}{dr} \Rightarrow \frac{RT}{\mu}\frac{d\rho}{dr}, \quad (9.12)$$

where, $dT/dr \rightarrow 0$ is used for isothermal wind. Also, considering the rate of mass loss as continuous, or constant, the equation of continuity (or the mass conservation law) gives:

$$\frac{dM}{dr} = 4\pi r^2 \rho v = constant \Rightarrow r^2\rho v = constant. \quad (9.13)$$

On differentiating Equation (9.13) with respect to r:

$$\frac{d(r^2\rho v)}{dr} = r^2\rho\frac{dv}{dr} + 2r\rho v + r^2 v\frac{d\rho}{dr} = 0 \Rightarrow \frac{1}{\rho}\frac{d\rho}{dr} = -\frac{1}{v}\frac{dv}{dr} - \frac{2}{r} \quad (9.14)$$

Substituting it in Equation (9.12) for dP/dr, and then replacing dP/dr in Equation (9.11), the following relationship is obtained

$$v\frac{dv}{dr} + \frac{RT}{\mu}\left(-\frac{1}{v}\frac{dv}{dr} - \frac{2}{r}\right) + \frac{GM_\odot}{r^2} = 0, \quad (9.15)$$

or,

$$\left(v - \frac{RT}{\mu v}\right)\frac{dv}{dr} - \frac{2RT}{\mu r} + \frac{GM_\odot}{r^2} = 0. \quad (9.16)$$

Equation (9.16) shows that there occurs a critical point when $dv/dr \rightarrow 0$, where $2RT/(\mu r) = GM_\odot/r^2$. Thus, setting $v_c = \sqrt{RT/\mu}$, and $r_c = GM_\odot/(2v_c^2)$, and rearranging Equation (9.16), the momentum equation is obtained in terms of the flow velocity

$$(v^2 - v_c^2)\frac{1}{v}\frac{dv}{dr} = 2\frac{v_c^2}{r^2}(r - r_c). \quad (9.17)$$

From this relation, it can be noted that if $r = r_c$, $dv/dr \rightarrow 0$ or $v = v_c$. Also, if $v = v_c$, $r = r_c$ or $dv/dr \rightarrow \infty$. The acceptable solution is when $r = r_c$, and $v = v_c$, which is the critical point. The solution to Equation (9.17) is found by integrating it, which provides Parker's solutions

$$\left(\frac{v^2}{v_c^2}\right) - ln\left(\frac{v^2}{v_c^2}\right) = 4ln\left(\frac{r}{r_c}\right) + 4\frac{r_c}{r} + C, \tag{9.18}$$

where C is a constant of integration. Equation (9.18) leads to five different solutions, depending on the value of C (Figure 9.2). The critical point fixes it as -3. Solutions I and II are double valued, and unphysical. Solution III has very high supersonic source speed everywhere. Solution IV is subsonic everywhere, and is physically valid. It is called the *solar breeze solution*, which provides $v << v_c$ at large distances. Here, the isothermal sound speed v_c defines the subsonic or supersonic speeds.

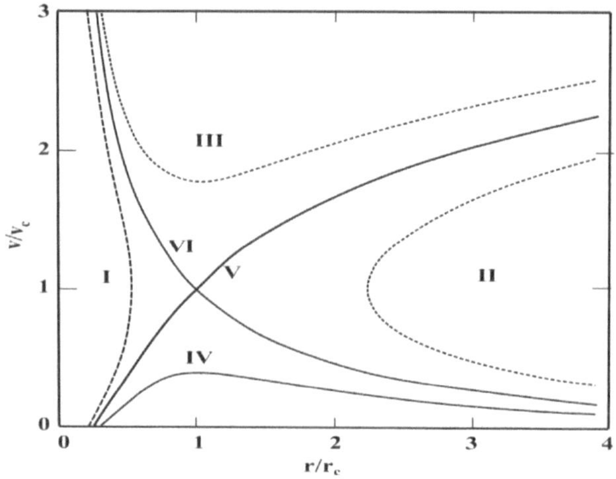

FIGURE 9.2: Parker's solution from Equation (9.18).

Only Solutions V and VI pass through the critical point $r = r_c$, and $v = v_c$, and are supersonic and subsonic at different regions. Of these, let us see Parker's solution V of $v >> v_c$ for large r, as considered by Parker for solar wind. It starts from subsonic near the solar surface, and then accelerates to high speeds. Parker's equation (9.18) can be written as

$$\left(\frac{v^2}{v_c^2}\right) \approx 4ln\left(\frac{r}{r_c}\right) \Rightarrow v \approx \left[ln\left(\frac{r}{r_c}\right)\right]^{1/2} \tag{9.19}$$

From $\rho r^2 v = c$ one gets the density $\rho = c/(r^2 v) = c/(r^2\sqrt{ln(r/r_c)})$, which implies that $\rho \to 0$ and also, $P \to 0$ as $r \to \infty$, as plasma is isothermal (*i.e.*, $T = const$), and ideal gas law is applicable between pressure and density. This solution matches with interstellar gas properties at large distances; hence it is a physically realistic model, and is called the *solar wind solution*.

Furthermore, Parker noticed that the weakening effect of gravity has the same effect on hydrodynamic flow as a *de Laval nozzle*: it incites a transition from subsonic to supersonic flow. Parker's papers generated a lot of controversy at that time, and were not generally accepted in the scientific community. In fact, the referees recommended rejection of the papers, but they were rescued by Subramaniam Chandrasekhar, then editor of the Astrophysical Journal. Later, Parker was vindicated in January 1959, when the Soviet spacecraft Luna 1 first directly observed the solar wind, and measured its strength, using hemispherical ion traps. This discovery, made by Konstantin Gringauz, was further verified by Luna 2, Luna 3, and by the more distant measurements of Venera 1, and the Mariner 2 spacecraft [283].

Parker's solution is too simplified for real solar wind, as it faces several difficulties due to its assumption of isothermal case. In fact, solar wind cools as it expands in space, therefore, thermal conduction and heating effects must be included to be more realistic. As the coronal plasma is very tenuous, it is collisionless, and hence, protons and electrons cool with different cooling rates. Also, the interaction of plasma with magnetic field fluctuations affect electron and proton expansion differently. Inspite of these complexities, Parker's solutions provided elegant and important insights about solar wind.

9.2.3 Heliospheric magnetic field: the Parker spiral

Solar wind carries magnetic field from the Sun to interplanetary space. This field is called the *interplanetary magnetic field* (IMF), which is maintained by electric currents in the expanding solar wind. The heliospheric magnetic field is dipolar in structure, in which magnetic fields remain attached to the Sun as the solar wind carries the magnetic field lines outward.

Let us consider a cylindrically symmetric case in the equatorial plane, and assume that the flow is radial, the angular speed of the solar rotation is Ω, and the angle between the radial direction and the magnetic field be ψ (Figure 9.3). Also, let us assume that the IMF is frozen into the expanding solar wind, such that close to the Sun the plasma rotates with the body, but in the radial expansion as the Sun rotates, the field lines attain spiral structure.

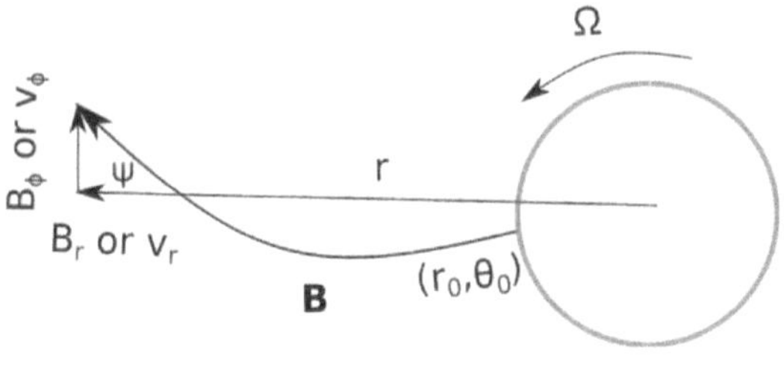

FIGURE 9.3: Parker's spiral.

Let v be the flow velocity, then its component perpendicular to the field is $v_\perp = v \sin\psi$, which can be imagined as the speed of the field line in this direction. The high plasma conductivity ties the field line to the surface of the

Sun. Thus, the speed of the field line perpendicular to the radial direction is $\Omega(r - r_\odot)$, and $v \sin\psi = \Omega(r - r_\odot)\cos\psi$. From this, one obtains

$$\tan\psi = \frac{\Omega(r - r_\odot)}{v}. \tag{9.20}$$

As r increases, this approaches an Archimedean spiral, called *Parker's spiral* for solar magnetic field. At 1 A.U., the equatorial spiral angle is about 45°, and at 10 A.U., it is 90°.

In spherical coordinates, magnetic field $\vec{B} \equiv (B_r, 0, B_\phi)$ is radial, and constant on the solar surface, and velocity $\vec{v} \equiv (v_r, 0, v_\phi)$ is determined by the rotation axis of the Sun, and the components of $\vec{B}$ and $\vec{v}$ are functions of r.

From $\nabla.\vec{B} = 0$ one has $B_r = B_0(r_\odot/r)^2$, *i.e.*, B_r decreases as $1/r^2$ with distance. The azimuthal component B_ϕ can be evaluated from the momentum conservation equation

$$\rho(\vec{v}.\nabla\vec{B})_\phi = (\vec{j} \times \vec{B})_\phi \Rightarrow r^2\rho v_r \frac{d}{dr}(rv_\phi) = \frac{1}{\mu} r^2 B_r \frac{d}{dr}(rB_\phi), \tag{9.21}$$

where $(\nabla p)_\phi$ is neglected, and use of the Ampere's law has been made. Then, both L.H.S. and R.H.S. are multiplied by r^3. Here, the mass flux $(r^2\rho v_r)$ and magnetic flux $(r^2 B_r)$ are constants. By integrating Equation (9.21), one obtains

$$L = rv_\phi - \frac{rB_rB_\phi}{\mu\rho v_r}, \tag{9.22}$$

where, L, the constant of integration is the angular momentum per unit mass. In a frame which rotates with the angular speed Ω, we have the velocity $\vec{v} \equiv (v_r, 0, v_\phi - r\Omega)$, which is parallel to $\vec{B}$. Therefore,

$$B_\phi = \frac{v_\phi - r\Omega}{v_r} B_r \Rightarrow B_\phi = \frac{v_\phi - r\Omega}{v_r} B_0 \left(\frac{r_\odot}{r}\right)^2. \tag{9.23}$$

This implies that at large distances, B_ϕ decreases as $1/r$, *i.e.*, it decreases more slowly than B_r. This behaviour explains the formation of the spiral magnetic field configuration.

The calculation of magnetic field out of the equatorial plane $\theta \neq \pi/2$ is more complex, where B_ϕ is given by

$$B_\phi \approx -\frac{B_0 r_\odot^2 \Omega \sin\theta}{rv_r}. \tag{9.24}$$

From these relations, one can infer that the magnitude of magnetic field B, far away from the Sun, in the equatorial plane decreases as $1/r$, while in the equatorial direction, it varies as $1/r^2$. The field is directed inward in one hemisphere, and outward in the other.

9.2.4 Heliospheric magnetic field and solar cycle

During the minimum phase of solar activity, Sun's magnetic field is nearly poloidal, and the coronal structure is dominated by two large polar regions with opposite magnetic polarities. Solar wind flows radially out from these regions of open fields (coronal holes), dragging the frozen-in magnetic field to form a heliospheric current sheet near the equatorial plane. During quiet periods, the current sheet can be nearly flat. Since the Sun rotates with a period of 27 days, the solar wind becomes a complex spiral of high and low speeds, and high and low densities.

As solar activity increases, the solar surface fills with active regions, coronal holes, and other complex structures, which modify the solar wind and the current sheet. The current sheet moves up and down, which led Alfvén to call this structure the ballerina's twirling skirt (Figure 9.4). As the Earth moves through this evolving ballerina skirt, it is sometimes within the heliospheric current sheet, sometimes above it and sometimes below it. Depending on whether the field is pointing mostly toward or away from the Sun, the Earth is either in the toward sector, or the away sector.

FIGURE 9.4: The heliospheric magnetic field - the ballerina's twirling skirt. The spiral shape results due to the effect of the Sun's rotating magnetic field on the interplanetary plasma. Credit: Werner Heil/NASA.

This sector structure, discovered in 1965 by N. Ness and J. Wilcox in data from the IMP I satellite, has been shown to be quite stable for years. When the magnetic field of the solar wind switches polarity, it is a strong indication that Earth has crossed the current sheet. The location of the Earth with respect to the current sheet is important because space weather impacts are highly dependent on the solar wind speed, the solar wind density, and the direction of the magnetic field embedded in the solar wind. Superposed to this large-scale structure, there are large variations in all components of the IMF. For further details, one may refer to [292].

Around solar maximum the solar magnetic field structure turns more complex. At this phase, polar coronal holes reduce in size, while at lower latitudes, closed field line structures associated with active regions start appearing. Temporal variations are more during this phase due to change in the active regions size, and explosive eruptive events.

9.3 Some properties of solar wind

The solar wind plasma consists mainly (about 95% by number) of protons and electrons, with the rest consisting almost entirely of doubly ionized helium nuclei (α-particles). The helium abundance varies significantly during energetic transient events. The total plasma density mostly lies between 3–20 particles/cm^{-3} with an average value of about 10 particles/cm^{-3}. The solar wind velocity is mostly in the range of 350–750 km/s, and changes a little with distance between 1-20 A.U.

The magnetic field intensity ranges between 2–10 γ (1 $\gamma = 10^{-5}$ G). The plasma temperature is found from the particle velocity dispersion in the frame of reference of the plasma bulk motion. The mean values of the temperatures of proton and electron components are both between 100,000 and 150,000 K, while the α-particles are 4–5 times hotter.

The radial variation in the main solar wind parameters, such as the density, mass flux, and bulk kinetic energy agree with an inverse square dependence, at least in the range $0.81 < r < 1.5$ A.U. Embedded within the solar-wind plasma is the interplanetary magnetic field, found to follow a $r^{-1.5}$ relation.

The total number of particles carried away from the Sun by the solar wind is about 1.3×10^{36} per second, or about $(1.3 - 1.9) \times 10^6$ tons/s, or the total mass loss per year is around $(2 - 3) \times 10^{-14} M_\odot$ [273]. Other stars may have even stronger stellar winds that result in significantly higher mass loss rates. For further details on solar wind properties, one may refer to [273].

Table 9.1 lists some of the solar wind parameters during different phases of solar activity. These are functions of solar latitude. For example, density is maximum, and speed is minimum around the equator. However, the hemispheres are not exactly symmetric.

Periodicities in the solar wind can be divided into the time scales of the solar processes, *i.e.*, rotation period of the Sun, and the orientation of Earth with respect to the Sun. It also varies according to the 11-year solar activity cycles, and the 1.3-year and 154-day cycles, apart from effects related to transient events on the Sun.

TABLE 9.1: Solar wind characteristics

Parameter	Minimum phase	Average	Maximum phase
Flux ($cm^{-2}s^{-1}$)	1	3	100
Velocity (km/s)	200	400	900
Density (cm^{-3})	0.4	6.5	100
Helium %	0	5	25
B (nT)	0.2	6	80

9.3.1 Fast and slow solar wind

The solar wind is highly variable and structured depending on factors governed by the presence of coronal holes, active regions, and solar transient eruptions. There are the streams of *slow solar wind*, and *fast solar wind*. In near-Earth space, the slow solar wind has a velocity of 300–500 km/s, temperature of 1.4–1.6 $\times$ 10^6 °K, and a composition that is a close match to the ambient corona. By contrast, the fast solar wind has a typical velocity of 750 km/s, and a temperature of 8×10^5 °K. It nearly matches the composition of the Sun's photosphere [134]. The slow solar wind is twice as dense, and more variable in nature, than the fast solar wind [357].

The solar equatorial plane is the region of slow speed solar wind, with speeds of about 400 km/s. This region is also known as the streamer belt, where coronal helmet streamers are produced by magnetic fields opening up at higher altitudes to the heliosphere, and draping over closed magnetic loops of active regions. This portion of the solar wind forms the equatorial current sheet. The exact coronal structures involved in slow solar wind formation and the release of plasmas is still debatable [15]. Observations of the Sun between 1996 and 2001 showed that slow solar wind emissions occurred at latitudes up to 30–35° during the solar minimum that expanded polewards as the solar cycle approached maximum [267].

Different regions on the Sun produce solar wind of different speeds and densities. Coronal holes produce fast solar wind of high speed in the range of 500–800 km/s [410]. The escape of fast solar wind streams varies with the presence or absence of coronal holes, and as they change in size, shape and location on the Sun. The north and south poles of the Sun are regions of large persistent coronal holes, which possess open field-lines. Fast streams are also observed from chromospheric magnetic networks [168].

Solar transients also eject out slow and fast plasma clouds, which can create shocks in the interplanetary medium. The plasma source is small magnetic fields created by convection cells in the solar atmosphere. These fields confine the plasma and transport it into the narrow necks of the coronal funnels, which are located only 20,000 kms above the photosphere. The plasma is released into the funnel when these magnetic field lines reconnect.

When the high-speed solar wind overtakes the slow-speed wind, it results in a corotating interaction region (CIR). These CIRs consist of solar wind with

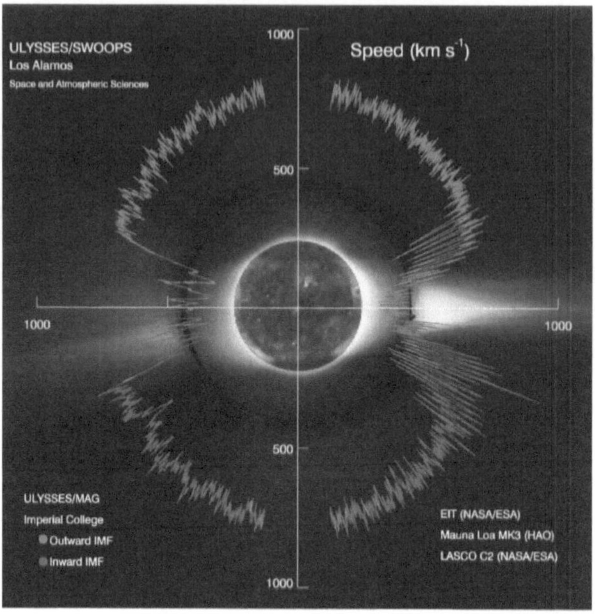

FIGURE 9.5: Ulysses' observations of solar wind speed as a function of helio latitude during solar minimum. Slow wind ($\sim$ 400 km/s) is confined to the equatorial regions, while fast wind ($\sim$750 km/s) is seen over the poles. Also shown are inward (top hemisphere)/outward (bottom hemisphere) polarities of the heliospheric magnetic field. With permission from D.J. McComas [266], also `https://swoops.lanl.gov/lasco_swoops.html`.

very high densities and strong magnetic fields, and can trigger geomagnetic storms when they interact with Earth's magnetosphere.

NASA's Ulysses mission, launched on October 6, 1990, studied the Sun at various latitudes, and measured the various properties of the solar wind over the course of nearly two decades (the mission ended in 2009) (Figure 9.5). Due to the large orbital velocity of the Earth (30 km/s), Ulysses had to be sent first to Jupiter to be deflected from the ecliptic plane using the planet's large gravitation field. Consequently, the final orbit of the spacecraft had its aphelion of 5.3 A.U. at the distance of Jupiter, and perihelion at 1.3 A.U., with the highest heliographic latitude of 80°. The Ulysses mission over the solar poles has greatly improved our knowledge of the three-dimensional structure of the heliosphere.

The Advanced Composition Explorer (ACE) satellite was launched in 1997 to the Lagrangian L1 point (see Section 10.6.1) to measure the solar wind and real-time constant flow of solar wind particles. It was planned initially for a five-year period, but has been operational to date. Spacecraft having their orbits beyond the Earth's magnetospheric cavity directly sample the charged particles and fields flowing out from the Sun. Such in situ measurements account for most of our understanding of the solar wind near the plane of the Earth's orbit.

The information about magnitude and orientation of plasma density fluctuations over scales of tens of kilometres at high solar latitudes is also obtained from interferometric studies of radio wave propagation from cosmic sources, such as the Crab nebula, as the line of sight to such a source moves through the outer corona. These fluctuations cause the interplanetary scintillations of the compact sources in an analogy to the optical scintillations of point sources, such as stars by Earth's atmosphere.

9.3.2 Acceleration of solar wind particles

In the late 1990s, the Ultraviolet Coronal Spectrometer (UVCS) instrument on board the SOHO spacecraft observed the acceleration region of the fast solar wind emanating from the poles of the Sun, and found that the wind accelerates much faster than can be accounted for by thermodynamic expansion alone. Parker's model predicted that the solar wind should make the transition to supersonic flow at an altitude of about 4 $R_\odot$ from the photosphere. But the transition (or sonic point) appears to be much lower, around 1 $R_\odot$ above the photosphere, suggesting that some additional mechanism accelerates the solar wind away from the Sun. The acceleration of the fast wind is still debatable, and cannot be fully explained by Parker's theory. The gravitational and electromagnetic explanation for this acceleration was suggested by Hannes Alfvén in 1970. Acceleration of the solar wind is also suggested to be due to the reconnection of open magnetic flux with coronal loops [120].

In 2018, NASA launched the Parker Solar Probe, named in the honour of Eugene Parker, on a mission to study the structure and dynamics of the solar corona. It will attempt to understand the mechanisms that cause particles to be heated and accelerated as solar wind. During its expected seven year mission, the probe will make twenty-four orbits of the Sun, passing further into the corona with each orbit's perihelion, ultimately passing within 0.04 A.U. of Sun's surface. It is the first NASA spacecraft named for a living person. Eugene Parker himself, at age 91, was on hand to observe the launch.

9.4 The interplanetary medium: quiet and disturbed

The interplanetary medium (I.P.M.) is essentially the hot solar wind plasma which fills the solar system, and in which all the solar system bodies, such as planets, dwarf planets, asteroids, and comets, are immersed. Apart from solar wind plasma, it also includes interplanetary dust and cosmic rays. The temperature of the interplanetary medium varies with distance from the Sun. For dust particles within the asteroid belt, typical temperatures range from 200°K at 2.2 A.U. down to 165°K at 3.2 A.U. [251]. The density of the interplanetary medium is very low, about 5×10^4 *particles*/m^3 in the vicinity of the Earth. However, the I.P.M. density is variable, and may be affected by magnetic fields and solar transient events such as coronal mass ejections, where it may increase to as high as 100 $\times10^4$ *particles*/m^3.

The plasma in the interplanetary medium is responsible for the strength of the Sun's magnetic field at the orbit of the Earth being much greater than originally expected. If space were a vacuum, then the Sun's 10^{-4} Tesla magnetic dipole field would reduce with the cube of the distance to about 10^{-11} Tesla. But satellite observations reveal that it is about 100 times greater at around 10^{-9} Tesla. Magnetohydrodynamic (MHD) theory predicts that the motion of a conducting fluid (*e.g.*, the interplanetary medium) in a magnetic field induces electric currents, which in turn generate magnetic fields, and in this respect it behaves like an MHD dynamo.

The interplanetary medium is responsible for several effects which can be seen from Earth. The *Zodiacal light* is a broad band of faint light, sometimes seen after sunset and before sunrise, stretched along the ecliptic and brightest near the horizon. It is caused by sunlight scattered off the dust particles in the interplanetary medium between the Earth and the Sun. A similar effect is the "*Gegenschein*", which is seen directly opposite to the Sun's position in the sky. It is much fainter than the Zodiacal light, and is caused by sunlight reflecting off dust particles outside the Earth's orbit.

9.4.1 Monitoring the I.P.M. from space

NASA's STEREO mission was launched on October 25, 2006, with the primary goal of stereoscopic measurements to study the Sun and space weather, particularly, the CMEs. One of its instruments, SWAVES, was an interplanetary radio burst tracker to trace the generation and evolution of traveling radio disturbances from the Sun to the orbit of Earth ([72], [54]). SWAVES carried several radio receivers for measuring: (i) radio wave intensity, source direction, and angular size in the frequency range of 16 MHz to 40 kHz, corresponding to source distances of about $1R_\odot$ to 1 A.U., (ii) radio and plasma waves near the electron plasma frequency at 1 A.U. (10-40 kHz), (iii) radio emissions at a fixed frequency of 50 MHz, at high time resolution, to complement ground-based radioheliograph measurements. (For details, `https://swaves.gsfc.nasa.gov/swaves_science.html`)

With radio probing and in situ observations from two identical stereoscopic spacecraft, the SWAVES experiment tracked the radio emission associated with CMEs in 3-D as they propagate from less than $1R_\odot$ above the photosphere out to 1 A.U. and beyond. It used radio imaging to study CMEs at medium frequency (300kHz - 3000 kHz). This instrument was intended to study generation, evolution, and interaction of CMEs with Earth's magnetosphere. Its aim was to measure interplanetary type II (associated with the propagation of CMEs in the corona and interplanetary medium), and type III radio bursts (associated with impulsively accelerated electrons streaming outward from the Sun on open field lines), both remotely and in situ ([142], [146]).

The type II radio signature provides a complementary view of the CME which can be related to coronagraph images and in situ measurements [314], [249]. The type III radio signature can be used to measure the density and interplanetary magnetic field (IMF) structure between the Sun and Earth. CME propagation direction and speed, size, and density structure are some of the primary factors in determining the *geoeffectiveness*, or terrestrial impact, of a CME [248].

9.5 Solar wind interaction with planetary magnetospheres and atmospheres

The solar wind is an excellent laboratory for ideal MHD in a variety of scale lengths. Due to the very large magnetic Reynolds number the solar wind exhibits well-developed turbulent structures. Related to turbulence and MHD waves, there are several types of shocks in the solar wind that occur due to high-speed CMEs, corotating interaction regions (CIRs) between sectors of fast and slow solar wind, encounter by magnetospheres of planets that cause obstacles to the solar wind flow, and eventually while meeting the interstellar plasma. The solar wind interacts with all the planets and other objects in the solar system as it moves outward from the Sun. NASA's New Horizons mission has been detecting these interactions as it traveled between Uranus and Pluto.

A typical magnetosphere forms when magnetized solar wind plasma flows around a magnetized object, such as a planet or a moon. Some of the planets and moons in the solar system have molten metallic (conducting) cores having strong convective overturning, added to the rotation of the planet, that develops a magnetic dynamo. The more rapid rotators turn into more powerful dynamos, such as Earth, Jupiter, Saturn, Neptune and Uranus.

9.5.1 Encounter with non-magnetic planets/satellites

Probes have found that Mercury, with slow rotation, has an extremely weak intrinsic field. Venus with its slow retrograde rotation has no dynamo. Mars having lost its dynamo has almost no measurable fields. The giant planets Jupiter, Saturn, Uranus, and Neptune, on the other hand, have magnetic fields much stronger than that of the Earth. Smaller objects, such as comets when they are active close to the Sun, and slowly spinning objects, like Mars and Venus, with low conducting core may not excite a magnetic dynamo. But, with a dense enough atmosphere, embedded within a strong flow of solar wind plasma, they form a structure that is called *induced magnetosphere*. Solar wind deflection is not due to their weak magnetospheres, but due to the inability of the solar wind plasma to penetrate through the ionized atmosphere (or, cometary gas).

The planets having comparatively weak or non-existent magnetospheres are subjected to atmospheric stripping by the solar wind. Mercury, the nearest planet to the Sun, bears the full brunt of the solar wind. Since its atmosphere is vestigial and transient, its surface is exposed to radiation. Mercury has a weak intrinsic magnetic field, so normally the solar wind cannot penetrate its magnetosphere, and particles only reach the surface in the cusp regions. During CMEs, however, the magnetopause may get pressed into the surface of this planet, and under these conditions, the solar wind may interact freely with the planetary surface. On the other hand, Venus which has a 100 times denser atmosphere, and little or no magnetic field of its own, develops a comet-like tail that extends all the way to Earth's orbit [154].

Solar wind has stripped away nearly a third of the original atmosphere of Mars. It has left only a layer that is 1/100th as dense as the Earth's. In 2015, NASA Mars Atmosphere and Volatile Evolution (MAVEN) measured the rate of atmospheric stripping caused by the magnetic field carried by the solar wind as it flows past Mars [198] to determine how the planet's atmosphere and water, presumed to have once been substantial, were lost over time. Its other objectives are to determine (i) the current state of the upper atmosphere, ionosphere, and interactions with the solar wind, (ii) the current rates of escape of neutral gases and ions to space and the processes controlling them, and (iii) the ratios of stable isotopes in the Martian atmosphere [196]. Data from the MAVEN orbiter confirmed that the solar wind is responsible for stripping away the atmosphere of Mars over the years, as the shielding effect of the global magnetic field was lost, along with the cooling of the planet's internal dynamo.

The planet's atmosphere can disappear in mainly two ways: (i) an electron is knocked off an atom in the upper atmosphere, and then the charged atom is accelerated away by the electric and magnetic fields of the solar wind, and (ii) atmospheric atoms can also get knocked into space through collisions with incoming solar wind particles. The Maven mission also observed the response of the Martian system due to the interaction of a CME that hit the planet on 8th March 2015 [197]. During the solar storm, the rate of charged atoms flying into space was found to be 10–20 times as high, or around 500 gm/s. It was estimated that the energy hitting the Martian atmosphere during the storm was equivalent to a million tons of TNT an hour, corresponding to a large nuclear weapon per hour. Such solar storms might have been more frequent and intense during the early phase of the solar system. Thus it may have implications for the loss of Mars' atmosphere to space.

Unlike Earth, Mars does not possess a global magnetic field, which channels the solar wind particles towards Earth's polar regions where terrestrial auroral displays are seen. Therefore, it was a surprise that Maven's instruments captured several episodes of ultraviolet auroras glowing in the Martian atmosphere during periods of intense solar wind in 2015. On the other hand, on Mars, without a global magnetic field, auroras can occur anywhere and everywhere during periods of large solar activities. Earth's global magnetic field actually prevents auroras from happening everywhere on the Earth.

The Earth's Moon has no atmosphere, and no intrinsic magnetic field. Thus, the lunar surface is continuously bombarded with the solar wind, and the impact of the high energy particles from the solar wind cause it to emit faintly in X-ray wavelengths. The Project Apollo missions deployed passive aluminium collectors in an attempt to sample the solar wind, and returned with lunar soil. It confirmed that the lunar regolith is enriched in atomic nuclei deposited from the solar wind. Over billions of years, the lunar regolith has acted as a collector for solar wind particles. Therefore, the studies of rocks from the Moon's surface is valuable in studies of the solar wind [348].

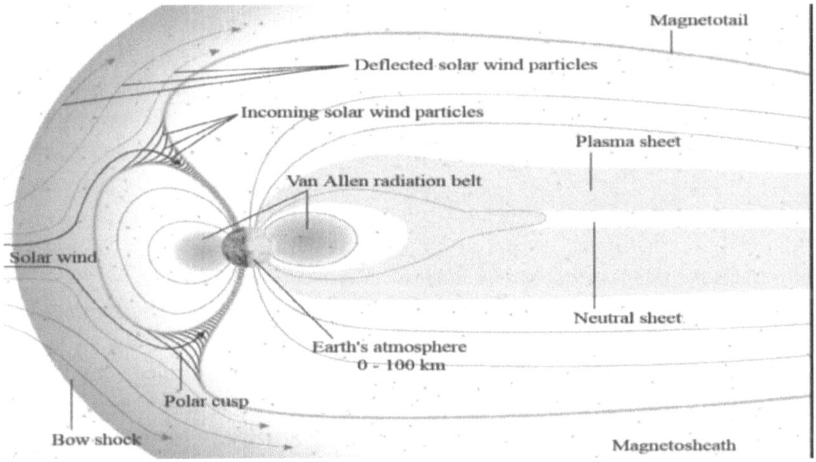

FIGURE 9.6: Schematic of Earth's magnetosphere and its interaction with solar wind. Credit: William Crochot `http://science.nasa.gov/newhome/headlines/guntersville98/images/mag_sketch_633.jpg`.

9.5.2 Encounter with magnetized planets

All magnetic planets (Earth, Jupiter, Saturn, Uranus, Neptune) are known to form a magnetosphere around them, which is essentially a magnetic cavity in the solar wind. Where the solar wind encounters a planet that has a well-developed magnetic field, the solar wind particles are deflected by the strong Lorentz force. This region, known as the *magnetosphere*, causes the particles to travel around the planet, rather than directly bombarding the atmosphere, or surface (Figure 9.6). This is very much like a stream of water flowing around a large obstacle in its path. The magnetosphere is shaped like a slightly compressed hemisphere on the side facing the Sun due to the ram pressure of the solar wind, and is drawn out in a long wake on the opposite side.

The boundary of this region is called the *magnetopause*. Some of the solar wind particles, however, are able to penetrate the magnetosphere through the magnetopause by partial reconnection of the magnetic field lines.

The Cluster mission of the European Space Agency (E.S.A.) suggests that solar wind particles may find it easier to infiltrate the magnetosphere than believed so far [123]. It showed that Kelvin-Helmholtz waves appear at the magnetopause that enable incoming solar wind to breach past it. These findings are also relevant to studies of magnetospheric progressions around other planetary bodies.

9.5.3 Drivers of space weather

Fluctuations in solar wind speed, density, direction, and entrained magnetic field strongly affect the planet's (Earth's) local space environment. The levels of ionising radiation, and radio interference, can vary by factors of hundreds to thousands; and the shape and location of the magnetopause and bow shock wave upstream of it can change by several Earth radii, exposing satellites in geosynchronous orbits to the direct solar wind. These phenomena are collectively called *space weather*.

There are several processes that play a role in the state of space weather. High-speed winds bring geomagnetic storms while slow-speed winds bring quiet space weather. Corotating interaction regions (CIRs) and to a lesser extent, current sheet crossings, can also cause geomagnetic disturbances.

CMEs inject large quantities of mass and magnetic flux into the heliosphere, causing transient disturbances, driving interplanetary shocks, solar energetic particles and act as major contributors to space weather. They are known as interplanetary CMEs, or ICMEs. The abrupt increase in solar wind density, velocity, and magnetic field due to solar transients, such as flares and CMEs lead to the enhancement of ion production and energisation, and an increase in the rate of escape of atmospheric gas to space.

When ejected towards the Earth, reaching as ICME, the shock wave causes a geomagnetic storm, compressing the magnetosphere on the day side and extending the night-side magnetic tail, which reconnects releasing terra-watt scale power directed back toward Earth's upper atmosphere. Solar energetic particles can cause aurorae in Earth's polar and high latitude regions. Energetic protons released by CME can cause an increase in the number of free electrons in the ionosphere, especially in the polar regions. This enhances radio wave absorption, especially within the D-region of the ionosphere, leading to Polar Cap Absorption (PCA) events. CMES and flares can disrupt radio transmissions, damage satellites, and electrical transmission lines resulting in massive power outages. Humans at high altitudes, as in airplanes or space stations, risk exposure to relatively intense solar particle events. Severe geomagnetic storms can disrupt large electric power grids, and even cause blackouts. While geomagnetic storms are less frequent, stronger events observed over several days from anywhere on the Earth, the substorms are more frequent, weaker events, observed over durations of hours near polar regions only.

Solar flares strongly influence local space weather in the vicinity of the Earth, as they are associated with the sudden intense enhancement of radiation, ejection of plasmas, and particles. The X-rays and UV radiation emitted by solar flares can affect Earth's ionosphere and disrupt long-range radio communications. Hard X-rays and energetic particles can damage the instruments on-board spacecrafts. The soft X-ray flux of flares increases the ionization of the upper atmosphere, and thus increase the drag on low orbiting satellites, leading to orbital decay. The enhanced ionization can interfere with short-wave radio communication. Direct radio emission at decimetric wavelengths may disturb the operation of GPS, radars and other devices that use those frequencies. The solar wind particles can affect Earth's satellites, and Global Positioning Systems (GPS). Powerful bursts can damage satellites, or can push GPS signals to be off by tens of metres.

Earth itself is largely protected from the solar wind by its magnetic field. However some of the charged particles are trapped in the *Van Allen radiation belts*. A smaller number of particles from the solar wind manage to travel to the Earth's upper atmosphere and ionosphere in the *auroral zones*. The only time the solar wind is observable on the Earth is when it is strong enough to produce the aurora and geomagnetic storms. Bright auroras strongly heat the ionosphere, causing its plasma to expand into the magnetosphere, increasing the size of the plasma geosphere, and injecting atmospheric matter into the solar wind. Geomagnetic storms result when the pressure of plasmas contained inside the magnetosphere is sufficiently large to inflate and thereby distort the geomagnetic field.

Apart from the interaction with magnetospheres, solar wind also affects the interaction of incoming cosmic rays with planetary atmospheres. Thus specifying and forecasting the solar wind is critical to developing forecasts of space weather and its impacts on Earth.

9.5.4 Detection of solar flares by ionospheric effects

An indirect means of observing solar flares is by the X-rays generated during the flash phase of flares, that cause the short wave fadeouts observed on Earth about 8 minutes later, and the slower moving shock wave and associated gas cloud which interacts with the terrestrial ionosphere 1 to 3 days later causing auroral displays.

When solar X-rays hit the Earth's ionosphere, it causes enhancement of ionization of the ionosphere, and thus disturbs the way the ionosphere reflects radio waves. At short wave frequencies, a dip in signal strength of distant stations can often be observed. At VLF frequencies, below 150 kHz, the opposite effect is observed, and the signal strength of a distant station will jump suddenly, and slowly decline. Using a VLF receiver tuned to a distant station is a reliable way to detect X-ray solar flares. The disturbances of the Earth's magnetic field, referred as a geomagnetic storm, can be observed with magnetometers. In modern times, the largest solar flare measured with instruments occurred on November 4, 2003, which saturated the GOES detectors. Extrapolating the GOES curve, it was estimated to be X28, but later analysis of the ionospheric effects amended this estimate to X45!

9.6 The heliosphere

The term *heliosphere* was first coined by Dessler in 1967. It is essentially the three-dimensional region containing the Sun, along with the solar system objects, filled with solar wind particles and the solar magnetic field [97] as shown in Figure (9.7). Although electrically neutral atoms from interstellar space can penetrate this bubble, virtually all the matter in the heliosphere emanates from the Sun itself. Solar wind plasma blown out from the Sun creates and maintains this bubble against the outside pressure of the interstellar medium comprised of hydrogen and helium.

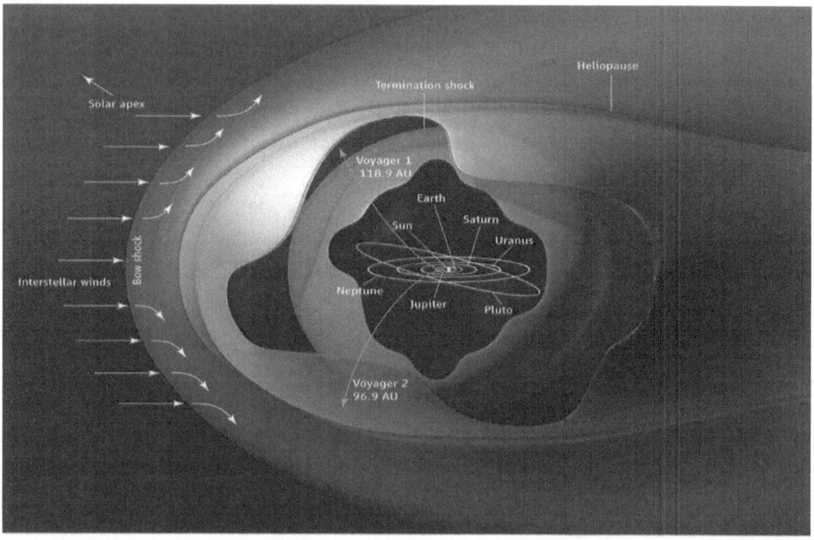

FIGURE 9.7: The heliosphere immersed in the interstellar medium travelling through the Milky Way galaxy. In this artist's concept, positions of the two Voyager spacecrafts in 2011 are shown along with their paths: Voyager I deflected by Saturn in 1980 above the planetary orbital plane, and Voyager II downward by Neptune. (By Brad Baxley, JILA, adopted from `https://physicsworld.com/a/voyager-sets-sights-on-milky-way/`)

The location where the solar wind becomes slower than the sound speed is called the *termination shock*. It is the region located around 80-200 A.U., where solar wind blowing outwards from the Sun slows down to subsonic speeds. As supersonic particles are ejected from the Sun, due to pressures caused by the interstellar medium, they are forced to slow down rather dramatically causing a shock wave-like effect. Within the bubble of the termination shock, the space is completely dominated by the Sun, solar wind and the interplanetary medium although cosmic rays and particles from the interstellar medium do penetrate into this region. The U.S. space probes Voyager 1 and 2 (launched in 1977) crossed the termination shock at a distance of 94 and 84 A.U. from the Sun in 2004 and 2007, respectively. Figure (9.7) shows their positions in 2011.

The solar wind continues to slow beyond the termination shock. It gradually becomes denser and hotter in the region called the *heliosheath*, where solar wind piles up against the inward approaching interstellar wind. The boundary between the solar wind and instellar wind is called the *heliopause* where the outward pressure of the solar wind balances the inward pressure of the interstellar wind. The observations of atoms in the heliosheath by the Cassini spacecraft orbiting Saturn, and the Interstellar Boundary Explorer orbiting the Earth have shown that the heliosphere is a spherical bubble rather than a comet-like shape.

Outside the heliosphere, one encounters the interstellar medium. This is basically the dust and gas that travels throughout our galaxy. Our Sun's heliosphere blends smoothly into the interstellar medium. As announced by NASA in 2013, Voyager 1 crossed the heliopause on August 25, 2012, and is the first of the two Voyagers to reach interstellar space, because the heliopause marks a boundary between the Sun's solar wind and the rest of the galaxy.

As reported in October 2018, NASA's Voyager 2 probe has detected an increase in cosmic rays that originate outside our solar system. A part of these cosmic rays is blocked by the heliosphere, so it is expected that Voyager 2 will measure an increase in the rate of cosmic rays as it approaches and crosses the boundary of the heliosphere. Since late August 2018, the Voyager probe's Cosmic Ray Subsystem instrument has measured about a 5% increase in the rate of cosmic rays hitting the spacecraft compared to that in early August 2018. The probe's Low-Energy Charged Particle instrument has also detected a similar increase in higher-energy cosmic rays. It is expected to continue transmitting weak signals until around 2025 as its electrics slowly fail.

It had been expected that the Sun would produce a shock wave as it travels within the interstellar medium, if the interstellar medium is moving supersonically toward the Sun. When the interstellar wind hits the heliosphere, it would slow down and create a region of turbulence. A bow shock was thought to possibly occur at about 230 A.U. However, in 2012 from the data obtained from Interstellar Boundary Explorer (IBEX), it was determined that the velocity of the Local Interstellar Medium relative to the Sun's was much lower to lead to a shock, and rather a 'bow wave' may be more appropriate to refer to this region [406].

9.7 Summary

From theoretical models to actual observations of solar winds, a lot of development has taken place during the last five decades. It has been noticed from monitoring the solar winds that their speeds have been decreasing in recent years. This affects the extent of heliosphere, and the decrease in its strength. As a consequence, the solar system becomes more vulnerable to energetic galactic cosmic rays. However, it is believed that this decreasing trend may not be permanent, and the solar winds are only going through a low period in solar cyclic activity.

There are several space missions that have, or continue to collect data related to the heliosphere. These are the Solar Anomalous and Magnetospheric Particle Explorer, Solar and Heliospheric Observatory (SOHO), Solar Dynamics Observatory (SDO), STEREO, Ulysses, Interstellar Boundary Explorer (IBEX), Cassini, *etc.* Large-scale, global models of the coupled solar wind-magnetosphere-ionosphere system have made tremendous progress in recent years. Due to advances in computational power, it has become possible to run MHD simulations nearly in real time. However, they have limitations due to missing physics especially in the ionosphere and in the inner magnetosphere. Also, the MHD approach is not adequate in describing the collisionless, multicomponent plasmas in the magnetosphere. These need to be improved for a better understanding of heliospheric physics.

The goal of space weather research is to be able to produce reliable forecasts and nowcasts of the space environment as well as assess the risks that are associated with a wide variety of technological applications. Observations of the near-Earth space environment are sparse and information on the global properties are often not available. Measurement in the solar wind, usually away from the Sun-Earth line, is not always sufficient to determine what impinges on the magnetopause an hour later. In the magnetosphere, the lack of global observations of the mass and energy circulation as well as the dynamics of the electromagnetic fields still limit our capabilities to evaluate the variety of physical processes that can be associated with the observed phenomena.

Chapter 10

Ongoing and Future Directions

Many institutes and observatories around the world are in the process of planning or establishing ambitious observational facilities from ground (in optical and radio wavelengths) and space (white-light coronagraph, UV, EUV, X- and γ - ray wavelengths). These projects require extensive planning, careful site survey (for ground-based instruments), and innovative instrumentation. Therefore, such endeavours usually take several years and even decades, as has been the case with the Advanced Technology Solar Telescope (ATST). In this chapter, we describe and discuss some of the ongoing and future efforts.

10.1 Ground-based and space-borne solar observatories

Space-borne observatories have the advantage of providing access to high-energy parts of the electromagnetic spectrum over the ground-based experiments. However, there are also some limitations in feasibility aspects, for example, in the IR and longer radio wavelengths. In these wavelengths, one requires large aperture telescopes, which are very expensive or impractical for space missions.

The sub-mm to metric radio bands provide valuable diagnostics of thermal structure in the chromosphere, and of non-thermal particle acceleration in flares. But, they require large radio antenna arrays in order to image the radio Sun interferometrically. In fact, for such long wavelength radio observations, space-based instruments would have no advantage, except in the very short sub-mm range, since the atmosphere is relatively transparent in the radio band, and atmospheric seeing is not an issue.

Another example of the advantage of ground-based observing is in solar neutrino detection. Neutrino detectors are usually too large in volume for capturing and detecting weakly-interacting neutrinos. In addition, they also require shielding from galactic cosmic rays by having several-km-deep underground experiments. These constraints rule out space orbiting platforms for neutrino detection projects.

For observing small-scale, magnetic features in the solar atmosphere, it is required to do spectro-polarimetry involving a large-aperture telescope to collect sufficient photon flux. This is essential for ensuring accurate measurements of the Stokes parameters over the spectral line profile at several positions. Although atmospheric seeing changes are no longer an issue when observing from space, the measurements still need to be done in a time short enough so that the solar conditions do not significantly change due to the dynamic plasma motions.

Another major advantage of ground-based facilities is that they can be easily maintained, repaired and upgraded over time. They can be supplied with state-of-the-art technology with the availability to new opportunities. For example, the GONG helioseismic ground-based network could be retrofitted over its 24-year period of operation to observe multiple spectral features. Upgrades with higher resolution cameras, faster data-storage devices, and computer resources, have helped in extending its life, and making new additions to its initial goal of inferring the solar interior using helioseismic observations.

Space weather has developed as a very important field in solar physics in recent years. It encompasses a wide range of time scales from short-time associated with transient activities to longer periods of variations associated with magnetic cycles. The study of long-term phenomena requires a consistent database of synoptic observations that is difficult to accomplish with space-based observations. Ground-based facilities have produced very long time data sets, such as the 400-year record of sunspot numbers, nearly 100 years of solar images from observatories, such as Kodaikanal Observatory, India, and Mt. Wilson, U.S.A. There are efforts to extend GONG and SOLIS projects to maintain a long-term full-disk Doppler and magnetic field imaging covering several solar cycles, which will have immense benefits in future solar research.

Space experiments, on the other hand, are usually of a one-off nature, and may be subjected to unrecoverable failures that can limit their useful lifetimes. The long lead times involved in designing, development, and fabrication of space experiments, and the flight-qualified hardware restrictions limit the advanced technology that can be incorporated. Upgrading is not much of an option once the experiment is launched to space.

From the many advantages discussed in this section, it is clear that ground-based solar observations remain as important as ever even while new, ambitious space-based instruments continue to evolve technologically, and in scope.

10.2 Ongoing, and proposed ground-based optical and IR projects

Current solar telescopes can observe with a spatial resolution of $\sim$300 kms on the Sun. However, there are scientific problems which require the capability of observing at much better resolutions of 30 km or even better.

These problems are related to thin magnetic flux tubes, twisted and tangled structures that can channel energy into the corona, and help in explaining the mystery of the million degrees hot corona. Large apertures are a primary need for improving the resolution, and also in increasing the photon flux. Although the Sun is extremely bright, the number of photons, or flux, becomes an issue while observing at very narrow wavelength passbands. The increased photon flux helps in avoiding long exposures, while observing solar transient phenomena, and reducing atmospheric seeing degradations. However, there is still a need for augmentation by active and adaptive optics capabilities to correct for these degradations.

10.2.1 Daniel K. Inouye Solar Telescope (DKIST)

NSF-DKIST, formerly the Advanced Technology Solar Telescope (ATST), is in the final stages of construction by the National Solar Observatory (N.S.O.), USA, at Haleakalä (alt. 3 km), on the Hawaiian island of Maui. Haleakalä means "House of the Sun" in the Hawaiian language. DKIST would be the world's largest solar telescope when it becomes operational, expected to be in early 2020 (Note: The first light image was released on January 29, 2020). With a resolution of 25 kilometres, the DKIST is expected to achieve a much sought improvement in high resolution requirements.

The 75-mm-thick f/2 primary mirror of DKIST is a 4.24-metre off-axis section of a 12-metre diameter, f/0.67 concave parabola. It was cast from Zerodur by Schott (Germany) and polished at the University of Arizona. The 0.65-metre secondary mirror, a concave ellipsoid with a focal length of 1 metre, was made from silicon carbide, and is mounted on a hexapod to compensate for thermal expansion, and bending of the telescope structure, keeping the mirror in its optimal position.

It is expected to have 5 back-end instruments: a Visible Broadband Imager by N.S.O., Visible Spectro-Polarimeter by High Altitude Observatory; Visible Tunable Filter by Kiepenheuer-Institut für Sonnenphysik; Diffraction-Limited Near-InfraRed Spectro-Polarimeter (DL-NIRSP), and Cryogenic Near-InfraRed Spectro-Polarimeter (Cryo-NIRSP) by the Institute for Astronomy, University of Hawaii (`https://dkist.nso.edu/`).

A 4-m all-reflecting IR optimised telescope was in fact proposed more than two decades ago by Livingston (1994) [250]. This dream is only now turning into reality with the DKIST [207]. The telescope and the AO system is designed to achieve 0.08 arc-second resolution at 1565 nm. The infrared instruments planned for the DKIST, the Cryo-NIRSP, and the DL-NIRSP ([239], [320]) will provide unprecedented observations from 1000 to 5000 nm, on the solar disk, at the limb, and into the solar corona. The DKIST, with these instruments, will deliver several advantages of observing in the IR solar spectrum at the highest spatial resolutions.

10.2.2 CrYogenic InfRAred Spectrograph (NST-CYRA)

New instruments in infrared solar physics are being developed, such as a cooled grating spectrograph spectro-polarimeter, or the CYRA instrument at the New Solar Telescope (NST), Big Bear Solar Observatory, now named as the Goode Solar Telescope (GST) [69]. Cyra has dual-beam polarimetric capability from 1000 to 5000 nm. The cryogenic spectrograph gives much lower background signals as compared to the existing IR telescopes operating at Sac-Peak Observatory, and at the Mees Solar Observatory.

A set of weak lines at 4135 nm have been observed, which may be a key target for the CYRA instrument. These lines are weak and have a strong dependence on temperature. They disappear at the low temperatures of a sunspot umbra. These have potential to provide critical diagnostics for the weak magnetic fields of the quiet Sun.

10.2.3 GREGOR Infrared Spectrograph (GRIS)

The 1.5-m GREGOR solar telescope at the Observatorio del Teide on Tenerife is home to the GRIS, or grating spectrograph instrument ([330], [88]). Although it can be used at visible wavelengths, the spectrograph is mostly used in combination with the IR detector of the Tenerife Infrared Polarimeter (TIP-II) in spectroscopic mode from 1000–2300 nm, as well as for spectro-polarimetric measurements from 1000–1800 nm. Initial intensity spectra from GRIS show that excellent image quality can be obtained with the system, promising interesting future polarimetric data.

10.2.4 European Solar Telescope (EST)

This is proposed to be a major 4-metre aperture telescope for studies of the magnetic coupling between the photosphere and chromosphere. This would require simultaneous multi-wavelength imaging, spectroscopy and spectro-polarimetry at high spatial and temporal resolutions. The goal is to resolve scales of ~ 10 km in the photosphere for observing physical processes at their intrinsic scale sizes, and the interaction of magnetic fields and plasma motions.

EST will be located in the Canary Islands. This is a project of the European Association for Solar Telescopes (EAST), founded in 2006 by a group of solar physicists from 17 European countries (`http://www.est-east.eu`). The expected time-line for EST first light is around 2028.

10.2.5 Chinese Large Solar Telescope (CLST)

The Chinese Large Solar Telescope (CLST) is being built in Gregorian configuration with a 1.8-m aperture telescope with an open structure, alt-azimuth mount, retractable dome, a large mechanical de-rotator, and adaptive optics to provide diffraction-limited observations. The CLST design and development began in 2011. The available literature does not provide a firm date for its commissioning. A multi-wavelength tomographic imaging system, ranging from visible to near-infrared, is planned as the first light scientific instrument [312].

10.2.6 National Large Solar Telescope (NLST)

It is a 2-m class optical and near infrared (NIR) observational facility, proposed to be installed at a high altitude lake site (~4500m) in Leh-Ladakh region of India. It is being designed to address several key scientific issues that are related to the origin and dynamics of solar magnetic fields at high spatial resolution of ~ 0.1-0.3 arc-second (`https://www.iiap.res.in/nlst/`).

NLST proposes to observe the photosphere and chromosphere at several wavelengths using narrow-band imagers at high spectral resolution, large spectral range, and fast change between wavelengths. A spectro-polarimeter would be the main backend instrument based on infrared and optical wavelengths to provide magnetic field maps of the Sun. It would be equipped with adaptive optics to correct for atmospheric distortions.

10.2.7 Solar Physics Research Integrated Network Group (SPRING)

This is a proposal for establishing a new ground-based network of identical 1m-class aperture telescopes for synoptic observations of the Sun. Planned under SOLARNET, it is a European initiative for high-resolution solar physics [https://doi.org/10.1117/12.2306555]. It plans to obtain a wide range of solar observations with better spatial, spectral and/or temporal resolution than the existing GONG, and SOLIS networks. Synoptic full-disk observations are crucial for addressing a variety of solar physics problems, such as global and local helioseismology, multi-wavelength imaging, and polarimetry, solar irradiance, *etc.* It proposes to have a three-, or six-site network, possibly located at the existing GONG network of observatories.

10.3 Future prospects for solar IR observations

The current studies with CoMP data show that imaging spectroscopy of the corona in the infrared has the potential to provide unique insights into the dynamics of the coronal plasma and magnetic field. Therefore, many attempts have been made in the field of infrared solar physics. However, it remains to be extended to newer wavelengths, with development of new instrumentation to open these windows. There is a mostly unexplored spectral region of IR in the range of 5000–10000 nm. The N.S.T. and McMath-Pierce, and D.K.I.S.T. are the telescopes that can explore this region. Of the two current facilities, the McMath-Pierce has the advantage of being a drier site.

10.3.1 Ground-based solar IR instrumentation

A large ground-based coronagraph for the Coronal Solar Magnetism Observatory (COSMO) at Mauna Loa Solar Observatory of the High Altitude Observatory (HAO) has been proposed to take continuous synoptic measurements of solar coronal and chromospheric magnetic fields, which has been a crucial but difficult problem. COSMO will consist of (i) Chromospheric and prominence Magnetometry (ChroMag), (ii) white light electron scattered K-corona (K-Cor), and (iii) the 1.5-m aperture large coronagraph (LC) to observe the faint coronal radiation and magnetic field in a number of visible and near-IR emission lines, (*e.g.*, FeXIV at 530.3 nm, Fe X at 637.4 nm, Fe XIII at 1074.7 and 1079.8 nm). These lines are faint, therefore require large apertures. Coronal magnetic fields of the Sun will be measured using the polarization of spectral lines [131], based on the previous prototype Coronal Multi-channel Polarimeter (COMP instrument). It would operate in a more synoptic manner than the DKIST Cryo-NIRSP is expected to do. In addition, it would have a larger field-of-view (COSMO project: `https://www2.hao.ucar.edu/cosmo`). The data products from COSMO will be available to the community via the Mauna Loa website: `http://mlso.hao.ucar.edu`.

10.3.2 Space-based solar IR instrumentation

From the experience of several IR space missions for night-time objects, through a variety of space missions from IRAS, WISE, Spitzer to Herschel, it is obvious that the required technology is available for space missions for the Sun in the infrared spectrum. The Japanese Aerospace Exploration Agency mission had proposed an infrared spectro-polarimeter. Prospects for infrared space-based solar observations have been mostly lacking, although proposals for space-based solar IR missions were made decades ago, including the SIRE mission ([96]).

While the spatial resolution of a space-based solar telescope operating in the IR is less than if it is operated at shorter wavelengths, the magnetic sensitivity gains promise to open new realms of possible science. The results from Hinode, SOHO and now SDO spacecrafts, demonstrate that a new solar mission with higher magnetic sensitivity would indeed be revolutionary [247].

10.4 Future prospects for radio solar observations

One spectral window that has remained relatively un-explored is the low-frequency (long wavelength) radio domain below a few hundred MHz, representing the lowest frequency extreme of the accessible spectrum. A first-order analysis suggested that a telescope with a collecting area of about one square kilometre was required. It has evolved to an ambitious project now known as the Square Kilometre Array (SKA; [115]).

Major progress is expected from ground-based solar dedicated radioheliographs designed to produce images at high spatial resolution and with high-dynamic range over a broad frequency range. There are several fronts on which efforts in long wavelength solar observations (Far IR to radio) would need further thrust: (i) imaging spectroscopy in existing domains, (ii) imaging spectroscopy in frequency ranges not imaged so far, (iii) opening new windows of radio observations toward sub-mm to far IR domain, and (iv) in space. Some recent and ongoing developments are given as follows:

- *The Atacama Large Millimeter Array (ALMA)*: It is an astronomical interferometer consisting of 66 radio telescopes in the Atacama Desert of northern Chile. The array has been constructed at a high elevation of over 5,000 m, and low humidity, crucial factors to reduce noise and decrease signal attenuation due to Earth's atmosphere [68]. This facility is for observing at millimetre and submillimetre wavelengths of the electromagnetic window. ALMA began scientific observations in 2011. The array has been fully operational since March 2013. `https://www.almaobservatory.org/en/home/`

- *Frequency Agile Solar Radio Telescope (FASR)*: It is a next-generation radio telescope for solar observation in the radio and microwave frequency range. In contrast to general-purpose radio telescopes, such as the Very Large Array (VLA), it is specifically designed for solar observations, such as quasi-continuous coverage in 30 GHz-100 MHz at high time resolution (less than 1 s). It has several goals including mapping of the coronal magnetic field in ARs, localisation of bursts in the 0.5-1.5 GHz range, energy release and acceleration region during flares (spectral imaging of the quiet, active and eruptive Sun, from the chromosphere to $1R_\odot$).

- *Low Frequency Array (LOFAR)*: It is a new-generation radio interferometer constructed in the north of the Netherlands and across Europe [373]. It utilises a novel phased-array design, and covers the largely unexplored low-frequency range from 10-240 MHz. With the longer European baselines in place, LOFAR can achieve sub-arc second angular resolution over most of its 30-240 MHz operating bandpass, limited primarily by atmospheric effects and scattering due to interplanetary scintillation (IPS).

LOFAR is expected to be very useful for studying transient processes in the high corona related with flares and CME.

- *Square-Km Array (SKA)*: It is proposed to be built as a collaborative project of 13 member countries: Australia, Canada, China, France, Germany, India, Italy, New Zealand, Spain, South Africa, Sweden, The Netherlands and the United Kingdom. It is to be located in the southern hemisphere, in Australia and South Africa (`https://www.skatelescope.org/`). It plans to build the world's largest radio telescope, with eventually over a square kilometre (one million square metres) of collecting area using thousands of dishes and up to a million low-frequency antennas. Its major science goals include radio transients and solar and heliospheric physics. With partial arrays, it is expected to start scientific observations in mid-2020.

10.5 Some recent and future space-borne missions

Observations from satellite missions have transformed the field of solar physics. High-resolution observations with near-continuous temporal coverage have greatly extended our capability for studying long-term and transient phenomena. The opening of new regions of the solar spectrum has made detailed investigation of the solar atmosphere possible.

There is a variety of orbits to which spacecrafts are launched to meet the goals of observing the Earth, the Sun, or other deep sky objects. We discuss some of these in the following.

- Low Earth-orbits (LEO) - an Earth-centred orbit with an altitude of 2,000 km or less, or with at least 11.25 periods per day (an orbital period of 128 minutes or less) and an eccentricity less than 0.25. Most of the manmade objects in space are in LEO.

- Polar (or near) orbits - having an inclination near 90 degrees, which allows the satellite to observe every part of the Earth as the Earth rotates underneath, taking around 90 minutes for the satellite to complete one orbit.

- Geosynchronous or geostationary orbits - in these orbits, satellites circle the Earth at the same rate as the Earth spins. According to Kepler's Laws of Planetary Motion, geosynchronous satellites should be located around 35,790 km above the Earth, usually near the equator so that the force of gravity is constant from all directions. They are used for observing a full hemisphere of the Earth, and study large-scale phenomena such as hurricanes, or cyclones, and also are used for communication.

- Sun-Synchronous orbits - where a satellite passes over a section of the Earth at the same time of day. For this, the satellite is made to shift its orbit by about a degree per day. These satellites orbit at an altitude of 700-800 km.
- Lagrange-L1 halo orbit - a region in space where the gravity of the Sun and the Earth balance each other out. This region is located at about 1.5 million kms from Earth on the Sun-Earth line.

Most of these are used for observing the same part of the Earth, or scanning various locations (remote sensing) over time. The International Space Station is located at LEO about 330 km to 420 km above Earth's surface, while the Hubble Space Telescope observes from LEO at 550 km. Spacecrafts have also been sent on various lunar, planetary, asteroidal, and cometary missions. There are also missions travelling across the heliosphere, such as Voyagers, and Pioneers.

Solar Dynamics Observatory (SDO) was placed in a circular, geosynchronous orbit at an altitude of 35,789 kms at 102° W longitude, inclined at 28.5°. SDO's inclined geosynchronous orbit was chosen to allow near-continuous observations of the Sun, and also allow for high data rate through the use of a single dedicated ground station. There are new missions which are planned to observe the Sun from very close quarters, in fact, immersed within the solar corona, such as the Parker Probe.

10.6 Solar physics from vantage positions in space

There exist five 'stationary' Lagrange points for two massive bodies rotating about their centre of mass, where the force acting on a third small body (say, a satellite) is nearly zero in the rotating reference frame. Such points exist for the Sun and any of its planets, or even for a planet and one of its moons. Each of these points is a vantage location in space providing benefits in observing the Sun and other objects.

10.6.1 Lagrangian points

There are three co-linear Lagrange points L1, L2 and L3 along a straight line joining the two massive bodies, with Lagrange points L4 and L5 (also referred to as Trojan points) as the apex points of two equilateral triangles with the main two masses of the system (Figure 10.1). The distances of L1, L2 from Earth (in the Sun-Earth system) are 1,491,557 km (0.009 970 A.U.), 1,501,557 km (0.010 037 A.U.), respectively, while L3 is located at the other side of the Sun at 149,599,737 km (0.999 998 A.U.) from the Sun. The L1 and L2 points are the sites of several spacecraft in the Sun-Earth system.

A spacecraft affords an uninterrupted view of the Sun from Lagrange L1, and can continuously monitor the Sun and space weather before the effect of a solar energetic event reaches the Earth. Thus, it can provide about one hour's alert. SOHO, ACE, and WIND are three such spacecrafts, which have halo orbits about L1 to keep it slightly away from a direct Sun-Earth line of sight. The proposed Indian space-mission dedicated to the Sun, Aditya-L1, plans to be located at L1 for these reasons. The Deep Space Climate Observatory, or DSCOVR, was launched on 12 February 2015 essentially to succeed NASA's Advanced Composition Explorer's (ACE) in supporting solar wind alerts and warnings from the L1 orbit.

On the other hand, some astronomical spacecrafts are placed at the L2 point (beyond the Earth away from the Sun) to partially shield them from the Sun. Examples are WMAP, Herschel and Planck. This helps them to have dark skies for observing deep space objects, keep sensors cool, and at the same time, provide the required sunlight to power the solar cells of the spacecraft.

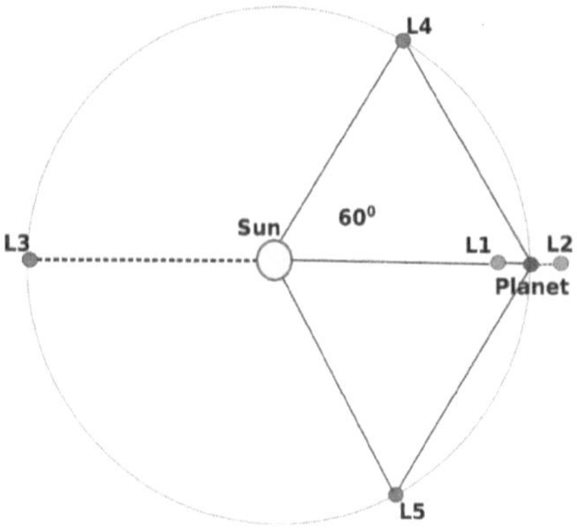

FIGURE 10.1: The five Lagrange points of a Sun-Earth system.

These five stationary points occur as a result of cancellation of the gravitational forces (per unit mass) exerted there due to both the Sun ($GM_\odot/R_\odot^2$) and the planet (GM_{planet}/R_{planet}^2) by the centrifugal force (per unit mass) ($r\omega^2$) at these points. Here r is the distance from the point to the center of mass, and ω is the angular velocity of the rotating reference frame, $M_\odot$ and M_{planet} are the respective masses of the Sun and the planet, and $R_\odot$ and R_{planet} the distances from the point to the Sun and planet, respectively. The mathematical problem, known as the 'General Three-Body Problem' was considered by Lagrange in 1772. For further reading, refer to `https://solarsystem.nasa.gov/resources/754/what-is-a-lagrange-point/`.

10.6.2 NASA-Parker Probe

It was launched on August 12, 2018, to explore the Sun's outer atmosphere or corona from within, approaching as close to the Sun as 8.5 $R_\odot$. Starting at a perihelion of 35.7 $R_\odot$ in November 2018 (already twice as close to the Sun as Helios ever went) it will reach its minimum perihelion of $9.86R_\odot$ in late 2024 [124]. It will observe in-situ the near-Sun environment to better understand coronal heating and the origin and evolution of the solar wind and several other frontier problems in heliophysics. Moreover, by making direct and close measurements of the energisation of solar energetic particles, Parker Solar Probe is expected to enhance the ability to characterise and forecast space weather.

In February 2019, Parker Solar Probe completed its first orbit of the Sun, reaching its aphelion, and then began the second of 24 planned orbits, for its second perihelion, or closest approach to the Sun, on April 4, 2019, to a distance of about 15 million miles from the Sun. The spacecraft's four instrument suites are designed to help in answering outstanding questions about the Sun's fundamental physics - including solar wind acceleration and coronal heating (`http://parkersolarprobe.jhuapl.edu/`).

10.6.3 Solar Orbiter (SolO)

It is expected to be launched in February 2020 as a joint ESA-NASA collaboration, with the aim to make in-situ measurements of the solar wind plasma, fields, waves, and energetic particles close enough to the Sun. (The spacecraft was launched from NASA's Kennedy Space Centre, Florida, on 10 February 2020.) At this location, the conditions are still relatively pristine and properties are not modified by subsequent transport and propagation processes. It will orbit the Sun between 0.28 and 0.7 A.U. Solar Orbiter will use multiple Venus gravity assist manoeuvres to move the inclination of its orbit to progressively higher heliolatitudes. A maximum inclination of 30°–34° out of the ecliptic is expected to occur by the end of its extended mission in 2029–2030. SolO aims to address a central question of heliophysics as to how the Sun creates and controls the heliosphere [279].

Solar Orbiter is the first mission of ESA's Cosmic Vision 2015-2025 programme. U.S. contributions are SoloHI (a wide-field heliospheric imager), and a Heavy Ion Sensor (part of the Solar Wind Analyser). ESA contributions are the spacecraft (built by Astrium, U.K.), and ten payloads by the European nations. The in-situ measurements would be traced back to their source regions on the Sun, requiring simultaneous, high-resolution imaging and spectroscopic observations of the Sun in and out of the ecliptic plane (`http://sci.esa.int/solar-orbiter/`).

10.6.4 Aditya - L1

Aditya-L1 is the first Indian mission to study the Sun from a halo orbit around the Lagrange L1. It is expected to be launched some time during 2020. The satellite would carry six payloads to provide observations of Sun's UV, soft and hard X-rays, corona in Visible (emission lines) and NIR. In addition, it will study the particle flux emanating from the Sun and reaching the L1 orbit, and the variation in magnetic field strength at its halo orbit location using a magnetometer. (`https://www.isro.gov.in/aditya-l1-first-indian-mission-to-study-sun`)

10.6.5 Observing from non-SEL Lagrangian points

Most solar space-based missions so far have been restricted to the Sun-Earth line (SEL), in orbit around the Earth, or from the Sun-Earth L1 Lagrangian point. As a result, observations from these satellites represent the same geometrical view of the Sun that is accessible from the Earth. A notable exception was STEREO, which began imaging the far-side corona of the Sun in UV since separating beyond $\sim 20°$ from Earth in early 2008, and achieved the full 360° corona in 2011, as the STEREO reached opposition. STEREO-A, since 2018, continues to image the far-side and provide the coverage of more than two-thirds of the corona, along with the SDO/AIA observations.

Extra Sun-Earth Line space mission may answer many outstanding problems of heliophysics, and also help in improving space-weather monitoring and prediction, as discussed in [137]. Most of our knowledge of CMEs has been derived from observations made from the Sun-Earth line. However, it is not the ideal vantage point to observe Earth-affecting CMEs.

A mission to Lagrange L5 or similar quadrature viewpoint seems to be the obvious next step after STEREO. The possibility has been widely explored ([389], [144], [380], [230]). This would help in more accurate forecasting, and speed measurements for Earth-directed CMEs, and corotating interaction regions (CIRs), important sources responsible for affecting space weather. There are several key benefits, such as: (1) it can provide near-Sun CME speed, an important input to models that forecast Earth-arrival time of CMEs, (2) it can distinguish the backside, and frontside CMEs, (3) it can identify preceding CMEs in the path of geo-effective CMEs, (4) it can provide significant lead time for alerts because CIRs reach the L5 point a few days before they arrive at the Earth, and (5) it can provide advance knowledge of coronal holes rotating to Earth's view, which are source regions of CME and CIRs [137].

10.6.6 Observing over the solar polar regions

Imaging of the solar poles has been a long-held interest of the solar community. Solar Polar Imager (Single and Constellation) mission concepts were proposed *e.g.*, POLARIS ([17], [16]), and the Chinese space program

SPORT [403]. The primary scientific objective of these missions is for measuring the evolution of the polar magnetic field, and sub-surface flows beneath the poles using helioseismology.

Ulysses (1990–2009) has been so far the only spacecraft to explore the heliosphere outside of the ecliptic plane. It achieved a high-inclination (80°) near-polar orbit with a Jupiter swing-by, and measured the solar wind plasma, electromagnetic fields, and composition in situ in between 1.3-5.4 A.U. from the Sun ([29], [28], [140]). Ulysses established the three-dimensional structure of the solar wind around solar minimum to be made up of a fast, uniform flow in the northern and solar polar regions, and slow-to-medium speed wind streams originating in a coronal streamer belt confined to a relatively narrow range of latitudes on either side of the heliographic equator [266].

A Jupiter gravity assist, as demonstrated for the Ulysses mission, remains the most practical method for going over the solar poles. The Ulysses-like mission results in a very long orbit with only a small fraction of time spent close to the Sun. However, new mission concepts continue to be developed due to the high scientific value and significant opportunity for new discoveries [167].

A new version of a solar polar mission [Solar Polar Explorer (SOLPEX)] is developed [284] using a combination of a Jupiter gravity assist, and solar electric propulsion [*e.g.*, NASA Evolutionary Xenon Thruster (NEXT) thrusters] to achieve a high inclination ($> 75°$) orbit with an eventual ~ 1 year period (final orbit is 0.5×1.5 A.U.). SOLPEX would contain a suite of remote sensing (Doppler magnetograph, EUV imager, coronagraph, and heliospheric imager) and in-situ instruments. It could complete three passes over each of the solar poles within a 9-year mission lifetime, with each pass lasting > 38 days above 60°.

10.6.7 Space-based observations in radio wavelengths?

Due to the large apertures required for observing in radio wavelengths, it is not very practical for space observations. However, the idea of carrying small aperture instruments on widely separated satellites, which might exploit the aperture synthesis approach, is quite attractive. Such missions would be advantageous for a variety of problems, such as (i) track and probe CME driven shocks from the corona to 1 A.U., (ii) map the structure of radio shocks and of flare electron beams, (iii) probe density and structure of the IP medium before and after a CME, (iv) understand radio emission and beam pattern of radio bursts, (v) measure electron density and temperature of filament material, *etc.* (`http://www-lep.gsfc.nasa.gov/swaves/swaves.html`). In a more distant future, space or lunar-based imaging instruments may also become feasible, and allow very low frequency imaging, and provide definitive protection against manmade interferences.

10.7 Summary

From humble beginnings in the early 1970s, the study of the Sun from space has blossomed into a mature field, helped by remarkable instrumental advances. Similar progress has occurred on the ground in high-resolution spectroscopy, photometry, and spectro-polarimetry at optical telescopes; near IR, radio imaging, and soon the mm/sub-mm astronomy. Solar physics has witnessed impressive strides with space missions such as SoHO, TRACE, STEREO, RHESSI, Ulysses, and now SDO, and Parker Probe; together with steady advances in ground-based solar imaging and spectro-polarimetry, especially with the development of modern adaptive optics systems. In the coming years, major 4-metre aperture telescopes, such as DKIST and EST, promise to unravel processes operating at unprecedented observable small scales on the Sun.

There are advantages and disadvantages of observing from ground and space. Therefore, efforts are not limited to space-based instruments. Major new ground-based instruments are being planned with new technology. There is, however, a pressing need to synergise the space-based observations with those from the ground. High-resolution, multi-wavelength observations from vantage locations in space need to be explored due to their many potential advantages. Further efforts in both ground-based and space missions are required to push the frontiers in heliosciences and solar-stellar insights [22].

Bibliography

[1] P. A. Gilman and P. A. Fox. Joint Instability of Latitudinal Differential Rotation and Toroidal Magnetic Fields below the Solar Convection Zone. *Astrophys. J.*, 484:439–454, 1997.

[2] M. Ackermann, M. Ajello, A. Albert, et al. High-energy Gamma-Ray Emission from Solar Flares: Summary of Fermi Large Area Telescope Detections and Analysis of Two M-class Flares. *Astrophys. J.*, 787:15, 2014.

[3] C. Aerts, J. Christensen-Dalsgaard, and D.W. Kurtz. *Asteroseismology.* Astronomy and Astrophysics Library, Springer, Dordrecht, 2010.

[4] G. Agnelli, A. Cacciani, and M. Fofi. The Magneto-Optical Filter I: Preliminary Observations in Na D Lines. *Solar Phys.*, 44:509–518, 1975.

[5] P. Ahnert. Der Komet 1942 g (Whipple-Fedtke). *Zeit.Astrophysik*, 22:288–309, 1943.

[6] M. Ajello, A. Albert, A. Allafort, et al. Impulsive and Long Duration High-Energy Gamma-Ray Emission from the Very Bright 2012 March 7 Solar Flares. *Astrophys. J.*, 789:20, 2014.

[7] C.E. Alissandrakis, F. Borgioli, F. Chiuderi Drago, M. Hagyard, and K. Shibasaki. Coronal Magnetic Fields from Microwave Polarization Observations. *Solar Phys.*, 167:167–179, 1996.

[8] C. W. Allen. *Astrophysical Quantities, 3rd edn.* Athlone, London, 1976.

[9] A. Ambastha, S. Basu, and H. M. Antia. Flare-Induced Excitation of Solar p-modes. *Solar Phys.*, 218:151–172, 2003.

[10] A. Ambastha and A. Bhatnagar. *Photographic Atlas of the Solar Chromosphere.* Udaipur Solar Observatory, Udaipur, 1985.

[11] A. Ambastha, M. J. Hagyard, and E. A. West. Evolutionary and Flare-Associated Magnetic Shear Variations Observed in a Complex, Flare-Productive Active Region. *Solar Phys.*, 148:277–299, 1993.

[12] H.M. Antia and S. Basu. Temporal Variations of the Solar Rotation Rate at High Latitudes. *Astrophys. J. Lett.*, 559:L67–L70, 2001.

[13] H.M. Antia and S. Basu. Local Helioseismology Using Ring Diagram Analysis. *Astron. Nachr.*, 328:257–263, 2007.

[14] H.M. Antia, S. Basu, and S.M. Chitre. Solar Internal Rotation Rate and the Latitudinal Variation of the Tachocline. *Mon. Not. R. Astron. Soc.*, 298:543–556, 1998.

[15] S. K. Antiochos, Z. Mikić, V. S. Titov, et al. A Model for the Sources of the Slow Solar Wind. *Astrophys. J.*, 731:112, 2011.

[16] T. Appourchaux, F. Auchère, and E. E. Antonucci. Solaris: Solar Sail Investigation of the Sun. In *Advances in Solar Sailing, Ed M. Macdonald*, volume 259, pages 279–288. Springer: Berlin; Heidelberg, 2014.

[17] T. Appourchaux, P. Liewer, M. Watt, et al. POLAR Investigation of the Sun - POLARIS. *Exp. Astron.*, 23:1079–1117, 2009.

[18] M. J. Aschwanden. *Physics of the Solar Corona - An Introduction.* Springer, Praxis, 2006.

[19] M.J. Aschwanden. What did Yohkoh and Compton Change in Our Perception of Particle Acceleration in Solar Flares? In *Observational Plasma Astrophysics : Five Years of Yohkoh and Beyond (Eds. Watanabe, T., Kosugi, T. and Sterling, A.C.*, volume 229, pages 285–294. Kluwer Academic Publishers, Boston, Mass., 1998.

[20] M.J. Aschwanden, H.S. Hudson, T. Kosugi, et al. Electron Time-of-Flight Measurements During the Masuda Flare, 1992 January 13. *Astrophys. J.*, 464:985–998, 1996.

[21] W. B. Atwood, A. A. Abdo, M. Ackermann, et al. The Large Area Telescope on the Fermi Gamma-Ray Space Telescope Mission. *Astrophys. J.*, 697:1071–1102, 2009.

[22] T. Ayres and D. Longcope. Ground-Based Solar Physics in the Era of Space Astronomy. In *A White Paper Submitted to the 2012 Heliophysics Decadal Survey*, 2012.

[23] H. W. Babcock. The Solar Magnetograph. *Astrophys. J.*, 118:387–396, 1953.

[24] H. W. Babcock. The Topology of the Sun's Magnetic Field and the 22-Year Cycle. *Astrophys. J.*, 133:572–587, 1961.

[25] J.N. Bahcall and H.A. Bethe. A Solution of the Solar Neutrino Problem. *Phys. Rev. Lett.*, 65:2233–2235, 1990.

[26] J.N. Bahcall, M.H. Pinsonneault, S. Basu, and J. Christensen-Dalsgaard. Are Standard Solar Models Reliable? *Phys. Rev. Lett.*, 78:171–174, 1997.

[27] S. Balachandra Rao. *Indian Mathematics and Astronomy: Some Landmarks.* Bhavan's Gandhi Center of Science and Human Values, Bharatiya Vidya Bhavan, Bangalore, 1994.

[28] A. Balogh, T. J. Beek, R. Forsyth, et al. The Magnetic Field Investigation on the Ulysses Mission Instrumentation and Preliminary Scientific Results. *Astron. Astrophys. Supp. Ser.*, 92:221–236, 1992.

[29] S. Bame, D. McComas, B. Barraclough, et al. The Ulysses Solar-Wind Plasma Experiment. *Astron. Astrophys. Supp. Ser.*, 92:237–265, 1992.

[30] M. K. V. Bappu. The Kodaikanal Observatory - A Historical Account. *J. Astrophys. Astron.*, 21:103–106, 2000.

[31] T. Baranyi. Comparison of Debrecen and Mount Wilson/Kodaikanal Sunspot Group Tilt Angles and the Joy's Law. *Monthly Not. Royal Astron. Soc.*, 447:1857–1865, 2015.

[32] S. Basu. Seismology of the Base of the Solar Convection Zone. *Mon. Not. R. Astron. Soc.*, 288:572–584, 1997.

[33] S. Basu. Global Seismology of the Sun. *Living Rev. Solar Phys.*, 13:2, 2016.

[34] S. Basu, W. J. Chaplin, Y. Elsworth, R. New, and A. M. Serenelli. Fresh Insights on the Structure of the Solar Core. *Astrophys. J.*, 699:1403–1417, 2009.

[35] S. Basu and A. Mandel. Does Solar Structure Vary with Solar Magnetic Activity? *Astrophys. J.*, 617:L155–L158, 2004.

[36] R. A. Bayanna, R. E. Louis, S. Chatterjee, et al. Membrane-Based Deformable Mirror: Intrinsic Aberrations and Alignment Issues. *Applied Optics*, 54:1727–1736, 2015.

[37] J. G. Beck. A Comparison of Differential Rotation Measurements. *Solar Phys.*, 191:47–70, 2000.

[38] J. M. Beckers. A Seeing Monitor for Solar and Other Extended Object Observations. *Experimental Astron.*, 12:1–36, 2001.

[39] J. M. Beckers, E. Leon, J. Mason, and L. Wilkins. The Caltech Solar Site Survey, 1965-1967. *Solar Phys.*, 176:23–36, 1997.

[40] J.M. Beckers and T. Brown. Solar Circulation Measurements: Consideration and Plans. In *Future Solar Optical Observations Needs and Constraints, Ed. G. Godoli.*, volume 106, pages 189. Pubblicazioni della Universita degli Studi di Firenze, 1978.

[41] J. Beer, A. Blinov, G. Bonani, et al. Use of 10Be in Polar Ice to Trace the 11-year Cycle of Solar Activity. *Nature*, 347:164–166, 1990.

[42] A. O. Benz. Flare Observations. *Living Rev. Solar Phys.*, 14:2, 2017.

[43] H. A. Bethe. Energy Production in Stars. *Phys. Rev.*, 55:434–456, 1939.

[44] C. Bethge, H. Peter, T. J. Kentischer, et al. The Chromospheric Telescope. *Astron. Astrophys.*, 534:12 pp., 2011.

[45] A. Bhatnagar, K. Jain, and S. C. Tripathy. GONG p-Mode Frequency Changes with Solar Activity. *Astrophys. J.*, 521:885–888, 1999.

[46] L. Biermann. Kometenschweife und Solare Korpuskularstrahlung. *Zeit.Astrophysik*, 29:274–286, 1951.

[47] D. E. Billings. *A Guide to Solar Corona.* Academic Press, New York, USA, 1966.

[48] V. M. Bogod and L. V. Yasnov. On the Comparison of Radio-Astronomical Measurements of the Height Structure of Magnetic Field with Results of Model Approximations. *Astrophys. Bull.*, 64:372–385, 2009.

[49] J.D. Bohlin, K.J. Frost, P.T. Burr, A.K. Guha, and G.L. Withbroe. Solar Maximum Mission. *Solar Phys.*, 65:5–14, 1980.

[50] V. Bommier, E. Landi Degl'Innocenti, J.-L. Leroy, and S. Sahal-Bréchot. Complete Determination of the Magnetic Field Vector and of the Electron Density in 14 Prominences from Linear Polarization Measurements in the He I D3 and H-α Lines. *Astron. Astrophys.*, 154:231–260, 1994.

[51] V. Bommier and S. Sahal-Bréchot. The Hanle Effect of the Coronal Lyman-α Line of Hydrogen: Theoretical Investigation. *Solar Phys.*, 78:157–178, 2004.

[52] V. Bommier, S. Sahal-Bréchot, and J. L. Leroy. Determination of the Complete Vector Magnetic Field in Solar Prominences, Using the Hanle Effect. *Astron. Astrophys.*, 100:231–240, 1981.

[53] U. Boser. Solar Circle. *Archaeology*, 59:22–34, 2006.

[54] J. L. K. Bougeret, Goetz M. L., S. D. Kaiser, et al. S/WAVES: The Radio and Plasma Wave Investigation on the STEREO Mission. *Space Sci. Rev.*, 136:487–528, 2008.

[55] D. C. Braun and Y. Fan. Helioseismic Measurements of the Subsurface Meridional Flow. *Astrophys. J.*, 508:L105–L108, 1998.

[56] D. C. Braun and C. Lindsey. Helioseismic Images of an Active Region Complex. *Astrophys. J.*, 513:L79–L82, 1999.

[57] D. C. Braun and C. Lindsey. Seismic Imaging of the Far Hemisphere of the Sun. *Astrophys. J.*, 560:L189–L192, 2001.

[58] D.C. Braun, T.L. Duvall, and B.J. Labonte. Acoustic Absorption by Sunspots. *Astrophys. J. Lett.*, 319:L27–L31, 2005.

[59] J.R. Brookes, G.R. Isaak, C. P. McLeod, et al. The Search for Solar Oscillations, 1974 to 1976. *Month. Not. Royal Astron. Soc.*, 184:759–767, 1978.

[60] J.R. Brookes, G.R. Isaak, and H.B. van der Raay. A Resonant-Scattering Solar Spectrometer. *Month. Not. Royal Astron. Soc.*, 185:1–18, 1978.

[61] J.W. Brosius. Coronal Diagnostics with Coordinated Radio and EUV/Soft X-Ray Observations. In *Solar and Space Weather Radiophysics—Current Status and Future Developments*, Gary, D.E., Keller, C.U. (Eds.), volume 314, pages 265–286. Kluwer, Dordrecht, 2004.

[62] J.W. Brosius, E. Landi, J.W. Cook, J.S. Newmark, N. Gopalswamy, and A. Lara. Measurements of Three-dimensional Coronal Magnetic Fields from Coordinated Extreme-Ultraviolet and Radio Observations of a Solar Active Region Sunspot. *Astrophys. J.*, 574:453–466, 2002.

[63] T. M. Brown and R. L. Gilliland. Asteroseismology. *Annual Rev. Astron. Astrophys.*, 32:37–82, 1994.

[64] T. M. Brown, R. L. Gilliland, R. W. Noyes, and L. W. Ramsey. Detection of Possible p-Mode Oscillations on Procyon. *Astrophys. J.*, 368:599–609, 1991.

[65] T. M. Brown, E. J. Kennelly, S. G. Korzennik, et al. A Radial Velocity Search for p-Mode Pulsations in ηBootis. *Astrophys. J.*, 475:322–327, 1997.

[66] A. S. Brun, S. Turck-Chièze, and J. P. Zahn. Standard Solar Models in the Light of New Helioseismic Constraints. II. Mixing below the Convective Zone. *Astrophys. J.*, 525:1032–1041, 1999.

[67] L. Burlaga, E. Sittler, F. Mariani, and R. Schwenn. Magnetic Loop Behind an Interplanetary Shock - Voyager, Helios, and IMP 8 Observations. *J. Geophys. Res.*, 86:6673–6684, 1981.

[68] R. Bustos, M. Rubio, A. Otárola, et al. Parque Astronómico de Atacama: An Ideal Site for Millimeter, Submillimeter, and Mid-Infrared Astronomy. *Publ. Astron. Soc. Pacific.*, 126:1126–1132, 2014.

[69] W. Cao, N. Gorceix, R. Coulter, K. Ahn, T.R. Rimmele, and P.R. Goode. Scientific Instrumentation for the 1.6 m New Solar Telescope in Big Bear. *Astron. Nachr.*, 331:636–639, 2010.

[70] F. Carrier, F. Bouchy, and P. Eggenberger. Asteroseismology on Solar-Type Stars, the Right Tool to Understand Stellar Interiors. *Recent Res. Develop. Astron. Astrophys.*, 1:219–244, 2003.

[71] A. Caspi, T.N. Woods, and H.P. Warren. New Observations of the Solar 0.5-5 keV Spectrum. *Astrophys. J. Lett.*, 802:1, 2015.

[72] B. Cecconi, X. Bonnin, S. Hoang, et al. STEREO/Waves Goniopolarimetry. *Space Sci. Rev.*, 136:549–563, 2008.

[73] S. Chandrasekhar. *Radiative transfer.* Dover, New York, 1960.

[74] W. J. Chaplin, Y. Elsworth, G. R. Isaak, et al. Recent Results from the Birmingham Solar-Oscillations Network (BiSON). In *American Astronomical Society, 188th AAS Meeting, id.69.04*, volume 28, page 936, 1996.

[75] W. J. Chaplin, Y. Elsworth, G. R. Isaak, et al. Variations in the Excitation and Damping of Low-l Solar p Modes Over the Solar Activity Cycle. *Monthly Not. Royal Astron. Soc.*, 313:32–42, 2000.

[76] S. Chapman. Notes on the Solar Corona and the Terrestrial Ionosphere. *Smithsonian Contr. Astrophys.*, 2:1–14, 1957.

[77] P. Charbonneau. Solar Dynamo Theory. *Annual Rev. Astron. Astrophys.*, 52:251–290, 2014.

[78] F. F. Chen. *Introduction to Plasma Physics and Controlled Fusion, 3rd Edition.* Springer International Publishing, 2016.

[79] D.-Y. Chou, M.-T. Sun, T.-Y. Huang, et al. Taiwan Oscillation Network (TON). *Solar Phys.*, 160:237–243, 1995.

[80] A. R. Choudhuri, P. Chatterjee, and J. Jiang. Predicting Solar Cycle 24 with a Solar Dynamo Model. *Phys. Rev. Lett.*, 98:131103, 2007.

[81] J. Christensen-Dalsgaard. Seismological Studies of the Sun and Other Stars. *Adv. Space Res.*, 2:11–19, 1982.

[82] J. Christensen-Dalsgaard. The 'Standard' Sun Modelling and Helioseismology. *Space Sci. Rev.*, 85:19–36, 1998.

[83] J. Christensen-Dalsgaard. Helioseismology. *Reviews Modern Phys.*, 74:1073–1129, 2002.

[84] J. Christensen-Dalsgaard. *Lecture Notes on Stellar Oscillations, Fifth Edition.* Institut for Fysik og Astronomi, Aarhus Universitet Teoretisk Astrofysik Center, Danmarks Grundforskningsfond, 2003.

[85] J. Christensen-Dalsgaard and S. Frandsen. Stellar 5 min Oscillations. *Solar Phys.*, 82:469–486, 1983.

[86] A. Claverie, G. R. Isaak, C. P. McLeod, et al. Solar Structure from Global Studies of the 5-minute Oscillation. *Nature*, 282:591–594, 1979.

[87] B. T. Cleveland, T. Daily, R. Davis, Jr., et al. Measurement of the Solar Electron Neutrino Flux with the Homestake Chlorine Detector. *Astrophys. J.*, 496:505–526, 1998.

[88] M. Collados, R. López, E. Páez, et al. GRIS: The GREGOR Infrared Spectrograph. *Astron. Nachr.*, 333:872–879, 2012.

[89] S. Couvidat, S.P. Rajaguru, R. Wachter, et al. Line-of-Sight Observables Algorithms for the Helioseismic and Magnetic Imager (HMI) Instrument Tested with Interferometric Bidimensional Spectrometer (IBIS) Observations. *Solar Phys.*, 278:217–240, 2012.

[90] T. G. Cowling. The Non-Radial Oscillations of Polytropic Stars. *Mon. Not. R. Astron. Soc.*, 101:367–375, 1941.

[91] R. Davis, Jr. Solar Neutrinos. II. Experimental. *Phys. Rev. Lett.*, 12:303–305, 1964.

[92] L. D'Azambuja. L'oeuvre de Bernard Lyot. *L'Astronomie*, 66:265–277, 1952.

[93] C. Debi Prasad, G. A. Gary, and A. Ambastha. Complex H_α Loop Activity in a Long Duration Flare. In *High Resolution Solar Physics: Theory, Observations, and Techniques*, volume 183, pages 523–530. Astron. Soc. Pacific, 1999.

[94] J.C. del Toro Iniesta. *Introduction to Spectropolarimetry.* Cambridge University Press, Cambridge, U.K., 2003.

[95] D. Deming, D. E. Jennings, J. Jefferies, and C. Lindsey. Physics of the Infrared Spectrum. In *Solar Interior and Atmosphere*, pages 933–963. University of Arizona Press, Tucson, AZ., U.S.A., 1991.

[96] D. Deming, T. Kostiuk, and D. Glenar. The Solar Infrared Explorer (SIRE): A Small Explorer Mission for Solar Physics. *Bull. Am. Astron. Soc.*, 23:1038, 1991.

[97] A. J. Dessler. Solar Wind and Interplanetary Magnetic Field. *Rev. Geophys. Space Phys.*, 5:1–41, 1967.

[98] F. L. Deubner. Observations of Low Wavenumber Nonradial Eigenmodes of the Sun. *Astron. Astrophys.*, 44:371–375, 1975.

[99] F. L. Deubner and D.O. Gough. Helioseismology: Oscillations as a Diagnostic of the Solar Interior. *Annual Rev. Astron. Astrophys.*, 22:593–619, 1984.

[100] M. Dikpati, G. de Toma, and P. A. Gilman. Predicting the Strength of Solar Cycle 24 Using a Flux-Transport Dynamo-Based Tool. *Geophys. Res. Lett.*, 33:L05102, 2006.

[101] M. Dikpati, P. A. Gilman, and G. de Toma. The Waldmeier Effect: An Artifact of the Definition of Wolf Sunspot Number? *Astrophys. J.*, 673:L99–L101, 2008.

[102] A.-C. Donea, D. C. Braun, and C. Lindsey. Seismic Images of a Solar Flare. *Astrophys. J.*, 513:L143–L146, 1999.

[103] A.-C. Donea and C. Lindsey. Seismic Emission from the Solar Flares of 2003 October 28 and 29. *Astrophys. J.*, 630:1168–1183, 2005.

[104] S. D'Silva and A. R. Choudhuri. A Theoretical Model for Tilts of Bipolar Magnetic Regions. *Astron. Astrophys.*, 272:621–633, 1993.

[105] T. L. Duvall, Jr., J. W. Harvey, S. M. Jefferies, and M. A. Pomerantz. Measurements of High-Frequency Solar Oscillation Modes. *Astrophys. J.*, 373:308–316, 1991.

[106] T. L. Duvall, Jr., S. M. Jefferies, J. W. Harvey, and M. A. Pomerantz. Time-Distance Helioseismology. *Nature*, 362:430–432, 1993.

[107] T. L. Duvall, Jr., A. G. Kosovichev, P. H. Scherrer, et al. Time-Distance Helioseismology with the MDI Instrument: Initial Results. *Solar Phys.*, 170:63–73, 1997.

[108] W. A. Dziembowski, D. O. Gough, G. Houdek, and R. Sienkiewicz. Oscillations of αUMa and other Red Giants. *Monthly Not. Royal Astron. Soc.*, 328:601–610, 2001.

[109] W.A. Dziembowski and P.R. Goode. The Toroidal Magnetic Field Inside the Sun. *Astrophys. J.*, 347:540–550, 1989.

[110] J.A. Eddy. A Nineteenth-Century Coronal Transient. *Astron. Astrophys.*, 34:235–240, 1974.

[111] J.A. Eddy. The Maunder Minimum. *Science*, 192:1189–1203, 1976.

[112] J.A. Eddy. Anomalous Solar Rotation in the Early 17th Century. *Science*, 198:824–829, 1977.

[113] J.A. Eddy. The Maunder Minimum: a Reappraisal. *Solar Phys.*, 89:195–207, 1983.

[114] Sh. Egamberdiev and Eric Fossat. International Research on the Interior of the Sun (IRIS). In *IRIS: 2nd Workshop Proc., Tashkent, Uzbek SSR, 1989*, volume 133, 1991.

[115] R.D. Ekers. The History of the Square Kilometre Array (SKA) - Born Global. In *Resolving The Sky - Radio Interferometry: Past, Present and Future*, page id7. Manchester, UK, 2012.

[116] S. Enome. Hinotori-A Japanese Satellite for Solar Flare Studies. *Advances Space Res.*, 29:201–202, 1982.

[117] J. W. Evans and R. Michard. Observational Study of Macroscopic Inhomogeneities in the Solar Atmosphere. I. Velocity Displacements of Fraunhofer Lines as a Function of Line Strength and Position on Disk. *Astrophys. J.*, 135:812–821, 1962.

[118] J. Evershed. Radial Movement in Sunspots. *Kodaikanal Obs. Bull.*, 2:63–66, 1909.

[119] J. Feynman and S. F. Martin. The Initiation of Coronal Mass Ejections by Newly Emerging Magnetic Flux. *J. Geophys. Res.*, 100:3355–3367, 1995.

[120] L. A. Fisk. Acceleration of the Solar Wind as a Result of the Reconnection of Open Magnetic Flux with Coronal Loops. *J. Geophys. Res.: Space Phys.*, 108:1157, 2003.

[121] E. Fossat, P. Boumier, T. Corbard, et al. Asymptotic g Modes: Evidence for a Rapid Rotation of the Solar Core. (id.A40). *Astron. Astrophys.*, 604:1–17, 2017.

[122] P. Foukal and J. Lean. An Empirical Model of Total Solar Irradiance Variation Between 1874 and 1988. *Science*, 247:556–558, 1990.

[123] K. C. Fox. *NASA Study Using Cluster Reveals New Insights Into Solar Wind.* N.A.S.A., 2012.

[124] N. J. Fox, M. C. Velli, S. D. Bale, et al. The Solar Probe Plus Mission: Humanity's First Visit to Our Star. *Space Sci. Rev.*, 204:7–48, 2016.

[125] G.W. Fraser. *X-ray Detectors in Astronomy.* Cambridge University Press, Cambridge, UK, 1989.

[126] H. Friedman, S. W. Lichtman, and E. T. Byram. Photon Counter Measurements of Solar X-rays and Extreme Ultraviolet Light. *Phys. Rev.*, 83:1025–1030, 1951.

[127] Y. Fukuda, T. Hayakawa, E. Ichihara, et al. Measurement of the Solar Neutrino Energy Spectrum Using Neutrino-Electron Scattering. *Phys. Rev. Lett.*, 82:2430–2434, 1999.

[128] Y. Fukuda, T. Hayakawa, K. Inoue, et al. Solar Neutrino Data Covering Solar Cycle 22. *Phys. Rev. Lett.*, 77:1683–1686, 1996.

[129] G. Galileo and S. Christoph. *On Sunspots.* University of Chicago Press, 2010.

[130] A.H. Gabriel and GOLF Team. Global Oscillations at Low Frequency from the SOHO Mission (GOLF). *Adv. Space Res.*, 11:103–112, 1991.

[131] D. Gallagher, S. Tomczyk, H. Zhang, and P.G. Nelson. Optical Design of the Cosmo Large Coronagraph. In *Ground-based and Airborne Telescopes IV: Stepp L.M., Gilmozzi R., Hall H.J. (Eds.)*. Bellingham, WA: SPIE Digital Library, 2012.

[132] D.E. Gary and C.U. Keller. *Solar and Space Weather Radiophysics-Current Status and Future Developments. Astrophy. Space Sci. Lib.*, volume 314. Kluwer, Dordrecht, 2004.

[133] S. Gburek, J. Sylwester, M. Kowalinski, et al. SphinX Soft X-ray Spectrophotometer: Science Objectives, Design and Performance. *Solar System Res.*, 45:189–199, 2012.

[134] J. Geiss, G. Gloeckler, and R. Von. Steiger. Origin of the Solar Wind from Composition Data. *Space Sci. Rev.*, 72:49–60, 1995.

[135] R. Giacconi, W. P. Reidy, T. Zehnpfennig, et al. Solar X-Ray Image Obtained Using Grazing-Incidence Optics. *Astrophys. J.*, 142:1274–1278, 1965.

[136] S. E. Gibson, T. A. Kucera, S. M. White, et al. FORWARD: a Toolset for Multiwavelength Coronal Magnetometry: doi: 10.3389/fspas.2016.00008. *Front. Astron. Space Sci.*, 3, 2016.

[137] Sara E. Gibson, A. Vourlidas, D. M. Hassler, et al. Solar Physics From Unconventional Viewpoints (`https://doi.org/10.3389/fspas.2018.00032`). *Front. Astron. Space Sci.*, 24:1–53, 2018.

[138] R. G. Giovanelli. The Relations Between Eruptions and Sunspots. *Astrophys. J.*, 89:555–567, 1939.

[139] L. Gizon and A.C. Birch. Local Helioseismology. *Living Rev. Solar Phys.*, 2(6):1, 2005.

[140] G. Gloeckler, J. Geiss, H. Balsiger, et al. The Ulysses Solar-Wind Plasma Experiment. *Astron. Astrophys. Supp. Ser.*, 92:267–289, 1992.

[141] G. Gonczi and F. Roddier. A Fourier Spectrum Analysis of Long Samples of Solar Line Oscillations. *Solar Phys.*, 8:255–259, 1969.

[142] N. Gopalswamy. Recent Advances in the Long-Wavelength Radio Physics of the Sun. *Planet. Space Sci.*, 52:1399–1413, 2001.

[143] N. Gopalswamy. Coronal Mass Ejections of Solar Cycle 23. *J. Astrophys. Astron.*, 27:243–254, 2006.

[144] N. Gopalswamy, J. M. Davila, O. C. St. Cyr, et al. Earth-affecting Solar Causes Observatory (EASCO): A Potential International Living with a Star Mission from Sun-Earth L5 (doi: 10.1016/j.jastp.2011.01.013). *J. Atmos. Solar Terrestr. Phys.*, 73:658–663, 2011.

[145] N. Gopalswamy, S. Hasan, and A. Ambastha. *Heliophysical Processes.* Springer, Berlin, 2012.

[146] N. Gopalswamy, M. L. Yashiro, R. A. Kaiser, et al. Radio Signatures of Coronal Mass Ejection Interaction: Mass Ejection Cannibalism? *Astrophys. J.*, 548:L91–L94, 2004.

[147] D. O. Gough, J. W. Leibacher, P. H. Scherrer, and J. Toomre. Perspectives in Helioseismology. *Science*, 272:1281–1283, 1996.

[148] D.O. Gough and M.J. Thompson. The Inversion Problem. In *Solar Interior and Atmosphere, Eds. Cox, A.N., Livingston, W.C. and Matthews, M.S.*, pages 519–561. University of Arizona Press, Tucson, 1991.

[149] D.O. Gough and J. Toomre. Seismic Observations of the Solar Interior. *Annual Rev. Astron. Astrophys.*, 29:627–685, 1991.

[150] G. Grec, E. Fossat, and M. Pomerantz. Solar Oscillations - Full Disk Observations from the Geographic South Pole. *Nature*, 288:541–544, 1980.

[151] N. Grevesse and A. Noels. Cosmic Abundances of the Elements. In *Origin and Evolution of the Elements, N. Prantzos, E. Vangioni-Flam and Cassé, M. (Eds.)*, pages 15–25. Cambridge Univ. Press, Cambridge, 1993.

[152] V.N. Gribov and B.M. Pontecorvo. Neutrino Astronomy and Lepton Charge. *Phys. Lett. B*, 28:493–496, 1969.

[153] D. B. Guenther, P. Demarque, Y.-C. Kim, and M. H. Pinsonneault. Standard Solar Model. *Astrophys. J.*, 387:372–393, 1992.

[154] H. Grünwaldt et al. Venus Tail Ray Observation Near Earth. *Geophys. Res. Lett.*, 24:1163–1166, 1997.

[155] G.E. Hale. On the Probable Existence of a Magnetic Field in Sunspots. *Astrophys. J.*, 92:315 343, 1908.

[156] G.E. Hale, F. Ellerman, S. B. Nicholson, and A. H. Joy. The Magnetic Polarity of Sunspots. *Astrophys. J.*, 49:153–178, 1919.

[157] W. Hampel, J. Handt, G. Heusser, et al. GALLEX Solar Neutrino Observations: Results for GALLEX IV. *Phys. Lett. B*, 447:127–133, 1999.

[158] W. Hanle. Über Magnetische Beeinflussung der Polarisation der Resonanzfluoreszenz. *Zeitschrift für Physik*, 30:93–105, 1924.

[159] I. G. Hannah, S. Christe, S. Krucker, et al. RHESSI Microflare Statistics. II. X-Ray Imaging, Spectroscopy, and Energy Distributions. *Astrophys. J.*, 677:704–718, 2008.

[160] J. Harvey. Trends in Measurement of Solar Vector Magnetic Fields Using the Zeeman Effect. In *Measurement of solar vector magnetic fields: Ed. Hagyard, M.*, volume 2374, pages 109–120. NASA, Washington D.C., 1985.

[161] J. Harvey. Helioseismology. *Physics Today*, 48:32–38, 1995.

[162] J. Harvey, K. Abdel-Gawad, W. Ball, et al. The Gong Instrument. In *Seismology of the Sun and Sun-Like Stars*, pages 203–208, 1988.

[163] J. Harvey and D. Hall. Magnetic Fields Measured with the 10830Å He I Line. In *Solar Magnetic Fields*, pages 279–288. Reidel, Dordrecht, 1971.

[164] J. Harvey and D. Hall. Magnetic Field Observations with Fe I 15648 Å. *Bull. Am. Astron. Soc.*, 7:459, 1975.

[165] J. Harvey, A. S. Krieger, A. F. Timothy, and G. S. Vaiana. Comparison of Skylab X-Ray and Ground-Based Helium Observations. In *Skylab Solar Workshop: Prelimary Results from the S-054 X- Ray Telescope and the Correlated Ground-Based Observations, (Ed.) Righini, G.*, volume 104, pages 50–58. Baccini and Chiappi, Florence, 1975.

[166] K.L. Harvey and C. Zwaan. Properties and Emergence Patterns of Bipolar Active Regions. *Solar Phys.*, 148:85–118, 1993.

[167] D. Hassler, J. Newmark, M. Velli, and N. Murphy. Solar Wind 15. In *Proceedings of a Conference (Brussels)*, pages 18–22, 2018.

[168] D. M. Hassler, I. E. Dammasch, P. Lemaire, et al. Solar Wind Outflow and the Chromospheric Magnetic Network. *Science*, 283:810–813, 1999.

[169] D. H. Hathaway. The Solar Cycle. *Living Rev. Solar Phys.*, 12(4):1–87, 2015.

[170] D. H. Hathaway, R. M. Wilson, and E. J. Reichmann. The Shape of the Sunspot Cycle. *Solar Phys.*, 151:177–190, 1994.

[171] D. H. Hathaway, R. M. Wilson, and E. J. Reichmann. A Synthesis of Solar Cycle Prediction Techniques. *J. Geophys. Res.*, 104:22375–22388, 1999.

[172] D. H. Hathaway, R. M. Wilson, and E. J. Reichmann. Group Sunspot Numbers: Sunspot Cycle Characteristics. *Sol. Phys.*, 211:357–370, 2002.

[173] W. Herschel. Experiments on the Refrangibility of the Invisible Rays of the Sun. *Philosophical Trans. Royal Soc. London*, 90:284–292, 1800.

[174] D. Hestroffer and C. Magnan. Wavelength Dependency of the Solar Limb Darkening. *Astron. Astrophys.*, 333:338–342, 1998.

[175] B. Hetherington. *A Chronicle of Pre-telescopic Astronomy.* John Wiley and Sons, 1996.

[176] F. Hill. Rings and Trumpets - Three-dimensional Power Spectra of Solar Oscillations. *Astrophys. J.*, 333:996–1013, 1988.

[177] F. Hill. Solar Oscillation Ring Diagrams and Large-Scale Flows. *Astrophys. J. Lett.*, 343:L69–L71, 1989.

[178] B. Hindman, D. Haber, J. Toomre, and R. Bogart. Local Fractional Frequency Shifts Used as Tracers of Magnetic Activity. *Solar Phys.*, 192:363–372, 2000.

[179] T. Hirayama. Theoretical Model of Flares and Prominences. I: Evaporating Flare Model. *Solar Phys.*, 34:323–338, 1974.

[180] G. Houdek, N. J. Balmforth, J. Christensen-Dalsgaard, and D. O. Gough. Amplitudes of Stochastically Excited Oscillations in Main-Sequence Stars. *Astron. Astrophys.*, 351:582–596, 1999.

[181] R. Howard. Solar Rotation. *Annual Rev. Astron. Astrophys.*, 22:131–155, 1984.

[182] R. Howard, P. I. Gilman, and P. A. Gilman. Rotation of the Sun Measured from Mount Wilson White-Light Images. *Astrophys. J.*, 283:373–384, 1984.

[183] R. Howard and B. J. Labonte. The Sun is Observed to be a Torsional Oscillator with a Period of 11 Years. *Astrophys. J.*, 239:L33–L36, 1980.

[184] R. F. Howard. Axial Tilt Angles of Sunspot Groups. *Solar Phys.*, 136:251–262, 1991.

[185] R. Howe. Solar Interior Rotation and its Variation. lrsp-2009-1. *Living Rev. Solar Phys.*, 6, 2009.

[186] D.V. Hoyt and K.H. Schatten. *The Role of the Sun in Climate Change.* Oxford University Press, 1997.

[187] H. S. Hudson, L. W. Acton, T. Hirayama, et al. White-Light Flares Observed by YOHKOH. *Publ. Astron. Soc. Japan*, 44:L77–L81, 1992.

[188] K. Hufbauer. *Exploring the Sun.* The Johns Hopkins University Press, USA, 1991.

[189] G. J. Hurford, S. Krucker, R.P. Lin, et al. Gamma-Ray Imaging of the 2003 October/November Solar Flares. *Astrophys. J.*, 644:L93–L96, 2006.

[190] G. J. Hurford, E.J. Schmahl, R.A. Schwartz, et al. The RHESSI Imaging Concept. *Solar Phys.*, 210:61–86, 2002.

[191] G. J. Hurford, R. A. Schwartz, S. Krucker, et al. First Gamma-Ray Images of a Solar Flare. *Astrophys. J.*, 595:L77–L80, 2003.

[192] J. Schou, J. Christensen-Dalsgaard, and M. J. Thompson. On Comparing Helioseismic Two-Dimensional Inversion Methods. *Astrophys. J.*, 433:389–416, 1994.

[193] S.A. Jaeggli, H. Lin, D.L. Mickey, et al. FIRS: A New Instrument for Photospheric and Chromospheric Studies at the DST. *Mem. S.A.It.*, 81:763–768, 2010.

[194] S.A. Jaeggli and A.A. Norton. The Magnetic Classification of Solar Active Regions 1992-2015. *Astrophys. J. Letters*, 820:L11, 2016.

[195] K. Jain, S. C. Tripathy, and A. Bhatnagar. Solar Cycle-induced Variations in GONG P-Mode Frequencies and Splittings. *Astrophys. J.*, 542:521–527, 2000.

[196] B. M. Jakosky, J. M. Grebowsky, J. G. Luhmann, and D. A. Brain. Initial Results from the MAVEN Mission to Mars. *Geophys. Res. Lett.*, 42:8791–8802, 2015.

[197] B. M. Jakosky, J. M. Grebowsky, J. G. Luhmann, et al. MAVEN Observations of the Response of Mars to an Interplanetary Coronal Mass Ejection. *Science*, 350:0201, 2015.

[198] B.M. Jakosky, R.P. Lin, J.M. Grebowsky, et al. The Mars Atmosphere and Volatile Evolution (MAVEN) Mission. *Space Sci. Rev.*, 195:3–48, 2015.

[199] S.M. Jefferies, T.L. Duvall Jr., J.W. Harvey, and M.A. Pomerantz. Helioseismology from the South Pole: Results from the 1987 campaign. In *Progress of Seismology of the Sun and Stars, Eds. Osaki, Y. and Shibahashi, H.*, pages 135–143. Springer-Verlag, 1990.

[200] J. M. Jensen. Time-Distance: What Does It Tell Us? In *SOHO 12/GONG+ 2002. Local and Global Helioseismology: The Present and Future (Ed. H. Sawaya-Lacoste)*, volume 517, pages 61–70. Noordwijk, Netherlands: ESA Publications Division, 2003.

[201] N. Kameshwar Rao. Aspects of Prehistoric Astronomy in India. *Bull. Astron. Soc. India*, 33:499–511, 2005.

[202] R. P. Kane. How Useful is the Waldmeier Effect for Prediction of a Sunspot Cycle? *J. Atmosph. Solar-Terres. Phys.*, 70:1533–1540, 2008.

[203] S.R. Kane. Impulsive (flash) Phase of Solar Flares: Hard X-ray Microwave, EUV and Optical Observations. In *Coronal Disturbances. Ed. G. A.Newkirk*, volume 57, pages 105–141. International Astronomical Union, Dordrecht; Boston: Reidel, 1974.

[204] S.R. Kane. Spatial Structure of High Energy Photon Sources in Solar Flares. *Solar Phys.*, 86:355–365, 1983.

[205] P. Kaufmann, S. M. White, S. L. Freeland, et al. A Bright Impulsive Solar Burst Detected at 30 THz: article id. 134. *Astrophys. J.*, 768:9pp, 2013.

[206] S. A. Kazantsev. Astrophysical and Laboratory Applications of Self-Alignment. *Sov. Phys. Usp.*, 26:328–351, 1983.

[207] S. L. Keil, T. R. Rimmele, C. U. Keller, and ATST Team. The Advanced Technology Solar Telescope. In *Advanced Solar Polarimetry – Theory, Observation, and Instrumentation: Ed. Sigwarth, M*, volume 236, pages 597–. Astronomical Society of the Pacific, San Francisco, 2001.

[208] A.B. Keith. *A History of Indian Literature.* Oxford, 1929.

[209] K.O. Kiepenheuer. Photoelectric Measurements of Solar Magnetic Fields. *Astrophys. J.*, 117:447–453, 1953.

[210] R. Kippenhahn and A. Schlüter. Eine Theorie der Solaren Filamente. *Zeit. Astrophysik*, 43:36–62, 1957.

[211] M.G. Kivelson and C.T. Russell (Eds.). *Introduction to Space Physics: Chap. 1.* Cambridge University Press, 1995.

[212] H. Kjeldsen, T. R. Bedding, M. Viskum, and S. Frandsen. Solar-Like Oscillations in ηBootis. *Astron. J.*, 109:1313–1319, 1995.

[213] H. Kjeldsen, T. R. Bedding, M. Viskum, and S. Frandsen. Confirmation of Solar-Like Oscillations in ηBootis. *Astron. J.*, 126:1483–1488, 2003.

[214] K.-L. Klein and A.L. MacKinnon (Eds.). *The High Energy Solar Corona: Waves, Eruptions and Particles.* Lecture Notes in Physics 725, Springer, Berlin Heidelberg, 2007.

[215] R. Kochhar. Pre-telescopic Astronomy in India. In *History of Indian Science, Technology and Culture, Ed. A Rahman.* Oxford University Press, 1999.

[216] R. A. Kopp and G. W. Pneuman. Magnetic Reconnection in the Corona and the Loop Prominence Phenomenon. *Solar Phys.*, 50:85–98, 1976.

[217] A. G. Kosovichev. Properties of Flares-Generated Seismic Waves on the Sun. *Solar Phys.*, 238:1–11, 2006.

[218] A. G. Kosovichev and J. O. Stenflo. Tilt of Emerging Bipolar Magnetic Regions on the Sun. *Astrophys. J. Let.*, 688:L115–L118, 2008.

[219] A. G. Kosovichev and V. V. Zharkova. X-ray Flare Sparks Quake Inside Sun. *Nature*, 393:317–318, 1998.

[220] A. G. Kosovichev and V. V. Zharkova. Variations of Photospheric Magnetic Field Associated with Flares and CMEs. *Solar Phys.*, 190:459–466, 1999.

[221] A.G. Kosovichev. Advances in Global and Local Helioseismology: An Introductory Review. *Lecture Notes in Physics*, 832:3, 2011.

[222] A.G. Kosovichev, T.L. Duvall, Jr., and P.H. Scherrer. Time-Distance Inversion Methods and Results (Invited Review). *Solar Phys.*, 192:159–176, 2000.

[223] T. Kosugi, K. Makishima, T. Murakami, et al. The Hard X-ray Telescope (HXT) for the SOLAR-A Mission. *Solar Phys.*, 136:17–36, 1991.

[224] M. R. Kundu. *Solar Radio Astronomy.* John Wiley & Sons Inc, New York, USA, 1965.

[225] M.R. Kundu and T.E. Gergely (Eds.). Radio Physics of the Sun. In *Radio Physics of the Sun.* Springer, 1980.

[226] M. Kuperus and M. A. Raadu. The Support of Prominences Formed in Neutral Sheets. *Astron. Astrophys.*, 31:189–193, 1957.

[227] E. Landi Degl'Innocenti. The Determination of Vector Magnetic Fields in Prominences from the Observations of the Stokes Profiles in the D3 Line of Helium. *Sol. Phys.*, 79:291–322, 1982.

[228] E. Landi Degl'Innocenti. The Physics of Polarization. In *Astrophysical spectro-polarimetry*, pages 1–53. Cambridge University Press, Cambridge, 2002.

[229] M. Landolfi and E. Landi Degl'Innocenti. Magneto-Optical Effects and the Determination of Vector Magnetic Fields from Stokes Profiles. *Solar Phys.*, 78:355–364, 1982.

[230] B. Lavraud, Y. Liu, K. Segura, et al. A Small Mission Concept to the Sun-Earth Lagrangian L5 point for Innovative Solar, Heliospheric and Space Weather Science (doi: 10.1016/j.jastp.2016.06.004). *J. Atmos. Solar Terrestr. Phys.*, 146:171–185, 2016.

[231] J. W. Leibacher. The Global Oscillation Network Group (GONG) project. *Adv. Space Res.*, 24:173–176, 1999.

[232] J. W. Leibacher and R. F. Stein. A New Description of the Solar Five-Minute Oscillation. *Astrophys. J. Lett.*, 7:191–192, 1971.

[233] R. B. Leighton. A Magneto-Kinematic Model of the Solar Cycle. *Astrophys. J.*, 156:1–26, 1969.

[234] R.B. Leighton, R.W. Noyes, and G.W. Simon. Velocity Fields in the Solar Atmosphere. I. Preliminary Report. *Astrophys. J.*, 135:474–499, 1962.

[235] J.R. Lemen, A.M. Title, D.J. Akin, et al. The Atmospheric Imaging Assembly (AIA) on the Solar Dynamics Observatory (SDO). *Solar Phys.*, 275:17–40, 2012.

[236] P. J. Léna. Infrared Observations of the Sun. In *Infrared and Submillimeter Astronomy: G. G. Fazio (Ed.)*, pages 97–107. D. Reidel Publishing Company, Dordrecht-Holland, 1977.

[237] J. L. Leroy. The Hanle Effect Applied in Magnetic Field Measurements. In *Measurements of Solar Vector Magnetic Fields*, volume 2374, pages 121–140. NASA. Marshall Space Flight Center, 1985.

[238] J. L. Leroy, V. Bommier, and S. Sahal-Brèchot. New Data on the Magnetic Structure of Quiescent Prominences. *Astron. Astrophys.*, 131:33–44, 1984.

[239] H. Lin. ATST near-IR spectropolarimeter. In *Innovative Telescopes and Instrumentation for Solar Astrophysics: Keil S.L., Avakyan S.V. (Eds.)*, pages 215–222. Bellingham, WA: SPIE Digital Library, 2003.

[240] H. Lin, J. R. Kuhn, and R. Coulter. Coronal Magnetic Field Measurements. *Astrophys. J.*, 613:L177–L180, 2008.

[241] H. Lin and R. Kuhn. An Imaging, Tunable Magneto-Optical Filter. *Solar Phys.*, 122:365–380, 1989.

[242] H. Lin, M.J. Penn, and J.R. Kuhn. He I 10830 Å Line Polarimetry: A New Tool to Probe the Filament Magnetic Fields. *Astrophys. J.*, 493:978–995, 1998.

[243] R.P. Lin, B.R. Dennis, G.J. Hurford, et al. The Reuven Ramaty High-Energy Solar Spectroscopic Imager (RHESSI). *Solar Phys.*, 210:3–32, 2002.

[244] C. Lindsey and D. C. Braun. Helioseismic Imaging of Sunspots at Their Antipodes. *Solar Phys.*, 126:101–115, 1990.

[245] C. Lindsey and D. C. Braun. Basic Principles of Solar Acoustic Holography. *Solar Phys.*, 192:261–284, 2000.

[246] C. Lindsey, D. C. Braun, and S. M. Jefferies. Local Helioseismology of Subsurface Structure. In *GONG 1992. Seismic Investigation of the Sun and Stars (Ed. Timothy M. Brown)*, volume 42, pages 81–84. Astron. Soc. Pacific, 1993.

[247] B.W. Lites, M. Kubo, H. Socas-Navarro, et al. The Horizontal Magnetic Flux of the Quiet-Sun Internetwork as Observed with the Hinode Spectro-Polarimeter. *Astrophys. J.*, 672:1237–1253, 2008.

[248] Y. Liu, J. A. Davies, J.G. Luhmann, et al. Geometric Triangulation of Imaging Observations to Track Coronal Mass Ejections Continuously Out to 1 AU. *Astrophys. J.*, 710:L82–L87, 2010.

[249] Y. Liu, J.G. Luhmann, S.D. Bale, et al. Relationship Between a Coronal Mass Ejection-Driven Shock and a Coronal Metric Type II Burst. *Astrophys. J.*, 691:L151–L155, 2009.

[250] W. Livingston. A 4-meter McMath Telescope for the Infrared. In *Infrared Solar Physics (IAU Symposium 154): Rabin D.M., Jefferies J.T., Lindsey C. (Eds.)*, pages 589–594. Kluwer Academic Publishers, 1994.

[251] F. J. Low et al. Infrared Cirrus – New Components of the Extended Infrared Emission. *Astrophys. J.*, 278:L19–L22, 1984.

[252] B. Lyot. Le Filtre Monochromatique Polarisant et ses Applications en Physique Solaire. *Annales d'Astrophysique*, 7:31–49, 1944.

[253] R.M. MacQueen, J.A. Eddy, J.T. Gosling, et al. The Outer Solar Corona as Observed from Skylab: Preliminary Results. *Astrophys. J.*, 187:L85–L88, 1974.

[254] P. K. Manoharan, S. Ananthakrishnan, M. Dryer, et al. Solar Wind Velocity and Normalized Scintillation Index from Single-Station IPS Observations. *Solar Phys.*, 156:377–393, 1995.

[255] P. K. Manoharan, M. Tokumaru, M. Pick, et al. Coronal Mass Ejection of 2000 July 14 Flare Event: Imaging from Near-Sun to Earth Environment. *Astrophys. J.*, 559:1180–1189, 2001.

[256] S. Masuda, T. Kosugi, H. Hara, S. Tsuneta, and Y. Ogawara. A Loop-Top Hard X-ray Source in a Compact Solar Flare as Evidence for Magnetic Reconnection. *Nature.*, 371:495–497, 1994.

[257] S. Mathew, A. Bhatnagar, C. Debi Prasad, and A. Ambastha. Fabry-Perot Filter Based Solar Video Magnetograph. *Astron. Astrophys. Suppl. Ser.*, 133:285–292, 1998.

[258] S. Mathew, A. Bhatnagar, C. Debi Prasad, and A. Ambastha. High Resolution Longitudinal Magnetic Field Measurements Using a Fabry-Perot Lithium Niobate Filter Based Video Magnetograph. In *High Resolution Solar Physics: Theory, Observations, and Techniques*, volume 183, pages 256–263. Astron. Soc. Pacific, 1999.

[259] S. K. Mathew, A. R. Bayanna, A.R. Tiwary, et al. First Observations from the Multi-Application Solar Telescope (MAST) Narrow-Band Imager. article id.106. *Solar Phys.*, 292:1–25, 2017.

[260] E. Maunder. Professor Spoerer's Researches on Sun-Spots. *Monthly Notices Royal Astron. Soc.*, 50:251–252, 1890.

[261] E. Maunder. A Prolonged Sunspot Minimum. *Knowledge*, 17:173–176, 1894.

[262] R. A. Maurya and A. Ambastha. Transient Magnetic and Doppler Features Related to the White-Light Flares in NOAA 10486. *Solar Phys.*, 258:31–52, 2009.

[263] R. A. Maurya and A. Ambastha. Sub-surface Meridional Flow, Vorticity, and the Lifetime of Solar Active Regions. *Astrophys. J. Lett.*, 714:L196–L201, 2010.

[264] R. A. Maurya, A. Ambastha, and S. C. Tripathy. Variations in p-Mode Parameters with Changing Onset Time of a Large Flare. *Astrophys. J. Lett.*, 706:L235–L239, 2009.

[265] B. H. McClintock and A. A. Norton. Recovering Joy's Law as a Function of Solar Cycle, Hemisphere, and Longitude. *Solar Phys.*, 287:215–227, 2013.

[266] D. J. McComas, S. J. Bame, B. L. Barraclough, et al. Ulysses' Return to the Slow Solar Wind (doi: 10.1029/97GL03444). *Geophys. Res. Lett.*, 25:1–4, 1998.

[267] D. J. McComas, H. A. Elliott, N. A. Schwadron, et al. The Three-Dimensional Solar Wind Around Solar Maximum. *Geophys. Res. Lett.*, 30:1517, 2003.

[268] P. S. McIntosh. The Classification of Sunspot Groups. *Solar Phys.*, 125:251–267, 1990.

[269] A.J. Meadows. *Early Solar Physics.* Pergamon Press, 1970.

[270] C. Meegan, G. Lichti, P. N. Bhat, et al. The Fermi Gamma-ray Burst Monitor. *Astrophys. J.*, 702:791–804, 2009.

[271] F. Meyer and H. U. Schmidt. Die Erzeugung von Schwingungen in der Sonnenatmosphäre durch einzelne Granula. *Zeitschrift Astrophysik.*, 65:274–298, 1967.

[272] F. Meyer and H. U. Schmidt. A Model for the Evershed Flow in Sunspots. *Mitteilungen Astron. Gesellschaft*, 25:194–196, 1968.

[273] N. Meyer-Vernet. *Basics of the Solar Wind.* Cambridge University Press, U.K., 2007.

[274] A. C. G. Mitchell and M. W. Zemansky. *Resonance Radiation and Excited Atoms.* Cambridge: University Press, 1934.

[275] W.M. Mitchell. The History of the Discovery of Solar Spots. *Popular Astron.*, 24:206–217, 1916.

[276] B. Montesinos and J. H. Thomas. The Evershed Effect in Sunspots as a Siphon Flow along a Magnetic Flux Tube. *Nature*, 390:485–487, 1997.

[277] G. Moruzzi and F. Strumia (Eds.). *The Hanle Effect and Level-Crossing Spectroscopy.* New York: Plenum Press, 1991.

[278] M. Mukerjee. Circles for Space. *Sci. American*, 289:22–34, 2003.

[279] D Müller, R.G. Marsden, O.C. St. Cyr, and H.R. Gilbert. Solar Orbiter: Exploring the Sun-Heliosphere Connection. *Solar Phys.*, 285:25–70, 2013.

[280] S. Musman and D.M. Rust. Vertical Velocities and Horizontal Wave Propagation. *Solar Phys.*, 13:261–268, 1970.

[281] V.M. Nakariakov. Coronal Oscillations. In *Dynamic Sun*, pages 314–324. Cambridge University Press, Cambridge, 2003.

[282] V.M. Nakariakov, L. Ofman, W. DeLuca, et al. Doppler Shift Oscillations of Hot Solar Coronal Plasma Seen by SUMER: A Signature of Loop Oscillations? *Astrophys. J.*, 574:L101–L104, 2002.

[283] M. Neugebauer and C. W. Snyder. Solar Plasma Experiment. *Science*, 138:1095–1097, 1962.

[284] J. Newmark. AGU Meeting. In *TESS Conference*, 2018.

[285] H. W. Newton and M. L. Nunn. The Sun's Rotation Derived from Sunspots 1934-1944 and Additional Results. *Month. Not. Royal Astron. Soc.*, 111:413–421, 1951.

[286] T. Nordgren. *Sun Moon Earth: The History of Solar Eclipses from Omens of Doom to Einstein and Exoplanets.* Basic Books, New York, 2016.

[287] Robert W. Noyes and Robert B. Leighton. Velocity Fields in the Solar Atmosphere. II. The Oscillatory Field. *Astrophys. J.*, 138:631–647, 1963.

[288] N. Oda. Morphological Study of the Solar Granulation. III - The Mesogranulation. *Solar Phys.*, 93:243–255, 1984.

[289] L. Ofman, V.M. Nakariakov, and C.E. DeForest. Slow Magnetosonic Waves in Coronal Plumes. *Astrophys. J.*, 514:441–447, 1999.

[290] Frank Q. Orrall. Observational Study of Macroscopic Inhomogeneities in the Solar Atmosphere.VI. Photospheric Oscillations and Chromospheric Structure. *Astrophys. J.*, 141:1131–1138, 1965.

[291] M. Ossendrijver. The Solar Dynamo. *Astron. Astrophys. Rev.*, 11:287–367, 2003.

[292] M. J. Owens and R. J. Forsyth. The Heliospheric Magnetic Field. *Living Rev. Sol. Phys.*, 10:1, 2013.

[293] T. Padmanabhan (Ed.). *Astronomy in India: A Historical Perspective.* Indian National Science Academy and Springer Nature, 2014.

[294] N. K. Panesar, D. E. Innes, S. K. Tiwari, and B. C. Low. A Solar Tornado Triggered by Flares? *Astron. Astrophys.*, 549:61–68, 2013.

[295] S. Parenti. Solar Prominences: Observations. *Living Rev. Solar Phys.*, 11:1–88, 2014.

[296] E. N. Parker. Dynamics of the Interplanetary Gas and Magnetic Fields. *Astrophys. J.*, 128:664–676, 1958.

[297] E. N. Parker. *Cosmical Magnetic Fields.* Clarendon Press, Oxford, 1979.

[298] E. N. Parker. Vector Structure of Active Magnetic Fields. In *Measurements of Solar Vector Magnetic Fields*, volume 2374, pages 7–16. NASA. Marshall Space Flight Center, 1985.

[299] E.N. Parker. The Formation of Sunspots from the Solar Toroidal Field. *Astrophys. J.*, 121:491– 507, 1955.

[300] M. Penn, S. Krucker, H. Hudson, et al. Spectral and Imaging Observations of a White-light Solar Flare in the Mid-infrared. article id. L30. *Astrophys. J.*, 819:1–5, 2016.

[301] M. J. Penn. Infrared Solar Physics. *Living Rev. Solar Phys.*, 11:2, 2014.

[302] W. D. Pesnell, B. J. Thompson, and P. C. Chamberlin. The Solar Dynamics Observatory (SDO). *Solar Phys.*, 275:3–15, 2012.

[303] L. E. Peterson and J. R. Winckler. Gamma-Ray Burst from a Solar Flare. *J. Geophys. Res.*, 64:697–707, 1959.

[304] K. Petrovay. Solar Cycle Prediction. *Living Rev. Sol. Phys.*, 7:6, 2010.

[305] M. Pick and N. Vilmer. Sixty-Five Years of Solar Radioastronomy: Flares, Coronal Mass Ejections and Sun-Earth Connection. *Astron. Astrophys. Rev.*, 16:1–153, 2008.

[306] A. K. Pierce. The Infrared Spectrum of the Sun. *Astron. Soc. Pacific Leaflets*, 9:217–224, 1965.

[307] E. R. Priest. *Dynamics and Structure of Quiescent Solar Prominences.* Kluwer Academic Publishers, Dordrecht, 1989.

[308] D.M. Rabin, J.T. Jefferies, and C. Lindsey. Infrared Solar Physics. In *Infrared Solar Physics.* Kluwer Academic Publishers, 1992.

[309] D.N. Rachkowsky. The Reduction for Anomalous Dispersion in the Theory of the Absorption Line Formation in a Magnetic Field (in Russian). *Izv. Krymsk. Astrofiz. Obs.*, 37:56–61, 1967.

[310] S. P. Rajaguru, Sarbani Basu, and H. M. Antia. Ring Diagram Analysis of the Characteristics of Solar Oscillation Modes in Active Regions. *Astrophys. J.*, 563:410–418, 2001.

[311] R. Ramaty, B. Kozlovsky, and R.E. Lingenfelter. Nuclear Gamma-Rays from Energetic Particle Interactions. *Astrophys. J. Supp. Ser.*, 40:487–526, 1979.

[312] C. Rao, N. Gu, L. Zhu, et al. 1.8-m Solar Telescope in China: Chinese Large Solar Telescope. *J. Astron. Teles. Instr. Sys.*, 1:1, 2015.

[313] N. E. Raouafi, P. Riley, S. Gibson, et al. Diagnostics of Coronal Magnetic Fields through the Hanle Effect in UV and IR Lines: doi: 10.3389/fspas.2016.00020. *Front. Astron. Space Sci.*, 3, 2016.

[314] M. J. Reiner, M. L. Kaiser, and J.-L. Bougeret. Coronal and Interplanetary Propagation of CME/Shocks from Radio, In Situ and White-Light Observations. *Astrophys. J.*, 663:1369–1385, 2007.

[315] F. Reines and C. L. Cowan. Free Antineutrino Absorption Cross Section. I. Measurement of the Free Antineutrino Absorption Cross Section by Protons. *Physical Rev.*, 113:273–279, 1959.

[316] E.J. Rhodes, R.K. Ulrich, and G.W. Simon. Observations of Nonradial p-Mode Oscillations on the Sun. *Astrophys. J.*, 218:901–919, 1977.

[317] J. C. Ribes and E. Nesme-Ribes. The Solar Sunspot Cycle in the Maunder Minimum AD1645 to AD1715. *Astron. Astrophys.*, 276:549–563, 1993.

[318] E. Rieger. Solar Flares - High-energy Radiation and Particles. *Solar Phys.*, 121:323–345, 1989.

[319] T. R. Rimmele and J. Marino. Solar Adaptive Optics. *Living Rev. Solar Phys.*, 8, 2011.

[320] T.R. Rimmele, K. Balasubramaniam, T. Berger, et al. First-Light Instrumentation for the Advanced Technology Solar Telescope. In *SP34A–03*, volume 86, pages 18–. EOS Trans. AGU., 2005.

[321] B. Roberts, P. M. Edwin, and A. O. Benz. On Coronal Oscillations. *Astrophys. J.*, 279:857–865, 1984.

[322] F. Roddier. Principle of Production of an Acoustic Hologram of the Solar Surface. *Comptes Rendus, Serie B - Sciences Physiques*, 281:93–95, 1975.

[323] F. Roddier. The Effects of Atmospheric Turbulence in Optical Astronomy. In *Progress in optics*, volume 19, pages 281–376. North-Holland Publishing Co. Amsterdam, 1981.

[324] Th. Roudier and R. Muller. Structure of the Solar Granulation. *Solar Phys.*, 107:11–26, 1986.

[325] G.B. Rybicki and A.P. Lightman. *Radiative Processes in Astrophysics.* Wiley-Interscience, New York, 1979.

[326] T. Sakurai et al. Construction of Long-Life Magneto-Optical Filters for Helioseismology Observations. In *Progress of Seismology of the Sun and the Stars, Eds. Y.Osaki and H. Shibahashi*, volume 367, pages 277–279. Springer Verlag, 1990.

[327] P. H. Scherrer, R. S. Bogart, R. I. Bush, et al. The Solar Oscillations Investigation - Michelson Doppler Imager. *Solar Phys.*, 162:129–188, 1995.

[328] H. U. Schmidt and J.B. Zirker. On the Oscillations of the Solar Atmosphere. *Astrophys. J.*, 138:1310–1311, 1963.

[329] H.U. Schmidt. Magnetohydrodynamics of an Active Region. In *Structure and Development of Solar Active Regions. Edited by K. O. Kiepenheuer*, volume 35, pages 95–107. Dordrecht, D. Reidel, 1968.

[330] W. Schmidt, O. von der Lühe, R. Volkmer, et al. The 1.5 Meter Solar Telescope GREGOR. *Astron. Nachr.*, 333:796–809, 2012.

[331] J. Schou, P.H. Scherrer, R.I. Bush, et al. Design and Ground Calibration of the Helioseismic and Magnetic Imager (HMI) Instrument on the Solar Dynamics Observatory (SDO). *Solar Phys.*, 275:229–259, 2012.

[332] H. Schwabe. Sonnen-Beobachtungen im Jahre 1843? *Astron. Nachr.*, 21:233–236, 1844.

[333] T. Sekii. Internal Solar Rotation. In *Sounding Solar and Stellar Interiors, Eds. Provost, J. and Schmider, F.X.*, volume 181, pages 189–202. Kluwer, Dordrecht, 1997.

[334] K. Shibata. Evidence of Magnetic Reconnection in Solar Flares and a Unified Model of Flares. *Astrophys. Space Sci.*, 264:129–144, 1998.

[335] K. Shibata and T. Magara. Solar Flares: Magnetohydrodynamic Processes. *Living Rev. Solar Phys.*, 8:6, 2011.

[336] P. J. A. Simões, G. S. Kerr, L. Fletcher, et al. Formation of the Thermal Infrared Continuum in Solar Flares. Id.A125. *Astron. Astrophys.*, 605:1–10, 2017.

[337] G.W. Simon and R.B. Leighton. Velocity Fields in the Solar Atmosphere. III. Large-Scale Motions, the Chromospheric Network, and Magnetic Fields. *Astrophys. J.*, 140:1120–1147, 1964.

[338] K.P. Singh. Techniques in X-ray Astronomy. *Resonance*, 10:8–20, 2005.

[339] K. R. Sivaraman, S. S. Gupta, and R. F. Howard. Measurement of Kodaikanal white-light images. I - A comparison of 35 years of Kodaikanal and Mount Wilson sunspot data. *Solar Phys.*, 146:24–47, 1993.

[340] K. R. Sivaraman, S. S. Gupta, and R. F. Howard. Measurement of Kodaikanal White-Light Images–IV. Axial Tilt Angles of Sunspot Groups. *Solar Phys.*, 189:69–83, 1999.

[341] H.B. Snodgrass. Separation of Large-Scale Photospheric Doppler Patterns. *Solar Phys.*, 94:13–31, 1984.

[342] D. Sobel. *Galileo's Daughter.* Walker & Co., New York, 1999.

[343] S. K. Solanki and N. A. Krivova. Solar Irradiance Variations: From Current Measurements to Long-Term Estimates. *Solar Phys.*, 224:197–208, 2004.

[344] S. K. Solanki, N. A. Krivova, M. Schüssler, and M. Fligge. Search for a Relationship between Solar Cycle Amplitude and Length. *Astron. Astrophys.*, 396:1029–1035, 2002.

[345] E. A. Spiegel and J.-P. Zahn. The Solar Tachocline. *Astron. Astrophys.*, 265:106–114, 1992.

[346] N. Srivastava, A. Ambastha, and A. Bhatnagar. Helically Twisted Prominence Eruption Event of 1979 March 11. *Solar Phys.*, 133:339–355, 1991.

[347] N. Srivastava, W. D. Gonzalez, A. L. C. Gonzalez, and S. Masuda. Solar Origins of Intense Geomagnetic Storms Observed During 6-11 March 1993. *Solar Phys.*, 183:419–434, 1998.

[348] L. V. Starukhina. Polar Regions of the Moon as a Potential Repository of Solar-Wind-Implanted Gases. *Adv. Space Res.*, 37:50–58, 2006.

[349] J. O. Stenflo. The Hanle Effect and the Diagnostics of Turbulent Magnetic Fields in the Solar Atmosphere. *Sol. Phys.*, 80:209–226, 1982.

[350] J.O. Stenflo. Applications of the Hanle Effect in Solar Physics. In *The Hanle Effect and Level-Crossing Spectroscopy.Physics of Atoms and Molecules*, pages 237–281. Springer, Boston, MA, 1991.

[351] J.O. Stenflo. *Solar Magnetic Fields, Polarized Radiation Diagnostics.* Dordrecht: Kluwer, 1994.

[352] J.O. Stenflo. Polarization Radiation Diagnostics of Solar Magnetic Fields. In *Astrophysical spectropolarimetry*, pages 55. Cambridge University Press, Cambridge, U.K., 2002.

[353] F. R. Stephenson. Pre-Telescopic Observations and Their Uses in Modern Astronomy. *Jour. British Astron. Assoc.*, 89:239–249, 1979.

[354] M. Stuiver and P. D. Quay. Changes in Atmospheric Carbon-14 Attributed to a Variable Sun. *Science*, 207:11–19, 1980.

[355] B. V. Subbarayappa and K. V. Sarma. *Indian Astronomy: A Source Book.* Nehru Centre, Mumbai, India, 1985.

[356] J. J. Sudol and J. W. Harvey. Longitudinal Magnetic Field Changes Accompanying Solar Flares. *Astrophys. J.*, 635:647–658, 2005.

[357] S. Suess. *Overview and Current Knowledge of the Solar Wind and the Corona: The Solar Probe.* NASA/Marshall Space Flight Center, 1999.

[358] B.J. Thompson, S.P. Plunkett, J.B. Gurman, et al. SOHO/EIT Observations of an Earth-directed Coronal Mass Ejection on May 12, 1997. *Geophys. Res. Lett.*, 25:2465–2468, 1998.

[359] M. J. Thompson, J. Christensen-Dalsgaard, M. S. Miesch, and J. Toomre. The Internal Rotation of the Sun. *Annual Rev. Astron. Astrophys.*, 41:599–643, 2003.

[360] W. T. Thompson. Coordinate Systems for Solar Image Data. *Astron. Astrophys.*, 449:791–803, 2006.

[361] W.T. Thompson, J.M. Davila, R.R. Fisher, et al. Cor1 Inner Coronagraph for Stereo-Secchi. In *Innovative Telescopes and Instrumentation for Solar Astrophysics*, volume 4853, pages 1–11. SPIE: The International Society for Photonics and Optics, 2003.

[362] A. M. Title, T. D. Tarbell, L. Acton, et al. White-light Movies of the Solar Photosphere from the SOUP Instrument on Spacelab. *Adv. Space Res.*, 6:253–262, 1986.

[363] A. R. Tiwary, S. K. Mathew, A. R. Bayanna, et al. Imaging Spectropolarimeter for the Multi-Application Solar Telescope at Udaipur Solar Observatory: Characterization of Polarimeter and Preliminary Observations. article id.49. *Solar Phys.*, 292:1–31, 2017.

[364] S. Tomczyk, A. Cacciani, and Jr. Rhodes, E. J. A Magneto-Optical Filter for Solar Oscillation Measurements. *Bull. American Astron. Soc.*, 19:701, 1975.

[365] S. Tomczyk, G.L. Card, T. Darnell, et al. An Instrument to Measure Coronal Emission Line Polarization. *Solar Phys.*, 247:411–428, 2008.

[366] S. Tomczyk and S. W. McIntosh. Time-Distance Seismology of the Solar Corona with CoMP. *Astrophys. J.*, 697:1384–1391, 2009.

[367] C. G. Toner, D. Haber, T. Corbard, et al. An Image Merge for GONG+. In *Proceedings of SOHO 12 / GONG+ 2002. Local and global helioseismology: the present and future (Ed. H. Sawaya-Lacoste)*, volume ESA SP-517, pages 405–407, 2003.

[368] S. Tsuneta, L. Acton, M. Bruner, et al. The Soft X-ray Telescope for the SOLAR-A Mission. *Solar Phys.*, 136:37–67, 1991.

[369] P. H. Ulmschneider. Possible Explanation of the 300-Second Type Oscillation in the Solar Chromosphere. *Astrophys. J.*, 152:349–350, 1968.

[370] Roger K. Ulrich. The Five-Minute Oscillations on the Solar Surface. *Astrophys. J.*, 162:993–1002, 1970.

[371] W. Unno and S. Kato. On the Generation of Acoustic Noise from the Turbulent Atmosphere, I. *Pub. Astron. Soc. Japan*, 14:417–430, 1962.

[372] I. G. Usoskin. A History of Solar Activity over Millennia. *Living Rev. Solar Phys.*, 10(1):1–94, 2014.

[373] M. P. van Haarlem, M. W. Wisel, A. W. Gunst, et al. LOFAR: The LOw-Frequency ARray: (DOI: 10.1051/0004-6361/201220873). *Astron. Astrophys.*, 556:1–53, 2013.

[374] A. Van Helden. Galileo and Scheiner on Sunspots. *Proc. Am. Phil. Soc.*, 140:338–342, 1996.

[375] J. M. Vaquero. Letter to the Editor: Sunspot Observations by Theophrastus Revisited. *Jour.British Astron. Assoc.*, 117:346, 2007.

[376] P. Vemareddy, A. Ambastha, and R. A. Maurya. On the Role of Rotating Sunspots in the Activity of Solar Active Region NOAA 11158. *Astrophys. J.*, 761:1–13, 2012.

[377] J. E. Vernazza, E. H. Avrett, and R. Loeser. Structure of the Solar Chromosphere. III - Models of the EUV Brightness Components of the Quiet-sun. *Astrophys. J. Suppl. Ser.*, 45:635–725, 1981.

[378] W.T. Vestrand and D.J. Forrest. Evidence for a Spatially Extended Component of Gamma Rays from Solar Flares. *Astrophys. J.*, 409:L69–L72, 1993.

[379] S. V. Vorontsov, J. Christensen-Dalsgaard, J. Schou, et al. Helioseismic Measurement of Solar Torsional Oscillations. *Science*, 296:101–103, 2002.

[380] A. Vourlidas. Mission to the Sun-Earth L5 Lagrangian Point: An Optimal Platform for Space Weather Research (doi: 10.1002/2015SW 001173). *Space Weather*, 13:197–201, 2015.

[381] M. Waldmeier. Neue Eigenschaften der Sonnenfleckenkurve. *Astron. Mitt. Eidgen. Sternwarte Zürich*, 14:105–136, 1935.

[382] J. Wang, Z. Shi, H. Wang, and Y. Lu. Flares and the Magnetic Nonpotentiality. *Astrophys. J.*, 456:861–878, 1996.

[383] T.J. Wang, S.K. Solanki, W. Curdt, et al. Doppler Shift Oscillations of Hot Solar Coronal Plasma Seen by SUMER: A Signature of Loop Oscillations? *Astrophys. J.*, 574:L101–L104, 2002.

[384] Y.-M. Wang, A. G. Nash, and N. R. Sheeley, Jr. Magnetic Flux Transport on the Sun. *Science*, 245:712–718, 1989.

[385] Y.-M. Wang and N. R. Sheeley, Jr. Understanding the Geomagnetic Precursor of the Solar Cycle. *Astrophys. J. Lett.*, 694:L11–L15, 2009.

[386] Y. M. Wang and N. R. Sheeley, Jr. Filament Eruptions near Emerging Bipoles. *Astrophys. J. Lett.*, 510:L157–L160, 1999.

[387] Y.M. Wang, R.C. Colaninno, T. Baranyi, and J. Li. Active-region Tilt Angles: Magnetic versus White-light Determinations of Joy's Law. article id. 50. *Astrophys. J.*, 798:14pp, 2015.

[388] H. P. Warren, A. R. Winebarger, and Brooks D. H. A Systematic Survey of High-temperature Emission in Solar Active Regions. article id. 141. *Astrophys. J.*, 759:1–13, 2012.

[389] D. F. Webb, D. A. Biesecker, N. Gopalswamy, et al. Using STEREO-B as an L5 Space Weather Pathfinder Mission (doi: 10.1016/j.srt.2010 .07.004). *Space Res. Today*, 178:1–10, 2010.

[390] S. M. White. Radio versus EUV/X-ray Observations of the Solar Atmosphere. *Solar Phys.*, 190:309–330, 1999.

[391] C. A. Whitney. Granulation and Oscillations of the Solar Atmosphere. *Smithsonian Contrib. Astrophys.*, 2:365–376, 1958.

[392] T. Wiegelmann, J. K. Thalmann, and S. K. Solanki. The Magnetic Field in the Solar Atmosphere. *Astron. Astrophys. Rev.*, 22:78, 2014.

[393] R. C. Willson and H. S. Hudson. Solar Luminosity Variations in Solar Cycle 21. *Nature*, 332:810–812, 1988.

[394] E.R. Wilson. *A Few Pre-Copernican Astronomers.* Popular Astronomy, 1917.

[395] L. Wolfenstein. Neutrino Oscillations in Matter. *Phys. Rev. D*, 17:2369–2374, 1978.

[396] W. H. Wollaston. A Method of Examining Refractive and Dispersive Powers by Prismatic Reflection. *Philosophical Trans. Royal Soc. London*, 92:365–380, 1802.

[397] H. Wolter. Glancing Incidence Mirror Systems as Imaging Optics for X-rays. *Annalen Physik*, 445:94–114, 1952.

[398] M.F. Woodard. Solar Subsurface Flow Inferred Directly from Frequency-Wavenumber Correlations in the Seismic Velocity Field. *Astrophys. J.*, 565:634–639, 2002.

[399] T. N. Woods, P.C. Chamberlin, J. W. Harder, et al. Solar Irradiance Reference Spectra (SIRS) for the 2008 Whole Heliosphere Interval (WHI). *Geophys. Res. Lett.*, 36:L01101, 2009.

[400] T. N. Woods, F. G. Eparvier, R. Hock, et al. Extreme Ultraviolet Variability Experiment (EVE) on the Solar Dynamics Observatory (SDO): Overview of Science Objectives, Instrument Design, Data Products, and Model Developments. *Solar Phys.*, 275:115–143, 2012.

[401] T. N. Woods, G. J. Rottman, S. M. Bailey, et al. Solar Extreme Ultraviolet Irradiance Measurements During Solar Cycle 22. *Solar Phys.*, 177:133–146, 1998.

[402] S. T. Wu, W. P. Guo, S. P. Plunkett, et al. Coronal Mass Ejections (CMEs) Initiation: Models and Observations. *J. Atmosph. Solar-Terres. Phys.*, 62:1489–1498, 2000.

[403] M. Xiong and Y. Liu. Taikong. In *ISSI-BJ Magazine*, volume 4, 2014.

[404] Y. Xu, W. Cao, C. Liu, et al. Near-Infrared Observations at 1.56 Microns of the 2003 October 29 X10 White-Light Flares. *Astrophys. J.*, 607:L131–L134, 2004.

[405] I. Ghezzi and C. Ruggles. Chankillo: A 2300-Year-Old Solar Observatory in Coastal Peru. *Science*, 315:1239–1243, 2007.

[406] G. P. Zank, J. Heerikhuisen, B. E. Wood, N. V. Pogorelov, et al. Heliospheric Structure: The Bow Wave and the Hydrogen Wall. *Astrophys. J.*, 763:1–13, 2013.

[407] W. Zessewitsch and W. Nikonow. Emission Lines in the Spectrum of the Solar Corona. *Nature*, 123:909–910, 1929.

[408] W. Zessewitsch and W. Nikonow. Die Deutung der Emissionslinien im Spektrum der Sonnenkorona. Mit 6 Abbildungen. *Zeitschrift Astrophysik*, 22:30–64, 1943.

[409] H. Zirin and J. M. Mosher. The Caltech Solar Site Survey, 1965-1967. *Solar Phys.*, 115:183–201, 1988.

[410] J. B. Zirker. Coronal Holes and High-Speed Wind Streams. *Rev. Geophys.*, 15:257–269, 1977.

Index